Ezra Stoller

Manufacturers Trust Company Building, New York City — a "glass-box" type of construction.

THE BOOK OF
POPULAR
SCIENCE

Volume 7

Grolier
INCORPORATED
New York

Distributed in the United States by
THE GROLIER SOCIETY INC.

Distributed in Canada by
THE GROLIER SOCIETY OF CANADA LIMITED

Cover photograph: all the colors of the rainbow are produced as natural sunlight is refracted through these prisms.

Herbert Matter, courtesy
Wyandotte Chemicals Corp.

13 ☼

CONTENTS OF VOLUME VII

Modern mammals of land, air and sea and extinct reptiles of past ages evolved from a common ancestor — the cotylosaur.

THE LIFE OF MAMMALS

History, Structure and Habits of Earth's Dominant Animals

BY IAN McT. COWAN

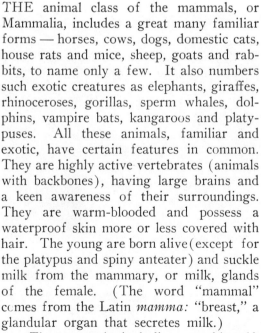

THE animal class of the mammals, or Mammalia, includes a great many familiar forms — horses, cows, dogs, domestic cats, house rats and mice, sheep, goats and rabbits, to name only a few. It also numbers such exotic creatures as elephants, giraffes, rhinoceroses, gorillas, sperm whales, dolphins, vampire bats, kangaroos and platypuses. All these animals, familiar and exotic, have certain features in common. They are highly active vertebrates (animals with backbones), having large brains and a keen awareness of their surroundings. They are warm-blooded and possess a waterproof skin more or less covered with hair. The young are born alive(except for the platypus and spiny anteater) and suckle milk from the mammary, or milk, glands of the female. (The word "mammal" comes from the Latin *mamma:* "breast," a glandular organ that secretes milk.)

The mammals, including man himself, are the dominant animals of the world. They live wherever life is possible — on Arctic ice sheets and in tropical rain forests, on mountain crags and a mile below the surface of the sea, in underground burrows and at the tops of the tallest trees, in fresh-water lakes and streams and also in the air. They vary in size from the minute pigmy shrew of North America — 2 inches in body length and weighing about as much as a dime — to the blue whale, a giant 100 feet long and weighing some 130 tons. Today there are roughly 3,500 distinct species of mammals; there are about 15,000 species and subspecies.

Mammals are comparative newcomers to the animal population of the earth. We can trace their descent back to the cotylosaurs, a reptile group that lived in the Carboniferous period, about 225,000,000 years ago. The cotylosaurs were sluggish and clumsy animals that crawled on their bellies with legs sprawled to the side. Many of them probably lived in or near water, but laid their tough-shelled eggs on dry land. Their bodily processes of digestion and respiration were slow. As a result, they released energy slowly from the foodstuffs they ate; they generated little body heat. They depended for warmth upon the temperature of the external surroundings.

When the temperature was high, the pace of life quickened somewhat. The cotylosaur moved more rapidly and ate more, bolting comparatively large pieces of food: insects and the flesh of amphibians and other reptiles. When temperatures were low, the life processes almost came to a standstill. Stored food, in the form of body fat, was utilized, but no new food was consumed.

of the mammallike reptiles, a gradual lengthening of the front and hind legs took place. The feet came more and more to point forward instead of to the side, and the legs were brought in under the body. Thus the body was lifted off the ground. The bones of the pelvis fused to form a single bony structure. These changes permitted the animals to move about more efficiently and rapidly.

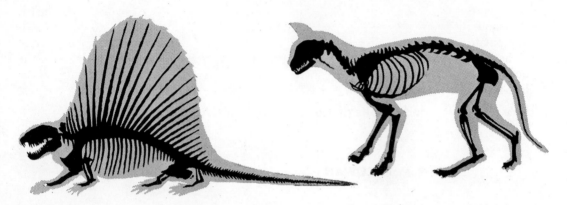

The primitive reptile *Dimetrodon* (left) and the domestic cat (right), a modern mammal, are both descended from the reptiles known as cotylosaurs. The darkened areas in the two drawings indicate skeletal structure.

From the cotylosaurs, directly or indirectly, evolved all of the reptiles: the modern forms — turtles, crocodiles, lizards and snakes — and those that we know about only from fossil remains. The cotylosaurs were also ancestors of the animals known as mammallike reptiles.

The mammallike reptiles began their evolution some 215,000,000 years ago. Some were carnivorous (flesh-eating); others, herbivorous (vegetation-eating). They all had one feature in common: a hollow in the skull behind the eye, which accommodated large jaw muscles. In time the mammallike reptiles became more and more like mammals. By about the beginning of the Jurassic period (150,000,000 years ago), a population of carnivorous mammallike reptiles seems to have given rise to small animals that clearly possessed the characteristics of mammals. The mammallike reptiles themselves became extinct.

How did a sluggish, sprawling reptile develop into an active, warmblooded mammal? We mentioned that the cotylosaurs crawled on their bellies. During the rise

In the mammallike reptiles certain bones hinged the lower jaw with the skull. By a remarkable transformation, these hinge bones came to have a different function in the mammals. They became part of a chain of bones in the middle ear — a chain that transmits vibrations from the eardrum to the inner ear. As a result of this and other changes, the lower jaw of mammals is now formed of only one bone.

Another change affected the nature of the teeth. From a row of simple, spikelike teeth, which the reptiles possessed, there developed the incisors, canine teeth and cheek teeth, or molars, of the mammals. These teeth were set in sockets of the jaw. On the crowns of the molars evolved several cusps, or points. The mammal's teeth could more efficiently nip off pieces of food, shear and grind it.

A secondary bony palate developed in the roof of the mammal's mouth. This structure, now found in all mammals, separates the nasal, or air, passages from the food passages. Consequently, the mammal can hold food in its mouth, chew it

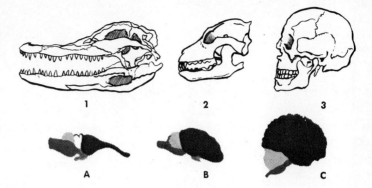

In this drawing we show the skull and brain of a typical reptile (an alligator) and two mammals (a dog and a man). 1: skull of alligator; 2: skull of dog; 3: skull of man; A: brain of alligator; B: brain of dog; C: brain of man. The brain case enclosing the brain has expanded in the evolution of the higher mammals, making possible greater brain development.

and still continue to breathe; it does not have to bolt its food. Moreover, the secondary palate makes suckling possible.

The bony case enclosing the brain expanded during the evolution from mammallike reptile to mammal. This permitted greater brain growth and development. Especially important was the development of the cerebral cortex, a covering of gray

A further change involved the heart, which came to be divided into four chambers. Through this development, blood carrying oxygen from the lungs to the heart is kept entirely separate from the blood pouring into the heart from the rest of the body. As a result, the tissues constantly receive a rich supply of oxygen, picked up in the lungs.

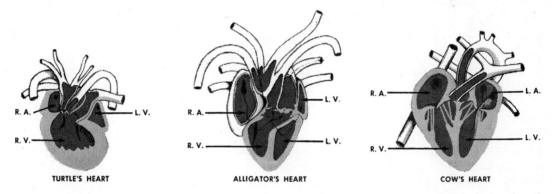

TURTLE'S HEART   ALLIGATOR'S HEART   COW'S HEART

Three types of hearts. The initial letters "L.," "R.," "A.," and "V." stand, respectively, for "left," "right," "auricle" and "ventricle." The turtle's heart has three chambers; the cow's, four chambers. The alligator's heart has only three chambers, but one of these — the left ventricle — is divided into two distinct sections.

matter over the cerebral hemispheres. The cerebral cortex is a complex area of nerve cells; it is concerned with higher processes such as thinking and reasoning.

The changeover from reptile to mammal involved important alterations in the internal structure. For one thing, a sheet of muscle appeared, separating the thoracic, or chest, cavity from the abdominal region. This muscular partition is the diaphragm of mammals; it allows these animals to supplement rib breathing with belly breathing. Rib breathing, which occurs in the reptiles, is produced by muscular movements of the rib cage; it is not so efficient as belly breathing.

Oxygen is necessary for converting food substances into energy. The food of mammals is broken into small pieces, through chewing, before being swallowed; consequently it can be quickly digested in the stomach and intestine. The digested food particles are rapidly transported to the tissues by the circulatory system.

With a greater supply of oxygen and food in its tissues, there is a greater output of energy. Some of this goes into muscular work, making the mammal more active. Some takes the form of heat, which, carried by the blood, warms the body and causes many of its chemical reactions to proceed at a quicker rate.

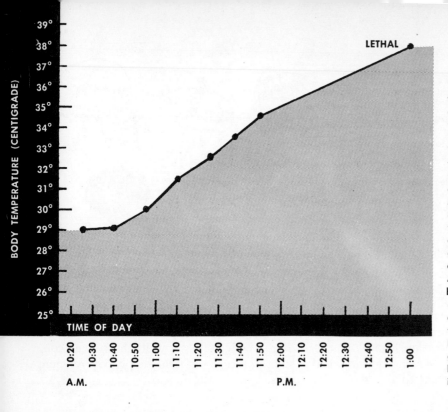

BODY TEMPERATURE (CENTIGRADE)

39°
38°
37°
36°
35°
34°
33°
32°
31°
30°
29°
28°
27°
26°
25°

LETHAL

TIME OF DAY

10:20 10:30 10:40 10:50 11:00 11:10 11:20 11:30 11:40 11:50 12:00 12:10 12:20 12:30 12:40 12:50 1:00

A.M.   P.M.

The body temperature of a typical reptile, such as an alligator, varies within wide limits, according to the external environment. This is strikingly shown in the above graph, adapted from one in the *Bulletin of the American Museum of Natural History*, Volume 86, Article 7. An alligator was tethered to a stake and subjected to the heat of direct sunlight for several hours (from about 10:20 A.M. to 1:00 P.M.). The body temperature gradually arose, as indicated, until a lethal, or fatal, point was reached (about 1:00 P.M.) In mammals there is generally little change in body temperature, except in hibernation (see Index).

As body temperature rises, the mammal's body requires more fuel (food), a greater consumption of oxygen and a higher production of carbon dioxide and other waste products. As a result, blood circulation is increased. Of course this requires increased filtration of the blood by the kidney.

The kidney of a mammal not only filters the blood of salts, sugar, water and wastes, such as urea, but also reabsorbs the sugar and most of the salt and water. Consequently a highly concentrated urine is produced, and water is conserved within the body.

The mammals developed a covering of hair, insulating the body against both excessive heat and cold. They also evolved an efficient control of their surface blood vessels. Because of this, blood can be turned, or shunted, away from the surface when body heat needs to be conserved during cold weather. In hot weather and during physical activity, the body temperature is kept down by dilating these same surface blood vessels. They then radiate heat; they also lose a watery fluid that escapes to the skin surface by way of the sweat glands. When a mammal sweats, the skin is moistened; evaporation of this moisture cools the body.

The milk glands of mammals secrete a rich, balanced, liquid diet for newborn offspring. Because this offspring depends for survival on its mother's milk, it is under parental care for some time.

The extended period of infant care clearly distinguishes the mammals from the reptiles, which do not give further attention to their young once the eggs are laid or the young are born. (Some reptiles bear living young.)

During the period of infant care, the young mammal has time to develop and learn a variety of responses. When it leaves the protection and guidance of its parent, it is fitted to modify its actions in the face of varying circumstances.

We pointed out that the first mammals probably arose about 150,000,000 years ago. These primitive mammals evolved into different types.

One group, the monotremes, has persisted down through the ages and is represented today by the platypus and spiny anteaters. They have fur and feed milk to their young, but they are very much like reptiles in many respects. For example, the waste products of the intestine, the urine and the sperm or eggs pass into a common chamber, called the cloaca, from which they are discharged. (The name "monotreme" means "single hole" in Greek.) This is the condition found in reptiles. The bones making up the shoulder and hip are very reptilian, as are certain features of the skull. Females lay reptilelike eggs.

Another group of primitive mammals, called pantotheres, were small creatures, omnivorous in habit: that is, they would eat almost anything. Sometime in the early Cretaceous period (120,000,000 years ago), the pantotheres gave rise to the marsupial mammals on the one hand and the placental mammals on the other.* These two groups changed but little during the long Cretaceous period, which lasted some 60,-000,000 years.

* The young of marsupial mammals are carried, from a very early stage of their development, in a marsupium, or pouch. Placental mammals retain their young within the body until development is quite complete. The name "placental" is derived from the placenta, a structure by means of which the fetus is nourished and discharges wastes in the uterus, or womb.

There are various differences between the skin of a mammal, such as man, and that of a reptile. Reptilian

During this time the reptiles, including the large and small carnivorous and herbivorous dinosaurs, ruled the land. The mammals continued to be small creatures; many of them probably lived in trees and were active only at night. The marsupials were omnivorous animals, the placentals were insect eaters.

By the end of the Cretaceous period (60,000,000 years ago), the dominant reptiles had dwindled in numbers or had become extinct. Cold and dry uplands had been created by the building of mountains and other geological changes. In these uplands the large reptiles were at a fatal disadvantage. The mammals, however, flourished under the changed conditions because they were warm-blooded creatures, and also because of their unrestricted diet, their exploratory activities and their more efficient reproduction. The extinction of

skin lacks glands and is drier than that of mammals. The epidermis is thinner; there is an absence of hair.

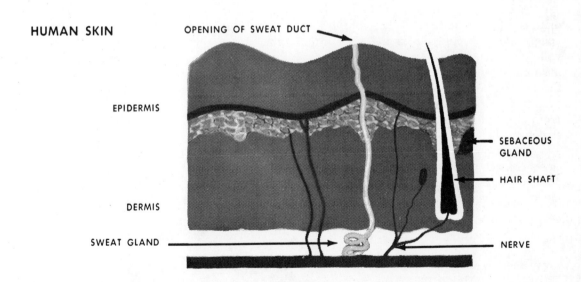

HUMAN SKIN

OPENING OF SWEAT DUCT

EPIDERMIS

SEBACEOUS GLAND

HAIR SHAFT

DERMIS

SWEAT GLAND

NERVE

SKIN OF REPTILE

EPIDERMIS

DERMIS

most of the reptiles left empty environments that could be filled by the mammals. The marsupials and placentals adapted to these new niches, changing more and more in form and habits until they occupied and exploited all the places where it was possible for them to live.

The female of the marsupial, as we have noted, rears her young in a special pouch on her belly. The marsupial embryo develops from a comparatively large-yolked egg cell in the female's uterus, or womb. While in the uterus, the embryo receives little or no nourishment from the mother; the young is born in a very undeveloped condition. (The newborn opossum is no larger than a bee.) It crawls into the shelter and warmth of the pouch, where the mammary glands are located, and hangs on to one of the nipples; milk is pumped into the young by the mother. The young remains in the pouch, attached to a nipple, until development is complete. Then it leaves the pouch to feed on its own, but frequently returns for protection.

Placental mammals differ in their manner of reproduction. The embryo develops from an egg cell having very little yolk. A membrane surrounding the embryo fuses with tissue lining the uterus of the female. This complex intergrowth of tissues from the wall of the uterus and from the embryo is the placenta. In this structure, blood vessels from the mother's and embryo's circulatory systems come into close contact. Though there is normally no exchange of maternal and embryonic blood, dissolved substances are transferred between the two systems. Food and oxygen diffuse from the mother's blood into her developing young. Carbon dioxide and wastes, such as urea, are given off by the embryo to the maternal blood stream. When the young placental mammal is born — that is, delivered from the mother's uterus — it is in an advanced stage of development.

The placentals also differ from marsupials by having a more expanded brain case and a greater development of the cerebral hemispheres. They lack the cloaca, or common chamber for wastes and sex products, found in all the monotremes and marsupials. They are more aggressive and efficient than the marsupials. When they compete directly with marsupials for living space or food, the pouched mammals are usually bested.

Mammals and their environments

Mammals do not live in a vacuum. They must cope with the many factors of their environment — temperature, amount of moisture, humidity, light, vegetation, other species of animals and members of their own species. Some mammals are less adaptable than others. For example, the giant panda (a bearlike mammal) can live only in bamboo forests, subsisting on bamboo shoots; the koala, which is an Australian marsupial resembling a teddy bear, survives only if the leaves of one or two kinds of eucalyptus trees are available to it for food. On the other hand, the brown rat can live almost anywhere and

The young of the opossum, a typical marsupial. The young opossums have been taken out of the pouch of the mother in order to show their undeveloped state at birth. Each is firmly attached to a teat of the mother.

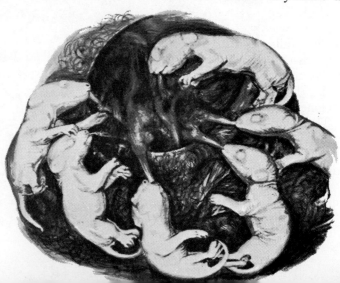

Water holes are vital to animals that feed on vegetation and to those that devour other animals. Not only do flesh-eaters need water themselves, but they prey on the herbivores living in the vicinity of water holes.

eat anything and has spread over most of the globe.

Though mammals are warm-blooded and generally control their internal temperature, the external temperature affects them. The white rat, for example, can maintain a nearly constant temperature when outside temperatures range from —13° to +104° F. If the outer temperature goes much beyond these extremes, the animal's own temperature regulation breaks down with fatal results.

Arctic foxes are active in winter even when the temperature drops to —60° F.; during blizzards the animals dig shelters in the snow. On the other hand, the armadillo, woodchuck, hamster and some other rodents hibernate during winter; their body temperature actually falls to a level only slightly above that of the environment. Bears also sleep in winter, but this is not true hibernation; for by burning a rich store of body fat their body processes continue at a rate that is only slightly lower than normal.

High humidity is important for small bats; their wing membranes dry out if the humidity falls much below 85 per cent. Moles also require dampness, and their burrows are usually very humid. They will die if forced from their burrows by a prolonged dry spell.

The abundance of water is very important, for many mammals drink a great deal. The large African herbivores, such as the antelopes, zebras, elephants, giraffes and buffalo, regulate their movements according to the water sources. These sources are also vitally important to animals such

as lions, leopards and hyenas. Not only do they need water themselves, but they find their prey among the herbivores that live in the general vicinity of water holes.

Some mammals drink dew or rain water running off foliage when there is no temporary or permanent water supply. Others get along with the water contained in their animal or vegetable food. This is true of many carnivores, rabbits and hoofed mammals, as well as certain desert rodents, such as pocket mice and kangaroo rats.

Most strictly land-living, or terrestrial, mammals are capable of swimming. Certain mammals spend much or all of their time in water; they include the platypus, water opossum, water shrew, beaver, muskrat, water civet and otter. These animals are usually sleek of body and have dense fur; most have elongated, pointed snouts. The feet are usually completely or partially webbed; the tail is often flattened horizontally or from side to side.

The most completely aquatic mammals are the seals, manatees and whales. They have streamlined spindle-shaped, or torpedolike, bodies with extremely short necks and no external ears. The front limbs are converted into paddles, and, in the case of seals, the hind legs too. Whales and manatees lack hind legs but have broad horizontally flattened tails; some whales, such as the fast-swimming killer whales, even possess a fleshy fin on the back.

Several factors permit the aquatic mammals to remain submerged for some time. For one thing, these animals renew the air in their lungs, when breathing at the surface, more completely than do terrestrial animals. In addition to the oxygen in the lungs, extra oxygen is taken up by blood that flows from the lungs into special networks of blood vessels. Once a dive is made, the heartbeat slows down; blood continues to flow in the brain and other vital centers but is shut off from the rest of the body. Thus the oxygen of the blood is not quickly exhausted. A dive may be prolonged without the animal being forced to come to the surface for air.

Light is another environmental factor that affects mammals. A great many species remain secluded during the day and become active only at sundown or after dark. This is true of numerous small mammals, such as bats, flying squirrels, badgers, shrews, rats, mice, anteaters and raccoons, to mention a few. Other species, including moles, monkeys, tree squirrels and ground squirrels, are most active by day. The large-hoofed mammals and the cats, foxes, coyotes and rabbits may be active day or night.

In addition to regulating daily activity, light also controls the time of mating in some mammals. For example, the reproductive glands of the ferret mature as the amount of daylight increases each day during the spring. On the other hand, with such mammals as goats and deer, a decrease in the amount of daylight per day, as in the fall, stimulates reproductive activity.

Light may also govern the molting cycle.* When daylight decreases day by day in the fall, weasels replace their summer pelage of brown with one of white. They change again to a brown coat when daylight increases in the spring. This is true of the snowshoe rabbit, which has a brown summer pelage and a white winter one.

When mammals such as sheep, dogs and white rats go from low to high altitudes, their red blood cells increase in number. At sea level, sheep, for example, have 10,500,000 red blood cells per cubic millimeter of blood. At an elevation of 16,000 feet their red blood cells increase to 16,-000,000. This increase is an adaptation

for the high altitudes, where the amount of oxygen in the atmosphere is reduced. Since the red blood cells carry oxygen, more of them are needed in order to supply the body tissues with sufficient oxygen. The llama, vicuña and viscacha (a rodent), which permanently live in the Andes at elevations of 12,000 to over 15,000 feet, have an even better adaptation. Their hemoglobin (pigment of the red blood cells) has a much greater capacity for carrying oxygen than does the hemoglobin of lowland-living species.

Certain mammals dig shelters in the soil; others live entirely underground. The platypus, various marsupials, many rodents and rabbits, armadillos and carnivores, such as foxes and skunks, dig simple shelters in which to sleep, hide and bring forth and care for their young. Voles, hamsters and various ground squirrels, including woodchucks and chipmunks, store food and hibernate, during winter, in their burrows. The truly subterranean mammals such as the marsupial mole, the true moles, the mole rats and the pocket gophers dig nesting chambers, chambers for storing food, connecting galleries and galleries in which they hunt food. They are completely adapted for an underground life, having thick, short hair and strong, oversized digging claws on the front feet. Their eyes are minute or missing; external ears are lacking or almost so. Moles hunt underground insects, insect larvae and earthworms; the mole rats and gophers feed on plant roots and tubers.

Plants serve mammals in many important ways. Grassy clumps, patches of weeds, bushes and the crowns of trees afford ideal places for small mammals to take shelter from the elements and escape the attack of carnivores and birds of prey. Kangaroos, deer, various antelopes, buffalo and elephants find the forest a good refuge.

Vegetation attracts hordes of insects, which provide food for numerous small mammals, including opossums, shrews, hedgehogs, bats, civets, skunks and certain ground squirrels and monkeys.

A number of mammals live in trees and only rarely descend to the ground.

---

* The pelage, or hair, of mammals is shed in periodic molts. Some mammals molt in spring and autumn; others, only in the autumn.

An accomplished acrobat — the spider monkey.

Among the most specialized of these animals are certain monkeys and apes. They have developed great dexterity with their hands and feet. The thumb and great toe can be opposed to the other fingers and toes of the hand and foot; this permits a very good grasp of tree branches. The whole hand and wrist are exceptionally mobile, and the long arms can move in almost any direction because of the great freedom in the shoulder joint. Gibbons and spider monkeys are very good at swinging by their arms from branch to branch. Some South American monkeys, such as spider and howling monkeys, possess a prehensile, or grasping, tail as an additional means for climbing and remaining secure in the trees.

Tree squirrels live only where there are trees. They cling to a trunk or branch by their sharp claws. A sloth spends its life upside down in trees, using enlarged, hooklike claws to hang from a branch; it travels on the ground only with great difficulty.

Some mammals glide, sometimes for considerable distances, from tree to tree. This is true of the flying lemurs, flying squirrels and the squirrellike marsupials known as flying and gliding phalangers. All have a furry membrane along the sides of the body extending between the front and hind limbs. When gliding, the membrane is stretched so that the animals almost float through the air like parachutes.

Plants serve numerous mammals as food. These animals are adapted in various ways for a vegetarian diet. Fruit-eating bats, for instance, have flattened

The howling monkey, an inhabitant of South America, is at home in the trees. It uses its prehensile, or grasping, tail in moving from branch to branch.

grinding teeth. Some bats even possess an elongated tongue for feeding on nectar. The monkeys and apes eat a considerable variety of foods; vegetation forms a good part of their diet. The stomachs of gue-rezas — monkeys of tropical Africa — are divided into a number of open sacs, which aid in digesting leafy food. The aquatic sea "cows" (manatee and dugong) are herbivorous mammals that use their lips to crop aquatic plants; these are digested in a complex stomach. Bears eat a good deal of plant food, including roots and berries.

Rodents and rabbits are well-known plant eaters. Their front teeth, or incisors, are formed like chisels and are used for gnawing; the molars are developed for grinding. Rabbits eat fresh, green food voraciously. Some of this is completely digested in the intestines; in this case the waste is excreted normally. Some of the green food, however, is fermented for a while in the caecum, which is a blind pouch off the large intestine. The food is then excreted, eaten again and passed through the digestive tract a second time. More energy can be obtained from a given quantity of food as a result.

Rodents often have storage places for food; hamsters, voles, squirrels and kangaroo rats store various seeds, bulbs, tubers, roots, nuts or mushrooms. The beaver provides itself with a winter supply of branches and trunks of poplar, aspen, cottonwood and willow, from which it gnaws the bark. The material is submerged near the animal's lodge sufficiently deep in the water to prevent its being caught in the ice.

Among the most specialized of the herbivores are the kangaroos and hoofed mammals. Kangaroos have a single pair of lower incisor teeth with sharp inner edges, perfect for cutting grass. The horses and their kin crop grass and other vegetation with gripping incisors in the upper and lower jaws. Deer, cattle and their relatives possess only lower incisors;

9

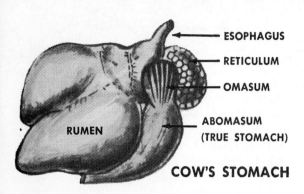

ESOPHAGUS

RETICULUM

OMASUM

ABOMASUM
(TRUE STOMACH)

RUMEN

COW'S STOMACH

Four-chambered stomach of the cow. Food first enters the rumen; from this it passes into the reticulum, where it is made into small masses, known as cuds. These are cast up to the mouth, where they are chewed at leisure. When swallowed again, the food passes to the omasum and from there to the abomasum (true stomach).

these work against a horny pad on the upper jaw for cropping. The tongue helps bring food to the mouth. It is covered with horny projections; it is large, mobile and pointed and can be considerably extended out of the mouth. The molar teeth of all these mammals have broad, grinding surfaces. The molars of the elephant are immense; this huge animal collects plant food with its trunk.

In the kangaroos, horses, deer, cattle and their kin, the digestive system includes a chamber or a sac in which bacteria break down the cellulose of plants so that the products can be used by the animal. Bacterial action takes place in the large intestine of horses. Deer, giraffes, antelopes, cattle, sheep and goats all have a complex, four-chambered stomach. Food is received into the first two stomachs when it is initially swallowed; here bacteria reduce the food to a cud (pulp). This is brought back into the mouth for chewing; later it is swallowed and further digested in the last two stomachs. Animals digesting food in this way are called ruminants.

Carnivores have teeth specialized for their flesh-eating habits. In general the incisors are developed for nipping and piercing, and the canines are elongated into daggers that stab and slash. Certain cheek teeth — called carnassials — in the upper

and lower jaws are formed into blades; as the jaws close, the carnassials shear past each other, like the blades of a pair of scissors and so cut or slice flesh and sinew. The hinder molars, not needed for grinding, are usually reduced in number and size; dogs and bears, however, have fairly well-developed molars that serve a grinding function. The hyenas possess massive bone-cracking teeth.

The land-living carnivores are strong and quick-reacting animals. Some can deliver powerful blows with their sharply clawed front feet. Many are strongly muscled in the limbs, allowing for climbing, leaping or the rapid pursuit of prey. The cheetah, or hunting leopard, the fastest of all terrestrial animals, can reach a speed of 65 to 70 miles an hour over a comparatively short distance.

Marine carnivores, such as the seals, lack carnassial teeth but have simple, pointed teeth used for catching and holding fish. The walrus' tusks are modified canine teeth that root up bottom-living crustaceans and mollusks; these are ground by the walrus' broad, crushing cheek teeth.

Certain whales, such as porpoises, sperm whales and killer whales, have simplified teeth, capable of grasping fish and squids or tearing apart the flesh of birds, seals and other whales. Other whales, such as the blue whale, possess a series of sheets of baleen — a horny substance — hanging down from the roof of the mouth. The baleen is frayed along the inner edge. It strains out plankton (see Index) from the water after the whale has engulfed a mouthful of water and animals.

The jaguar, a big land-living carnivore, can deliver powerful blows with its sharply clawed front feet.

The giant armadillo is protected by bony plates. Its front claws are sometimes used as defensive weapons.

Many mammals feed on insects. The winged mammals called bats catch insects while on the wing by opening the very large mouth. It is believed that the species that have a membrane between the hind legs may sometimes curve this membrane, in flight, so as to form a cup. The cup dips insects out of the air; the bat bends down its head, while still in flight, and eats its prey out of this pocket.

Certain mammals eat ants or termites almost exclusively; they include the spiny anteaters, the marsupial anteater, the true anteaters, the scaly anteaters (or pangolins) and the aardvark. Though they are not related to one another, these mammals have several features in common. The front feet possess stout digging claws for ripping apart termite and ant nests. The snout is elongated, and from it can be extended a long, thin tongue, sticky from the secretions of well-developed salivary glands. The teeth are reduced in size and number.

How mammals protect themselves from enemies

Mammals can protect themselves from their natural enemies in various ways. One of the most obvious protective measures is to run away. Kangaroos and hares, for instance, have long hind legs that give them good speed in a bounding escape flight. The hoofed mammals have slender, elongated, running-type legs that lift the body high off the ground; the animals run on their toenails, which are modified into hoofs.

An escaping mammal, if pressed too closely by an attacker, may seek to defend itself. Various methods are employed. The anteaters and sloths lash out with their long, strong claws; rodents and opossums bite their enemies. Hoofed mammals kick out with their feet, or use their tusks, horns or antlers.

Sometimes, instead of running away, many mammals may freeze on the spot. This is common with young fawns; their lack of movement and their dappled coat generally cause them to be overlooked by a predator. Hares will remain motionless until one comes close to them; then they bound away in a zigzag flight. The opossum "plays dead," lying limp with eyes closed and tongue hanging out of its partly opened mouth.

Many of the smaller mammals move about only at night or under the cover of vegetation or under fallen leaves and other debris of the forest floor. The large cud-chewing hoofed mammals come out into the open to feed rapidly. They then quickly return to cover, where they chew the cud and digest their food at leisure and in security.

Armadillos and pangolins are armored with horny plates or scales. When danger threatens they roll up so that only the protected surfaces of the body are exposed to an attacker. Hedgehogs, porcupines and spiny anteaters are protected by hairs that are modified into sharp spines or quills.

Civets, mongooses and skunks spray an attacker with an evil-smelling liquid — a secretion from modified sebaceous glands (see Index) in the region of the anus.

Many mammals are patterned or colored in such a way that they blend well with the background. If they refrain from moving, it is usually all but impossible to make them out. For example, the white winter coat of many weasels, foxes, hares and lemmings blends with the snow. Many desert mammals are pale buff or tawny, "melting in" beautifully with the desert soils.

The social life of mammals

Apart from the relations between a mother mammal and her young, a host of social relations exist among the adults and young of a population. Among the most conspicuous of these is dominance, in which one individual ranks over one or more of its fellows. The more dominant animal eats first; it displays aggressive behavior toward its inferiors. In a given mammal

Herd of elephants. These huge animals, like many other mammals, are social creatures and usually travel about in large groups.

population, male 1, say, dominates males 2, 3 and 4; male 2 dominates males 3 and 4; and male 3 dominates male 4. The same order may follow among females and among the young. The males dominate the females; the adults dominate the young.

Family groups are common among certain mammals. For example, the American beaver lives in a group consisting of two adults, the young of the current year and often the young of the year before. When the young reach maturity they are driven away. Howling monkeys form clans consisting of several adult males, a greater number of females and a variable number of young.

Fur seals and sea lions live in herds for the greater part of the year. During the breeding season, the adult males form harems of several females. At this time immature individuals and adult bulls that have not been able to gather any cows make up a separate group. A number of other species of seals live in herds but never form harems. The males and females breed at random.

Elephants, members of the horse family, hippopotamuses, many of the piglike mammals, giraffes, vicuñas, a majority of the deer and most of the cattle, antelopes, sheep and goats are social creatures. Generally the males associate with one or several females only during mating time. After breeding, the males abandon their mates or harems and establish herds of males only or else lead a solitary life. The females and young collect in a herd of their own, "controlled" by an adult female leader.

Certain mammals are not social. The chipmunk is intolerant of any others of its species. Only for mating do the sexes come together; and immediately afterward the male and female become antagonistic toward one another. The female associates with her young only as long as they remain in the nest. Bears, too, join company only to mate. During the rest of the year, the males live by themselves; the females, however, are followed by their cubs for a fairly long time.

Most mammals do not roam about at random; they restrict their explorations and food-searching activities to a home range. The mammal's lodging place — nest, burrow, thicket or other shelter — is within this range. There are also various places where the animal can attend to its comfort. For example, deer include in their home range wallowing places in which

Spiny anteater: order Monotremata.

Duck-billed platypus: order Monotremata.

to take mud baths. The badger has a particular spot for sunning itself. Zebras, bears and bison use termite nests or trees as "rubbing posts." In general, the size of the range of a particular mammal depends upon the amount of food available for its needs. Tigers, lions, wolves and bears have an extensive home range. At the other extreme there is the woolly opossum, which may stay in a single tree for several months.

Though many mammals spend the whole of their lives within a home range, some species migrate; that is, they travel from one area to another and back again. We discuss these migrations in the article Animal Migrations, in Volume 9.

Classification
of the mammals

There are eighteen orders of living mammals in the class Mammalia. The monotremes, belonging to the first order, are reptilelike and only remotely related to the rest of the mammals. The second order is made up of the marsupials. The animals grouped together in the remaining sixteen orders are placental mammals. Following is a brief account of the different orders.

Monotremata: duck-billed platypus, spiny anteaters. The monotremes are like reptiles in certain features of their skeleton and their egg-laying habit. They burrow and make nests to help protect the young. When adult, monotremes lack teeth. The platypus, of Australia and Tasmania, has a flattened bill covered with soft skin, and uses it to stir the mud of streams and lake bottoms as it seeks insect larvae, worms, mussels and snails. Its tail is flattened and its feet are webbed. Spiny anteaters, which possess a mixture of spines and hair on the back, have elongated faces and tongues adapted for eating ants. Spiny anteaters live in Australia and New Guinea.

Marsupialia: opossums, pouched mice, native cats, Tasmanian devil, Tasmanian wolf, banded anteater, pouched mole, bandicoots, opossum-rats, phalangers, koala, wombats, rat kangaroos, wallabies, kangaroos. These pouched mammals are found mainly in Australia and surrounding islands; the opossum-rats and opossums occur in South, Central and North America. Female marsupials usually have a pouch, or fur-lined pocket — the marsupium — in which the larvalike young complete their development. The marsupial's brain is small.

There are pouched mammals that resemble rats, mice, rabbits, woodchucks, squirrels, flying squirrels, moles, anteaters, small bears, otters, cats and wolves. The largest living marsupial is the giant red kangaroo, weighing two hundred pounds and standing seven feet high when raised up on its powerful tail and hind legs. The smallest is the shrew opossum of Brazil; it weighs less than a half dollar.

Insectivora: solenodon, tenrecs, golden mole, hedgehogs, elephant shrews, shrews, moles. These are small, secretive, mostly ground-living and nocturnal animals that feed on insects or are omnivorous. The snout is elongated; the cerebral hemispheres are small. There are five toes on the front and hind feet. Insectivores have a full complement of teeth — incisors, canines, premolars and molars. Some species still possess a cloaca; and the regulation of body temperature in many insectivores seems to be imperfect.

The solenodon, from Cuba and Haiti, and the tenrecs, from Madagascar, are somewhat like hedgehogs. The golden mole and elephant shrews are found in Africa; hedgehogs, in the Old World. Moles occur in Asia, Europe and North America; one kind, the desman of Europe, is an aquatic form with webbed feet. The mouselike shrews are found almost everywhere; some are aquatic in habit.

Dermoptera: colugos, or flying lemurs. The colugos, from the East Indies and the Philippines, are much like insectivores except that they are adapted for gliding. Broad, furry skin folds extend, at the side of the body, between the front and hind legs and onto the tail; when the colugo stretches its legs, the animal becomes a furry kite, gliding long distances from one tree to another. Colugos are the size of squirrels. They have foxlike faces; they eat leaves, flowers and fruit. When sleeping, they hang upside down from a branch, with all four feet grasping it.

Wallaby: order Marsupialia.

Flying lemur: order Dermoptera.

Elephant shrew: order Insectivora.

Bat: order
Chiroptera.

Giant anteater: order Edentata.

Chiroptera: bats. These are the only mammals that can truly fly, doing so by flapping their wings. The bats are essentially insectivores. They have greatly elongated bones in the front limbs and fingers for supporting the wing membranes. The hind legs are weak, making the animals almost helpless on the ground; they cannot walk, but can crawl and climb. When resting, bats hang upside down clutching a support with the clawed toes of the hind feet.

Bats are found on all continents, except Antarctica, and most oceanic islands. Large fruit-eating bats live in the tropics and subtropics of the Old World; the largest of these, called flying foxes, have a five-foot wing span. Most of the smaller species feed mainly on insects, which they catch on the wing. Several bats catch fish by hovering over the water's surface and seizing the quarry with their hind feet. Vampire bats, of Central and South America, feed on blood; they pierce the skin of large mammals with special pointed incisor teeth and lap the blood oozing from the small wound.

Edentata: anteaters, sloths, armadillos. The edentates are close relatives of the insectivores but are specialized for climbing and digging. Well-developed, long and stout claws arm the feet. The diet is somewhat restricted; hence, the teeth are reduced in number and are simplified. Additional joints in the backbone make edentates very strong. The brain is comparatively small, having poorly developed cerebral hemispheres. Edentates occur in South and Central America; one species, the nine-banded armadillo, is found in the southern part of the United States.

The anteaters have an elongated, tube-shaped snout, which is used to probe anthills and termite nests. They lack teeth. Sloths, which have a round head and a short face, are leaf eaters. They have no front teeth; the back teeth are simplified and lack a covering of enamel. Armadillos are covered by armor; shields of small flat bones in the skin, topped by horny plates, cover the shoulder and hip areas. Between these bony shields are bands of horny plates giving flexibility to the back. The top of the head and the tail are protected also by horny plates. As with the sloths, there are no front teeth; the teeth along the sides of the jaw are simple pegs without enamel. Armadillos feed on insects, carrion and whatever else they can find on the ground. They are excellent burrowers.

Pholidota: pangolins, or scaly anteaters. The pangolins are like anteaters but have evolved independently from an insectivore ancestor. They have an elongated snout, large claws and no teeth. Flattened, leaflike, horny scales cover the back and sides of the body from the nose to the tip of the long tail. These scales, which are composed of fused hairs, overlap like the scales of a pine cone. Pangolins live in tropical Africa and Asia; they are nocturnal, and several species can climb trees very well.

Primates: tree shrews, lemurs, lorises, galagos, tarsiers, New World monkeys, Old World monkeys, apes, man. The primates are closely related to the insectivores and apparently directly evolved from them. They have five digits, with nails instead of claws, on each hand and foot. The teeth are much like those of the insectivores. The diet is usually unrestricted, though vegetable matter makes up a good part of it.

The primates are adapted for living in trees and have developed great dexterity with their fingers and toes, permitting the efficient grasping of tree branches and the manipulation of objects. These animals have developed their vision to a high degree. Because the eyes look forward, making the fields of vision overlap, primates enjoy vision with depth. A flexible neck permits the eyes to be turned in any direction. The primates surpass all other animals in mental development. The brain, especially the cerebral hemispheres, is large. After birth, the young primate takes an unusually long time to develop and depends on parental guidance during much of its immature life. Family life is well established.

Primates are mostly forest-living animals of the Old and New World tropics and subtropics. Man, with the aid of his inventions, lives in almost all environments anywhere on the earth.

Rodentia: mountain beaver, tree squirrels, ground squirrels (including marmots, woodchucks, prairie dogs, chipmunks), flying squirrels, pocket gophers, pocket mice, kangaroo rats, beaver, New World mice (including wood, pack and cotton rats), hamsters, lemmings, voles (including muskrats, meadow mice), antelope rats, mole rats, bamboo rats, Old World rats and mice (introduced everywhere and including house rats and mice),

Gorilla: order
Primates.

Sloth: order Edentata.

Pangolin: order Pholidota.

African tree mice, dormice, jumping mice, jerboas, porcupines, Guinea pigs, capybaras, pacas, chinchillas, coypus and many others. There are more species of rodents than all other species of mammals combined.

Rodents have just four front teeth, two in the upper jaw and two in the lower one; these form large, sharp-edged chisels used for gnawing. Ridges on the cheek teeth permit the thorough grinding of hard grain and other plant material. Rodents are usually herbivorous; but they often eat insects and some species are omnivorous.

The skull of the rodent is long and low. The forelimbs are flexible enough for climbing, running and handling food. Usually there are five toes on each foot. Some rodents that are adapted for hopping have elongated, powerful hind legs and comparatively short front limbs.

Rodents have a great capacity for reproduction; this and their generally small size account for their great success in the world. They occur almost everywhere. Rodents would be the dominant mammals of the earth except for one thing — they are not so intelligent as we are.

Lagomorpha: pikas, or conies, hares, rabbits. These superficially resemble rodents but are not closely related to them. They have chisellike front teeth, but unlike rodents, the lagomorphs have a second, smaller pair of incisor teeth in the upper jaw. The cheek teeth cut vegetable food rather than grind or crush it.

Pikas are small, compact, short-legged mammals with short, rounded ears; they live in mountainous country of Eurasia and North America. Rabbits and hares are short-tailed, long-eared mammals adapted for hopping and running swiftly; the elongated hind legs give the power for long leaps, while the front limbs take up the shock of landing. Rabbits and hares occur in the Northern Hemisphere and have been introduced in Australia.

Cetacea: toothed whales (including river dolphins, beaked whales, sperm whales, narwhal, dolphins, killer whales, porpoises), whale-bone (or baleen) whales (including rorquals, humpbacked whales, blue whales, right whales). Cetaceans are highly specialized for an aquatic life. There is little or no hair on the body. Beneath the skin is a thick layer of oil-impregnated fibrous

tissue — the blubber; it may insulate the body or serve as a food-storage tissue.

The nostrils of whales (except those of the sperm whale) lie on the top of the head; the nasal passages go directly to the lungs by crossing the back of the throat in a species of closed tube. The brain of the whales is large and is highly developed; there is no sense of smell, but whales are keenly sensitive to water vibrations. The newborn whale is fed milk pumped into it by the mother. The young one seizes the mother's teats; these lie in a partially closed pouch that prevents the baby from swallowing too much sea water as it nurses.

The toothed whales have few or numerous simple, peglike teeth. The male narwhal has a single upper tooth that projects straight out of the front of the head as a long tusk. In the whalebone whales sheets of baleen, hanging down from the roof of the mouth, strain tiny organisms out of the water. Whales vary in size from small, five-foot species to giants one hundred feet long. They live in all oceans; some species frequent the fresh waters of large tropical rivers.

Carnivora: dogs, wolves, foxes, hunting dogs, bears, racoons, pandas, weasels (including ferrets, minks, wolverines, badgers, skunks, otters), civets, mongooses, hyenas, cats (from the lion to the domestic tabby), ear seals (including fur seals, sea lions), walrus, true seals (including common, harbor and harp seals, elephant seal). These mammals are highly specialized for hunting medium-sized to large prey. They usually have lithe and powerful bodies. The jaws are strong. Generally there are four or five toes, armed with sharp claws, on each foot. Carnivores have large, well-developed brains; they are extremely alert mammals with quick and well-co-ordinated movements. One or more of the senses of smell, vision and hearing are keen. Not all carnivores, by any means, eat flesh entirely; some species, such as the pandas, are vegetarians.

Land-living carnivores are either adapted for climbing or for running with great speed, though normally only over short distances. The seals and walrus are marine carnivores, streamlined of body; the feet have been transformed into paddles with webbing between the toes. Carnivores are native to all continents or their surrounding waters, and have also been introduced in Australia.

Jack rabbit: order Lagomorpha.

Walrus: order Carnivora.

Right whale: order Cetacea.

Aardvark: order
Tubulidentata.

African elephant:
order Proboscidea.

Tubulidentata: aardvarks. These short-legged, nocturnal, African mammals are the size of a small pig. The head is long and pulled out into an elongated tubular snout. Long, slender ears surmount the head. The tail is heavy and tapering. With its powerful front legs and sharp, flat nails, the aardvark digs burrows and tears up the nests of termites upon which it feeds. The cheek teeth are small cylinders lacking enamel; there are no incisors or canine teeth.

Proboscidea: Asiatic and African elephants. These are the largest of land mammals. The flexible muscular trunk is formed by the elongation of the nose and upper lip. The trunk's tip is a sensitive grasping organ. Tusks are protruding upper incisor teeth, which continue to grow long after adulthood is reached. Vegetable food is ground up by the enormous, tall, cross-ridged, cheek teeth. Elephants walk almost on the tips of the five toes of each foot. The brain is large.

Hyracoidea: hyraxes. The hyraxes, of Africa and Asia Minor, resemble rabbits somewhat in appearance and habit, though they have short ears and some live in trees. Two upper front, chisel-shaped incisors oppose four incisor teeth in the lower jaw. The cheek teeth are large and have grinding surfaces for crushing vegetable food. The four toes on the front feet and the three on the hind feet end in small, hooflike nails.

Sirenia: dugong (or sea cow), manatees. The naked-skinned sirenians are related to the hoofed mammals but are modified for an aquatic mode of life. The body is streamlined, and the head is low, ending in a blunt, bristly snout. The brain is small. Manatees live along the Atlantic coasts of Africa and the tropical Americas, sometimes going far up the larger rivers; the dugong inhabits the coasts of the Red Sea, Indian Ocean and western Pacific.

Perissodactyla: horses (including asses and zebras), tapirs, rhinoceroses. These are the hoofed mammals with odd-numbered toes. In these animals the axis of the foot passes through the mid-dle, or third, toe, which is best developed. The horse has one functional toe on each foot; the tapirs and rhinos, three. The toes end in hoofs. Horses have elongated limbs specialized for swift running; the tapir's legs and feet are short and stocky. The odd-toed hoofed mammals have a full set of incisor teeth for cropping plants.

Wild horses occur in Asia and Africa; but domestic forms are world-wide in distribution. The rhinoceroses, of Africa and Asia, carry horns that are formed of clumped masses of hair without a bony core. Tapirs, which live in the East Indies and Central and South America, have the nose elongated into a short, flexible trunk.

Artiodactyla: pigs (including water hog, wart hog, babirussa, peccaries), hippopotamuses, camels (including llama, vicuña), chevrotains, deer (including musk deer, muntjak, fallow deer, elk, moose, caribou), giraffe, pronghorn, cattlelike mammals, antelopelike mammals, goatlike mammals. These are the hoofed animals with even-numbered toes; the axis of the foot passes between the third and fourth toes. There are either two or four functional toes, usually bearing hoofs, on each foot. In many of these mammals the incisor teeth are replaced by a horny pad against which the lower incisors bite when cropping plants. The canine teeth of some species are elongated into tusks used for fighting; in other species the canines are small or lacking. The rear cheek teeth have excellent grinding surfaces. The brain is moderately well-developed, and the senses of smell, sight and hearing are usually very good. Those forms called ruminants — such as deer, giraffes, cattle and their kin — have the stomach divided into four compartments. The horns of cattle and antelopes are formed of a horny material over a bony core. Deer antlers are bony outgrowths, which are shed periodically. The even-toed hoofed mammals are native to all continents except Australia, where domestic forms have been introduced, and Antarctica.

See also Vol. 10, p. 275: "Mammals."

Tapir: order Perissodactyla.

Wart hog: order Artiodactyla.

Dugong: order Sirenia.

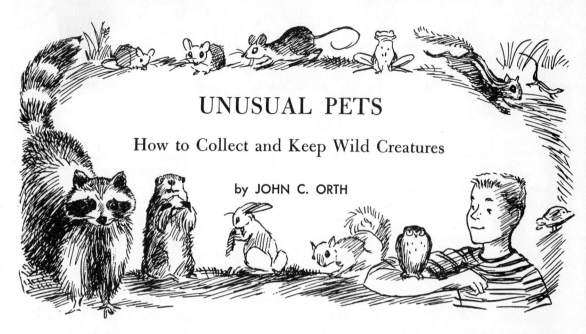

UNUSUAL PETS

How to Collect and Keep Wild Creatures

by JOHN C. ORTH

PROPERLY cared for and treated kindly, certain native wild animals make wonderful pets. You must always remember, however, that from the moment you capture and cage a wild creature, you take on certain responsibilities; you become its provider and its protector. It goes without saying that it is inhuman to make an animal a captive and then to neglect or mistreat it.

Various states in the United States and provinces in Canada have laws against collecting and keeping certain forms of wildlife. This is particularly true in the case of mammals and birds, although some regions also protect amphibians and reptiles of certain species. However, there are many forms of animal life that you can keep without a permit if you are sincerely interested in collecting, studying and exhibiting them in a home zoo. Before you attempt to keep any wild creature as a pet — especially a bird or mammal — find out what the law is where you live. You can get information from your wildlife or game commissioner, a game warden or perhaps an Audubon club.

Some creatures, such as foxes, raccoons, woodchucks and certain kinds of birds and reptiles, do not take to captivity readily when they are caught as adults. Do not try to keep them, because fierce animals in captivity will give you no particular satisfaction. On the other hand, many of these animals often make fine pets if caught when young.

It is advisable to keep birds and also the larger mammals, such as raccoons, foxes and opossums out of doors. All of our small native mammals and birds, except those that hibernate or migrate, can be left outdoors in winter provided they have good shelter boxes to protect them from the elements. Small mammals, reptiles and amphibians will do better if kept indoors.

Mammals are particularly responsive, but many other forms of animal life make satisfying pets. By observing the habits and actions of all these animals closely, you will learn many interesting things; you may even unearth new facts about them.

If there is a zoo in your vicinity, by all means visit it. Study the cages of small animals and the methods used to care for the animals. The keepers will probably be very glad to give you useful hints, if you ask questions.

This article will deal mainly with the kinds of animal life that need to be kept in

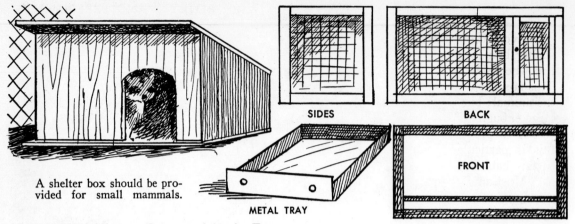

A shelter box should be provided for small mammals.

SIDES

BACK

METAL TRAY

FRONT

Cage for small mammals. The metal tray fits into a space left in front.

some sort of cage. Other articles in THE BOOK OF POPULAR SCIENCE tell how to keep insects and other small creatures in terrariums and also how to start and maintain an ant colony.

How to build a home for a small mammal

It may seem a bit odd to discuss cages before we take up the matter of collecting animals. You must keep in mind, however, that once an animal is captured, an adequate cage must be ready to receive it, or one must be built immediately. It is cruel to confine an animal in a small space where it cannot be properly cared for. When building a cage for a mammal or bird, remember that this creature has been used to unlimited freedom of movement; therefore the larger the cage the better. Of course, no matter how roomy it is, it still confines the animal to some extent. Strangely enough, however, captive animals often get used to their cages after a short period of time. At the Bear Mountain Trailside Museums, animals actually look to their cages for protection. Many times, when one of these animals has been given its freedom, it has returned to its cage and, if the door has been left open, has entered, apparently feeling that it was at home again.

Cages for squirrels, raccoons and other climbing animals should be at least 6 feet high, and no less than 4 feet square. Ground-dwelling animals, such as the woodchuck, skunk and fox, do not need a high cage, but do require more room for movement on the ground. The minimum size of the bottom of the cage for woodchucks and skunks is 4 feet by 6 feet; for more active animals, such as foxes, it should be at least 4 feet by 10 feet. Every part of the floor must be within easy reach for cleaning. For the framework, 2-inch by 4-inch wood construction is best; but 2-inch by 2-inch pieces, if properly braced, will prove quite satisfactory.

The floor can be made of wood planking; however, concrete or heavily painted sheet iron is best for good cleaning. A cage set directly on the ground is unsatisfactory, unless it can be moved from one spot to another without too much effort. The reason for this is that the ground becomes fouled quickly and the cage bottom cannot be thoroughly cleaned. Another disadvantage is that unless adequate underground guards are provided, the animal will probably be able to dig its way out with ease.

Wire should be attached to the framework on all four sides. It should be nailed on from the inside; otherwise the animals, whatever their species, will be very likely to chew the framework. The size of the wire you will use will depend on the kind of animal you want to keep. For an all-purpose cage, 1-inch by 1-inch welded wire is best. It is advisable to roof over a tall cage entirely; a pitch toward the back of about one inch to the foot will be enough to carry off the rain. In the case of long, low cages, only one third or one half should be roofed. For the door, use a heavy, well-braced wooden frame, covered with wire cloth of the same size used on the cage.

Every cage should be provided with a shelter box into which the animal can retire when sleeping or during bad weather. This box should furnish adequate space for the animal; but the opening should be just large enough so that it can enter. The object is to protect the animal from drafts; this is particularly important in the case of tiny babies.

Most animals require a certain amount of sunlight if they are to thrive; the cage should be placed accordingly. In areas where temperatures soar, select a cool, well-shaded situation where the cage will receive a certain amount of morning or late-afternoon sun. In more temperate regions, locations more exposed to the sun are permissible; however, the cage should never be placed in a location where the heat will become intolerable to the animal. When the temperature becomes high, hose down the area around the cage, as well as the cage itself. Always bear in mind that some sunlight is essential but that too much can be fatal.

Some animal fanciers prefer to keep an inch or two of soft wood shavings or other absorbent material on the bottom of cages. This practice has certain advantages as far as cleaning is concerned. The great disadvantage is that the shavings often become mixed with food and water and may cause your animals to suffer from indigestion. The best way to keep a large cage clean is to put nothing on the floor and to hose it out at regular intervals; in good weather, the cages dry rapidly. It is often advisable to apply some mild disinfectant to the bottom of the cage before using the hose.

Simple drop trap for small animals. The stick is pulled out by the trapper when the animal is under the trap.

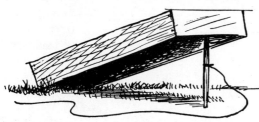

For mammals of chipmunk size or smaller, you can buy excellent cages at a pet shop; these are of all-metal construction for the most part. If you are handy with tools, it is a simple matter to build small cages to meet any requirements.

Collecting and feeding small mammals

Before you can keep a wild animal pet, you must first catch it. Although you can buy excellent live traps, it is part of the fun of this hobby to build your own traps. Many of them consist of a small cage with a door, or doors, that will close when the animal either steps on a treadle or tugs on the bait. There are various ways in which the tripping mechanism can be constructed; one of them is illustrated on this page. The traps should be built in such a way that the animal will not be injured either by the springing of the trap or by its frightened efforts to escape. Once it has been trapped, the best way to get an animal to its cage is to carry it home in the trap. Set the trap in the cage and open it, leaving the animal free to get out into its larger

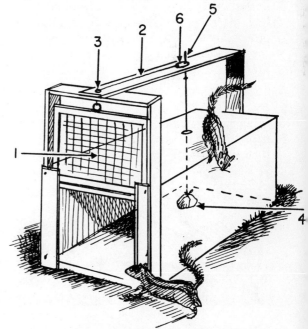

Bait trap for mammals. Door (1) is held open by cord (2), passing through hole in frame (3), and by ring (6), held in place by pin (5). When animal pulls bait (4), pin slips out of ring, releasing door, which then falls shut.

quarters. If you cannot put the trap in the cage, hold it up to the open door of the cage and let the animal out.

There is another kind of trap, in which no tripping mechanism is used. It consists of a large open receptacle, such as a garbage can. Set bait at the bottom of the can and tilt the can at an angle which will lead an animal to believe that it can get out. To make it easier for the animal to get at the bait, a plank is set against the can so that it forms a ramp from the ground to the edge of the receptacle. It is easy enough for the animal to enter the can, but impossible for it to get out, for the inside has a light coating of Vaseline or machine oil. A large-mouthed gallon jar set into the ground so that the top is level with the surrounding earth can be used in the same way to trap small mammals, such as mice.

Those who find it difficult to do live trapping can always enlist the services of a farmer's boy. He is generally wise in the ways of nature, and for a small fee he will keep you supplied with an excellent variety of animals for your home zoo.

All but very young mammals should be fed once a day. For nocturnal animals, late afternoon is the best time; for the others almost any time will do. Feed at the same time each day; it will be interesting to note how impatient the animals become when feeding time is near. Young

animals should be fed several times a day. Babies that have not been weaned require frequent meals of warm milk, fresh or canned; they must be fed slowly with a small nursing bottle or a medicine dropper. It is very difficult to raise sucklings; it is best not to try to do so unless the babies have been abandoned or are lost.

You can learn only by experience how much food to give an animal at each feeding. It should always be a little on the hungry side after feeding time; this will keep it alert and active. Your pets should never look fat and lazy. This is a sure sign of overfeeding, which may result in ill health. Cod-liver oil should be mixed with the food two or three times a week, and fresh water should always be present.

Some mammals that can be kept as pets

Opossums are among the easiest animals to live-trap. In fact an opossum will often trap itself by getting into a garbage can from which it cannot withdraw. Food is no problem, as table scraps (cooked vegetables and meat), supplemented with prepared dog food and occasional morsels of raw meat, will do nicely.

Moles and shrews are among the hardest animals to keep in captivity. Moles are difficult to capture with live traps; but quick work with a spade on a lawn that they frequent usually brings good results. You will have to have patience and a knowledge of their habits if you are to succeed in trapping them. Shrews can usually be

A simple but effective trap. Bait is set at the bottom of the can, which is tilted at an angle.

You can feed baby animals with a small nursing bottle.

caught in small live traps by using meat or peanut butter as bait. A fair-sized aquarium with from 4 to 6 inches of soft, damp (not moist) earth is satisfactory for moles; shrews will do best in a terrarium. Both moles and shrews require almost constant feeding; they will relish insects, worms and raw meat.

Bats should not be kept in captivity for more than a day or two because you will find it very hard to give them enough room for flight. When they do feed in captivity, which is all too seldom, it is usually on some form of insect life. Some species, however, will take bananas, cottage cheese or even chopped hard-boiled egg in milk.

Raccoons can easily be live-trapped by using meat or fresh fish for bait. If adults are caught, they should immediately be released as they seldom make good pets. Young ones are inclined to be a bit quarrelsome at first, but with careful handling they will become very tame. Patience and a pair of heavy leather gloves are necessary. These animals usually accept table scraps, supplemented with dog food, raw meat and various fruits. Of course, there should always be a large pan of fresh water in the cage.

Skunks are among the easiest of animals to keep in captivity; they make ideal pets if captured when young. It is not absolutely necessary but often advisable to have captive skunks descented by a veterinarian. Table scraps, meat and prepared dog food are acceptable to skunks. These foods should not be given in large quantities, as the animals tend to become very fat and sluggish when they are overfed.

Foxes should not be kept in captivity unless you have ample acreage. They are inclined to be smelly, no matter how well their cages are disinfected and cleaned; the aroma of fox wafted by the breezes may be quite offensive to a neighbor living nearby. The pups are usually very tractable, but as they grow older they become quite snappy; as a result, adults make poor pets. It is best to release them when they are grown up. Dog food and raw meat with occasional table scraps usually suffice for these animals.

Woodchucks make good pets only when they are young. Adults are savage and can cause serious damage to cage and captor alike with their long chisellike incisor teeth. Young that have wandered too far from their home burrow may often be run down and caught with a long-handled net. All sorts of vegetables (carrots are a must), apples and many wild, broad-leaved plants form an adequate diet. As the woodchucks are true hibernators, they should not be left out of doors over the winter.

Much that we just said about woodchucks applies to ground squirrels and chipmunks. Adults should be released if trapped; only the young should be kept as pets. Fruits, nuts and various grains are good foods for them. They are easily trapped in small live traps baited with peanut butter or the foods they ordinarily eat.

Gray squirrel.

Woodchuck.

Squirrels tend to remain wild if captured as adults, with the exception of the flying squirrel. Young that have just left the nest are often easy to catch and make excellent pets. As squirrels are very active they must have large, roomy cages; an old tree with several branches should be set into the floor. Nuts, assorted fruits and vegetables and occasional treats of insects and meat are a fine diet for squirrels.

Mice and rats of many species can be kept in captivity and most of them will do well. Small live traps, baited with peanut butter, bacon, nuts or fruit will usually catch them. The sunken gallon jar trap, already mentioned, will catch some of the smaller kinds. Dried grasses and strips of cloth should be placed in the cage so that a nest can be built. Don't forget that mice and rats (particularly rats) can bite; handle them with care until you are certain they can be trusted. Carrots, lettuce, fruit, nuts, grain, prepared pellets and meat (occasionally) should be fed to them.

Porcupines are easy to capture and often come into the hands of collectors. If a porcupine refuses all food after several days in captivity, it is best to release the animal, if possible somewhere near the place where it was captured. Sometimes if a porcupine is tempted with a large variety of foods, the animal may select one of them. As the animal adjusts, it may eat other foods. Porcupines often eat vegetables, fruits, dry bread and the tips of hemlock, elm and aspen branches.

Wild rabbits usually do not make good pets. Those that have been captured when young are apt to be shy and nervous even when they reach maturity; any undue commotion may literally frighten them to death. If you want to keep rabbits in captivity, there are many domestic varieties that are extremely hardy and make excellent pets. A variety of vegetables and prepared rabbit pellets are the best foods.

Armadillos are most interesting pets; they usually can be caught quite easily. Although they are mostly nocturnal, they occasionally move about during the daytime. Soft fruits, raw chopped meat and cooked eggs are excellent food for these animals.

Fine pets. Above: raccoon; center: lively chipmunk group; below: crow.

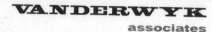
Few wild birds may
be kept without a permit

Because of federal and local restrictions, only a very few kinds of birds may be kept in captivity without a permit. This is generally difficult to obtain.

Wild geese and ducks are carefully protected by law; you should not make any plans for keeping these birds without first getting in touch with the wildlife authorities. There are, however, a number of domesticated species that may be kept without a permit. Because a fairly large investment is required, you should be sure that you know how to keep these birds properly before you purchase any. You will find ample information on this subject in books such as Reginald Appleyard's DUCKS: BREEDING, REARING, MANAGEMENT (Poultry World, London, 4th edition) and GEESE; BREEDING, REARING, GENERAL MANAGEMENT (Poultry World, London, 3rd edition), and F. J. S. Chatterton's DUCKS AND GEESE AND HOW TO KEEP THEM (Cassell & Co., London, 8th edition).

Hawks and owls are sometimes kept as pets if local conservation regulations permit. The young of certain species, particularly red-shouldered, red-tailed, broad-winged and sparrow hawks, and barred, great horned and screech owls, are most likely to adjust themselves to living in a cage. Adult owls may be caged with fairly good results; it is much more difficult to keep hawks in captivity. Adult-hawk-trapping should be left to the professional. Young owls and hawks are best fed freshly killed mice, but, as they get older, chunks of horsemeat (and an occasional mouse) are usually adequate. An excellent source of natural food may be found in animals freshly killed by automobiles.

Crows, jays and magpies are related, and their cage and food requirements are the same. They need a large cage with an inch or two of sand in the bottom and a number of perches. Only young birds should be caged, for adults caught wild seldom become adjusted to a cage. As the young grow to maturity they make ideal pets. The birds will do very nicely on a diet of shelled nuts, mixed grain, sunflower seed, chopped meat, chopped carrot, lettuce and assorted fruits.

Quails and pheasants are considered game birds, and they are protected by law. A permit for keeping and raising them may usually be obtained without too much trouble from your local conservation department. Books such as Jean T. Delacour's PHEASANT BREEDING AND CARE (All-Pets Magazine, Fond du Lac, Wis., 1959) and David B. Greenberg's RAISING CAGE BIRDS IN CAPTIVITY (D. Van Nostrand Co., New York, 1949) will supply much useful material on raising these birds.

Song birds cannot be legally caged without a permit. However, you can attract them to your property by means of various kinds of feeding stations. (See Index, under Feeding devices.)

Turtles and other reptile pets

Turtles are among the easiest animals to keep in captivity. They should be kept out of doors only during the warmer months of the year. Indoors or outdoors they should never be constantly in sunlight. A large aquarium, with from 5 to 6 inches of water and a few flat rocks rising above the surface, is ideal for most varieties (but not for box, wood and gopher turtles). Generally speaking, the turtles that will thrive in the aquarium eat only food that has been dropped into the water; some will take food from your hands but will then have to get the food wet before eating it. Pieces of fish, raw meat, aquatic plants, earthworms and insects form about the best food. As the water in the aquarium will become fouled quickly, it should be changed daily.

Box, wood and gopher turtles are essentially terrestrial and, therefore, require cages with only a small pool of water. In captivity they do well on melons, bananas and soft fruits, and will also occasionally eat earthworms. As the gopher tortoise is a powerful digger, be sure that the floor of the cage is well constructed.

Alligators are protected by law in some areas, but they may still be collected in

Two baby turtles, kept in a bowl.

Aquatic species are best collected by means of a seine or hand net.

others. They may also be obtained in pet shops. Only small specimens up to a foot in length should be kept by an amateur. Larger ones can be extremely dangerous. They are best collected by means of a hand net or noose on a long pole and should be handled by holding behind the head. Their food and cage requirements are about the same as for the aquatic turtles, except that they do not feed on vegetation of any kind.

Lizards generally get along best in a desert terrarium. They should be exposed to sunlight or to the rays of a sun lamp for a short time each day. In keeping lizards such as the chameleon, fence lizard and skinks, try to reproduce as closely as possible the habitat where they were caught. These little animals should be fed on meal worms and other forms of insect life. Water should be provided by spraying down the cage. The little droplets that will collect on the rocks and plants will furnish the necessary water for these animals.

Snakes may be purchased from animal dealers, but it is more fun to hunt them. One word of caution to the amateur — be sure you know the poisonous snakes of your region before you go snake collecting. A good knowledge of the habits and habitat of each species, such as you can acquire from the FIELD BOOK OF SNAKES OF THE UNITED STATES AND CANADA (G. P. Putnam's Sons, New York, 1941) by Karl P. Schmidt and D. D. Davis or the FIELD BOOK OF NORTH AMERICAN SNAKES (Doubleday & Co., New York, 1939) by Raymond L. Ditmars, usually makes for more successful collecting. You will be most likely to find snakes in or under rocks, old boards and logs, in sheltered brush piles, along the edges of streams, under the loose bark of trees and along old stone walls and foundations. A broom handle with a 6-inch angle iron screwed to one end is a good thing to use for pinning down snakes, as well as for turning over old boards and logs. Caution: if you are in poisonous-snake country, never put your fingers under rocks or logs or similar hiding places of snakes.

When pinning down a snake, it is always best to pin down the top of the head and then to pick up the animal just behind the head. There is a widespread belief that the best way to catch a snake is to pin it down just *behind* the head. The trouble is that if you hold down the snake in this way, it may be able to move its head just enough to bite you when you try to pick it up. Use heavy leather gloves in handling the larger nonpoisonous species. There is nothing better than a cloth bag or sack for transporting snakes. *Do not try to collect poisonous snakes; leave them for the professionals.*

Most species of snakes are not too difficult to keep in captivity. A solid cage with a glass front and a screen top is easily constructed. The cage should open from the top for cleaning and feeding; the size will depend on the size of the snakes that are to be kept. Fine gravel or coarse sand and a few large rough stones are to the liking of most of these reptiles. Place a broad-bottom water dish that cannot be easily tipped over in the cage; provide

To pin down snakes, you can use a broom-stick with an angle iron screwed to it.

A cage with a glass front and a screen top for snakes.

branches or small logs for the species that like to climb.

Highly nervous species, such as the racers and whip snakes, often do not do well in captivity. Watch these animals closely; if they do not quiet down and accept food, it is best to release them before they fret themselves to death.

A good feeding of live food once a week is all that a snake requires. The animal should be allowed to take all it wants. A good indication that it is not getting enough food is the formation of slight folds of loose skin toward the tail. When feeding or cleaning the cages of aggressive snakes, wear a leather glove. This is adequate protection from the strike of our native nonpoisonous species.

Ring-necked snakes like a diet of earthworms, but will take small toads, salamanders and various soft-bodied insects.

Hog-nosed snakes are difficult to keep for any length of time unless a quantity of toads can be collected and kept on hand for food.

Green snakes are insect eaters exclusively. It is useless to try to keep them unless a good supply of insect life is readily available.

Racers and whip snakes, as we have pointed out, are hard to keep because of their nervous nature. Young ones that will accept frogs, mice and insects sometimes do quite well, however.

Indigo snakes are excellent for handling because of their gentle disposition. They will readily accept frogs, fish, mice and other foods.

Rat, fox, chicken, bull, gopher, corn and pilot black snakes, which are among the largest of our native species, can also be handled very easily. Their favorite foods are live mice and rats.

King snakes feed on snakes in a state of nature; many, however, will accept mice in captivity.

Milk snakes are popularly supposed to enter barns where cows are kept in order to obtain milk from them! This is an old wives' tale; they actually are searching for the mice found in such places. To keep them happy in captivity, give them plenty of mice.

Water snakes are aggressive and very smelly creatures and, therefore, are not ideal pets. They are easily kept in captivity, however, on a diet of fresh fish (either whole or in pieces) and frogs.

Brown and red-bellied snakes are small, inoffensive creatures; they can be fed on earthworms and soft-bodied insects.

Garter and ribbon snakes are easily kept on a diet of frogs and toads; small individuals may prefer earthworms and soft-bodied insects. Ribbon snakes seem to have a special fondness for fresh fish, either whole or cut into strips.

Salamanders, frogs and other amphibians

Salamanders may be divided into three general groups: those that are aquatic, those that occasionally frequent water (some only during the breeding and egg-laying season) and those that are terrestrial. In order to collect and keep sala-

Above: frogs can be kept in an aquarium. Right: certain animals can be collected at night as they cross unfrequented roads.

manders, their habits should be carefully studied.

Aquatic species are best collected with a seine or hand net. Use a hand net to catch salamanders found in or under rotted logs, old boards and flat rocks. If you use a car for collecting, cruise slowly along unfrequented roads on warm, rainy nights; you will sometimes see hundreds of these animals crossing the road. In spring, certain of the larger species congregate in large numbers in small pools and ponds to mate and lay their eggs. This is an excellent time to collect the species that are usually hard to find.

Use an aquarium for the aquatic species of salamanders and a terrarium with a pool of water for the others. Remember that these creatures like cool, moist conditions; extreme dryness or heat will kill them quickly. Change the water in the aquarium or terrarium when any signs of fouling appear. Since many salamanders are able to climb the vertical glass sides of a terrarium, a piece of glass should be used as a cover. The glass should not rest tightly on the top, but should be raised to a distance of about one sixteenth of an inch to allow for some ventilation.

Hellbenders, mud puppies, spotted newts and axolotls can be kept in an aquarium with coarse gravel or flat rocks in the bottom. The aquarium should be filled to within an inch or two of the top and should be kept as cool as possible. A glass or wire top is necessary, as these animals may otherwise escape. The food should consist of tadpoles, earthworms and small fish.

For terrestrial species and for red efts (immature spotted newts), a cool terrarium with a pool of water is most satisfactory. Earthworms, meal worms and other soft-bodied insects are excellent foods.

Frogs and toads are found in many different places. Some frogs are most easily collected by hand or in a hand net along the borders of lakes, ponds and streams; others, in wet meadows; still others, in wooded areas. Toads, for the most part, favor open areas, though they are occasionally found elsewhere. All native frogs and toads come to water, whether it be a puddle in a wagon rut or a lake, in order to mate and lay their eggs. By collecting the eggs, you can learn a great deal about the life history of these animals. When the eggs hatch, keep only a dozen or so of the tadpoles; release the rest of them in some nearby pond. A simple aquarium with plenty of aquatic plants will suffice for the eggs as well as for the tadpoles. The tadpoles will feed on bits of chopped liver, algae and crushed worms.

Spring peepers, toads, tree frogs and wood frogs can best be kept in a terrarium with a small pool of water. The other varieties do well in an aquarium with several inches of water. Several flat rocks should be so placed that the frogs can emerge from the water occasionally. All sorts of insect life and earthworms are excellent foods for these animals. The water in the aquarium should be replaced daily.

SCIENCE AND PROGRESS

(1815-95) V

BY JUSTUS SCHIFFERES

NEW VOYAGES OF DISCOVERY

THE daring sea voyages of the fifteenth and sixteenth centuries were undertaken primarily to discover new lands and found new empires. With certain notable exceptions, the overseas and overland expeditions of the seventeenth and eighteenth centuries were sent out to establish colonies or to convert native peoples or to find new sources of raw materials. In the nineteenth century, a number of exploring expeditions were launched, after careful preparation, with the express aim of adding to man's knowledge of the globe he inhabits. Vast areas of North and South America and Africa, to say nothing of the

The man who touched off this great movement of exploration was America's famed scientist-president Thomas Jefferson. On April 30, 1803, the United States had acquired by purchase from the Emperor Napoleon a vast region called Louisiana, extending from the Mississippi to the Rocky Mountains and from Mexico to the Lake of the Woods. President Jefferson decided to send out an expedition to explore this vast new addition to the fledgling republic.

As leader of the expedition, he selected his one-time private secretary, Captain Meriwether Lewis (1774–1809), who

Sacajawea, an Indian squaw, guiding the Lewis and Clark Expedition.

sea bottom and the Arctic and Antarctic regions, were revealed to mankind. These expeditions, undertaken without thought of immediate gain, often opened up unexpected natural resources, which later were fully exploited.

During the nineteenth century, the continent of North America was more thoroughly explored than any other; by the end of the nineteenth century, its vast expanses had been quite accurately mapped.

hailed from Albemarle County, Virginia. Lewis chose as his second-in-command Lieutenant William Clark (1770–1838), also a native Virginian but then residing in Kentucky. These two men organized the Lewis and Clark Expedition whose purpose was, in Jefferson's words, "to trace the Missouri to its source, to cross the highlands [that is, the Rocky Mountains] and follow the best water communication from thence to the Pacific Coast."

The trip down the Ohio, up the Mississippi and then up the Missouri to its Great Falls was accomplished by means of river boats and canoes with much labor but little danger. In April 1805, Lewis and Clark sent sixteen of their men back down the Missouri by keelboat with specimens, gifts for the President and careful accounts of what they had already seen and what they had learned from the Indians.

The party that went on included, in addition to its leaders, twenty-six soldiers and rivermen; Clark's Negro servant York, an amazing sight to the Indians, who had never seen a black man before; two French hunters and the Indian wife and half-breed papoose of one of them. The Indian squaw, Sacajawea, a remarkably intelligent woman who acted as guide and interpreter, was as much responsible as anyone for the success of the expedition. She was familiar with the territory between the Missouri and Columbia rivers, for she had been a member of a Shoshone tribe that had dwelt in this area.

The expedition passed the Three Forks of the Missouri, above Great Falls; then the explorers obtained horses from the Indians and crossed over the Great Divide. At last they came to the Columbia River, teeming with salmon. They sailed down the river and in November 1805 reached the Pacific Ocean where the Columbia empties its waters. The return journey was made by the same route the next year. The information that Lewis and Clark brought back about the Great Northwest encouraged fur trading and then settling in this region. Within a single generation the Oregon Trail rivaled the Santa Fe Trail as a road to the West.

Baron Alexander von Humboldt, famous German naturalist, traveler and statesman, exploring the Orinoco River in South America.

Lewis and Clark were not the first white men to reach the Pacific by the land route; they had been anticipated by a young fur trader, Alexander Mackenzie (1755?–1820), in the employ of the Northwest Fur Company. This venturesome young man had journeyed by canoe in 1789 from Fort Chippewyan along the Great Slave Lake and down the river that now bears his name to the Arctic Ocean. In 1792–93, Mackenzie made an overland journey from Fort Chippewyan to the Pacific coast near Cape Menzies. At the top of a large rock bluff overlooking Vancouver Bay, Mackenzie wrote in red grease paint an inscription, since chiseled in the rock, which summed up his achievement:

"Alexander Mackenzie, from Canada, by land, the twenty-second of July, one thousand seven hundred and ninety-three."

The interior of North America was explored by fur traders like the famous Missouri Legion, by Indian scouts like Kit Carson and by religious exiles like the Mormons; it was mapped and surveyed by railroaders and geologists. A journey "unequaled in the annals of geographical exploration for the courage and daring displayed in its execution" was made in the summer of 1869 by an American geologist, John Wesley Powell (1834–1902), who later became the director of the United States Geological Survey. He termed his magnificent adventure a "modest boat trip down the Colorado [River]"; as a matter of fact, it was a hazardous feat that has seldom been repeated. Powell explored the cavernous depths of the Grand Canyon. He reported that two great natural agencies had sculptured the face of the valley of the Colorado and the Southwest in general; these agencies were the upheaval of land masses and the erosion of their surfaces. Control of erosion, particularly of topsoil, remains a vital problem today.

There were also notable explorations of South America in the nineteenth century. No man did more to reveal the mysteries of the continent than the German naturalist, traveler and statesman Baron Alexander von Humboldt (1769–1859).

This versatile man, a Prussian by birth, was educated at Goettingen, Freiberg, Frankfurt and Berlin. His early travels took him to Belgium, Holland, England and France. They served as a fitting prologue to his notable five-year trip of exploration to South America.

In the year 1799 he set out on this famous trip aboard the Spanish frigate Pizzaro, with the French botanist Aimé-Jacques-Alexandre Bonpland as his companion. In the course of the five years that followed, Humboldt thoroughly explored South America: its rivers, its mountains, its plants, its skies, its people. The luxuriant vegetation fascinated him; the strange animals and sea creatures fired his imagination. "Even the crabs are sky-blue and gold!" he exclaimed. He followed the course of the Orinoco through tropical jungles, and he established its connection with the upper Amazon. He crossed the Andes four times, exploring the interior of Ecuador and Peru. At length he returned to Europe by way of Mexico and the United States. He was welcomed back to Europe with joy and enthusiasm. He had been gone so long that his friends had given him up for dead.

Upon his return Humboldt settled in Paris and began a notable career as a writer and lecturer on scientific subjects. He soon became one of the most famous men in Europe. He penned a narrative in French of his journey up the Orinoco under the title of VOYAGE TO THE EQUINOC-TIAL REGIONS OF THE NEW CONTINENT; he also wrote a series of technical books on climate, temperature and plants, mostly in French or Latin. In 1827, he reluctantly returned to Germany at the bidding of his royal master, the King of Prussia. Two years later, at the age of sixty, he launched upon his last great journey — a nine-thousand-mile jaunt through the Ural and Altai mountains of Asiatic Russia.

Humboldt spent the last years of his life in Berlin, receiving visitors, exchanging correspondence with the great of the world and writing his most ambitious work, called THE COSMOS. This work, in five volumes, was published from 1845 to 1862 (the last volume appeared after his death). It sought to provide an accurate and complete description and a unified conception of the entire physical world. Understandably it has become outmoded in many respects; yet it remains a superb monument to a great man.

Humboldt's contributions to science were many and lasting. In the course of his travels he accumulated a great mass of geographical data. He discovered the cool current that sweeps up the west coast of South America from the Antarctic Ocean and that now bears his name. He climbed mountains to prove that temperature decreases with height above sea level. He pointed out that the very nature of plants and animals is controlled by the climate and terrain they inhabit. He studied the origin of tropical storms and the changes in the

The good ship *Beagle* in the Strait of Magellan.

earth's magnetic force from the equator to the poles; he also devoted much attention to "magnetic storms," a name that he himself coined. Meteorology became a much more exact science in his hands. He made an intensive study of volcanoes; he advanced the theory, still held, that zones of volcanic activity follow a definite line of cracks in the earth's crust. To quote Goethe, Humboldt was "without a rival in extent of information and acquaintance with existing sciences."

Humboldt's writings provoked a renewed interest in scientific exploration. Inspired by his example, many naturalists now betook themselves to the vast plains and tall mountains of South America. For example, a British expedition, sent out in 1831 in the small ship Beagle, of the British Navy, did important work in surveying various areas of the continent: Patagonia, Tierra del Fuego, Chile and Peru. The Beagle also explored many islands of the Pacific and Atlantic. Young Charles Darwin, who, as we shall see, was destined to stir up the greatest scientific controversy of the nineteenth century, traveled aboard the Beagle as "naturalist."

After the close of the Napoleonic Wars, Britain turned her attention to the exploration of Africa, called the Dark Continent because comparatively little was known about it. British explorers discovered Lake Chad and the mouth of the Niger, and explored the regions between Lake Chad and Timbuctoo. Protestant missions were organized on the Guinea coast in South Africa and in the dominions of Zanzibar. The missionary David Livingstone (1813–73) crossed the Kalahari Desert from north to south in 1849. In the years that followed he traversed the African continent from the west to the east; in the course of these wanderings he discovered Victoria Falls (1855).

In 1866 he set out on his last journey of discovery. After a time the world lost all trace of him. James Gordon Bennett, the fabulous managing editor of the New York HERALD, interested himself in the matter; he sent out the British-born reporter Henry Morton Stanley (1841–

In the Galapagos Islands, Charles Darwin discovered these lizard species: left, *Tropidurus pectinatus*; right, *Tropidurus multimaculatus*.

1904), born John Rowlands, to find the famous missionary-explorer. Stanley came upon Livingstone in Ujiji in 1871 and uttered the memorable salutation, "Dr. Livingstone, I presume." Livingstone continued his explorations in Stanley's company for a time; he went on alone after the newspaperman's departure. At last, worn out by disease and fatigue, he succumbed in 1873. In the following year, Stanley explored the Congo River.

Other Englishmen made notable discoveries in Africa. In 1862 John Hanning Speke (1827–64) proceeded along the river that flows from Victoria Nyanza (Lake Victoria) and, following it into Egypt, answered once and for all the age-old question, "Where does the Nile obtain its waters?" Another explorer of the mighty Egyptian river was Samuel Baker (1821–93), who discovered Albert Nyanza, the principal western reservoir of the river. Portuguese, Germans, French and Italians also made notable explorations in Africa. By the end of the century much of the mystery of the Dark Continent had been dissipated.

In the nineteenth century, the British made many surveys of their vast possessions in India. One of these surveying

STANLEY AND LIVINGSTONE

expeditions resulted in the most important development in geology since the time of Hutton. This was the theory of isostasy. It resulted from a survey undertaken in 1823 by Sir George Everest (1790–1866), the man for whom the tallest peak in the world — Mount Everest — is named.

Like all surveyors, Everest used a plumb line to level his instruments. According to the theory of gravitation, the plumb line, wherever it is used, should point directly to the center of the earth. It was soon observed, however, that the direction of the line in northern India was different from its direction in southern India. This deflection was attributed to the gravitational attraction of the nearby mass of the Himalaya Mountains. Joseph Pratt (1811–71), a British clergyman and mathematician who spent many years in India as Archdeacon of Calcutta, set out to examine the deviation of the plumb line. He found that the line was attracted far less by the overwhelming mass of the Himalayas than had been expected.

The explanation of this phenomenon was offered by George Biddell Airy (1801–92), English Astronomer Royal and President of the Royal Society. He held that the Himalayas are made up of comparatively light rock, while the plain of the Ganges, to the south, has a denser rock structure. Hence the deviation of the plumb line in the direction of the Himalayas is not so great as it would be if the mountains and the plain had the same density. Airy also advanced the idea that the tablelands and mountains of the earth rest upon a mass of lava. "It appears to me," he said, "that the state of the earth's crust lying upon the lava may be compared with perfect correctness to the state of a raft of timber floating upon water."

According to Airy, the mountains of the Himalayan range and the plains of the Ganges, as well as heights and depressions all over the world, press down upon the underlying sea of lava. They are in balance because the lower-lying lands are heavier than the peaks and mountain ranges. This is the theory of isostasy, or general equilibrium in the earth's crust. The name "isostasy" was first applied in 1889 by the American geologist and army officer Clarence Edward Dutton.

The British took the lead in exploring the depths of the ocean with the famous Challenger Expedition. The Challenger, a wooden corvette of 2,306 tons, was fitted out by the British Navy, "at a cost no more than that of keeping the vessel in commission," for the purpose of exploring the open ocean in every quarter of the globe. Much of the credit for inspiring the voyage of the Challenger belongs to the scientific director of the expedition, Professor Charles Wyville Thomson (1830–82) of the University of Edinburgh.

The Challenger, Captain G. S. Nares commanding, left Portsmouth jetty on December 21, 1872, with a full complement of scientists on board. The route was south to Madeira, thence to the Canaries, the West Indies and north as far as Nova Scotia; south again to Cape Verde, to the Cape of Good Hope, Australia, the China Sea, Japan, along the west coast of South America, through the Strait of Magellan and back to England. When the ship finally dropped anchor at Spithead, England, on May 24, 1876, it had logged 68,890 nautical miles.

The scientists aboard the ship — "philosophers," the bored naval officers called them — were particularly concerned with what they might find at the bottom of the sea. As a result, they laid the foundations of the modern science of oceanography, which seeks to solve the riddles of the deep. The birthday of this science, according to J. Y. Buchanan, one of the Challenger's scientists, was February 15, 1873, when, forty miles south of Tenerife, the ship took the first of its 362 official "stations." A dredging bucket, suspended on ropes, was lowered into 1,890 fathoms of water to see what could be dragged up from the bottom of the ocean. The temperature of the water at different depths was also recorded. The mud from the bottom of the sea proved cold enough to chill a bottle of champagne.

The Challenger Expedition was fruit-

The Challenger Expedition of 1872–76 was the first to collect the following species of deep-sea fish (top to bottom): *Ceratias carunculatus*, *Ceratias uranoscopus*, *Melanocetus murrayi*, *Bathypterois longicauda*.
From "Challenger Reports"

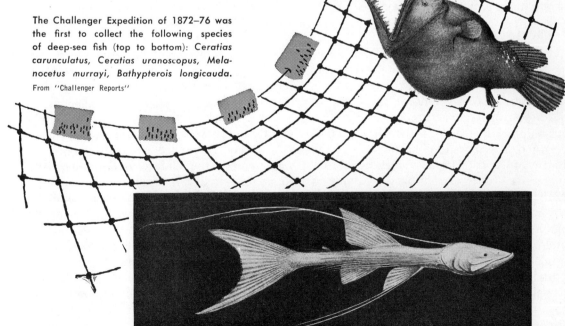

ful indeed. Through numerous analyses of samples, it established the chemical composition of sea water throughout the globe. It made the first comprehensive analysis of the sea bottom. It also exploded a number of myths. It disproved the belief that bathybius, a slimy substance found in the sea, represented vast masses of floating protoplasm. Buchanan showed that this so-called protoplasm is simply a precipitate of calcium sulfate in sea water. The Challenger's scientists also proved that Huxley and others were wrong in supposing that chalk was constantly being laid down on the ocean floor by the shells of tiny sea creatures. Finally, they showed that the "lost continent of Atlantis" — Plato to the contrary — was just not there.

In the year 1880, Thomson issued the first of fifty volumes of Challenger RePORTS, containing an account of the expedition and its findings. Upon his death, in 1882, his brilliant young assistant, Canadian-born John Murray (1841–1914), carried on the publication of the REPORTS, completed in 1895. These reports are still "the indispensable starting points for the study of marine biology." Since the oceans cover so large a proportion of the earth's surface, it is truly amazing that the serious study of its composition and its inhabitants should have been so long delayed.

Explorers of the
Arctic and Antarctic

The barren wastes of the Arctic and Antarctic also attracted explorers in the nineteenth century. One of the foremost of these sturdy adventurers was Sir John Franklin (1786–1847), a British naval officer who had served at Trafalgar and the Battle of New Orleans. He headed four successive Arctic expeditions, beginning in 1818; the last one, which set out in 1845, never returned. Between 1850 and 1857 his heartbroken wife Jane (nee Griffin) fitted out five ships to search for her husband. At last indisputable evidence was found that Sir John's ship had been deserted and that he and the rest of his party had been lost. Jane Griffin Franklin was honored for her services to geographical

science by receiving, in 1860, the gold medal of the Royal Geographical Society.

In the last decades of the nineteenth century it became a point of honor among nations to be the first to reach the North and South Poles. England, Norway, Sweden, Austria and the United States all entered the race; many different routes to each pole were explored.

Traveling first by ship and then by dog sled through bitter icy wastes, explorers edged closer to the North Pole, representing the northernmost point of the earth — 90 degrees north latitude. By the end of the century Fridtjof Nansen of Norway and Captain Umberto Cagni, of the Italian Duke of Abruzzi's expedition, had gone beyond the 86th parallel north. It was not until April 6, 1909, that the Pole was finally conquered by the determined American naval officer Robert E. Peary, who had made five previous attempts to reach the goal. The Norwegian Roald Amundsen was the first to reach the South Pole (December, 1911). One month later (January 18, 1912) valiant Captain Robert F. Scott of the British Navy also attained the goal; on the return trip he and his companions lost their lives.

In time the importance of these continuing expeditions became apparent. The northern polar route for air traffic between Europe and America has the most startling possibilities in both peace and war. The weather stations that have been set up in the far north have made weather forecasts more accurate than before. As for the Antarctic regions, they hold immense mineral treasures that some day will add greatly to the world's resources.

Today there are comparatively few regions upon the surface of the globe still to be explored. But science has set up new goals for modern explorers. Armed with microscopes, telescopes, cyclotrons, Geiger-Mueller counters, they seek to ferret out the secrets of the physical universe — the mysteries of the life that fills the earth, of the atoms, of the visible heavenly bodies, of the infinite space that is as yet beyond our ken.

Continued on page 156.

PROGRESS IN CHEMURGY

New Riches down on the Farm

BY WHEELER McMILLEN

A FIELD of ripening oats in July makes a pretty picture. The tall, slender stalks bow to the slightest breeze; a few long leaves reach out from the stalks. The sheathed grains clustering at the tips shine like green and gold bells.

When man has extracted the kernel for food and gathered the straw to bed down his livestock, has he made full use of the oat plant? Not very long ago the answer to this question would have been "Yes." Now another story can be told.

A manufacturer of breakfast foods looked out one day upon the mountain of oat hulls that was rising near his factory. A small part could be ground up for livestock feed; but the market was small. The rest of the hulls were fit only to be tossed into the boiler fires; that is, unless someone found a new use for them.

A chemist did. He found that oat hulls yield a chemical compound called furfural. At first there was no known use for furfural, but laboratory workers kept experimenting with it until their curiosity began to pay dividends. They discovered that in the refinement of lubricating oils furfural would dissolve out the substances that made these oils gummy. Soon the petroleum industry was buying entire tank carloads of furfural.

Then some one found synthetic resin adhesives are even more adhesive when furfural is added. Modern grinding tools and abrasives are made by cementing millions of tiny particles of great hardness. They are made to stick together by adhesives that are tough and resistant to heat and water.

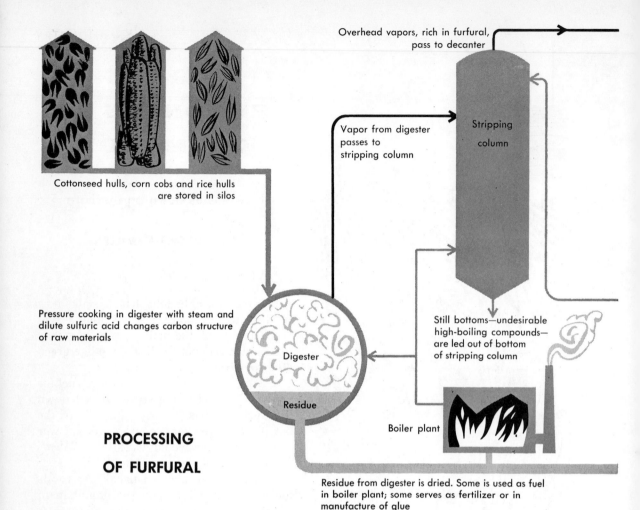

Overhead vapors, rich in furfural, pass to decanter

Vapor from digester passes to stripping column

Stripping column

Cottonseed hulls, corn cobs and rice hulls are stored in silos

Pressure cooking in digester with steam and dilute sulfuric acid changes carbon structure of raw materials

Digester

Residue

Still bottoms—undesirable high-boiling compounds—are led out of bottom of stripping column

Boiler plant

PROCESSING

OF FURFURAL

Residue from digester is dried. Some is used as fuel in boiler plant; some serves as fertilizer or in manufacture of glue

Furfural now plays an important part in the cementing process.*

The demand for furfural grew so much that supplies of oat hulls were not enough. Then another farm product found a new market. Corncobs became the raw material for furfural, and nowadays most of this versatile chemical is made from them. Nor was this the end of the story. The spent corncobs, after the furfural has been extracted from them, are formed into charcoal briquettes and sold as fuel for outdoor cooking.

This story of the modern uses of oats and corn shows how science takes materials that used to be thrown away and creates useful products that add to man's wealth and promote his welfare. We give the name of chemurgy to such industrial uses of plant materials. The word "chemurgy" was coined in the twenties of the present century from the Egyptian root word *chem* and the Greek word *ergon*. Literally it means "putting chemistry to work."

Chemistry — more specifically, organic chemistry — is indeed the key science in chemurgy. Organic chemistry has created a wide range of useful products — fuels, dyes, explosives, plastics, rayon and nylon, among others. Its raw materials are cellulose, proteins, starches, sugars and other natural compounds — substances that are readily available in plant life. It is the organic chemist's endless search for such materials that has made the rise of chemurgy possible.

A few pioneers began to promote the idea of chemurgy in the 1920's, when farm surpluses in the United States were a national problem. They pointed out that nearly half the tonnage grown by farmers consisted

* For a number of years, furfural was used extensively in the manufacture of nylon. It has now been replaced by cheaper materials, which are used in the production of nylon intermediates.

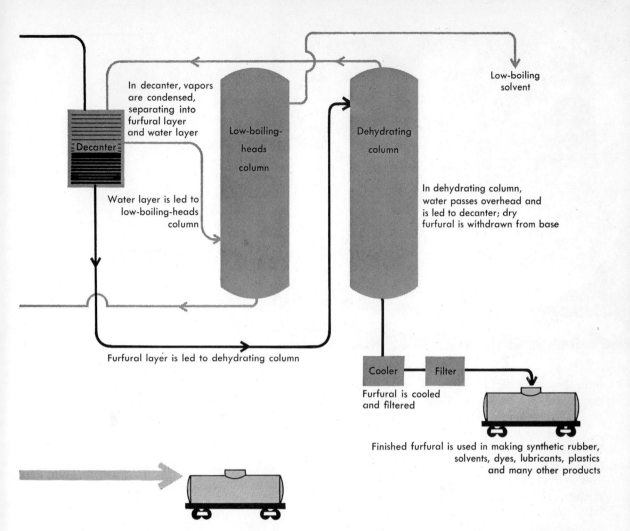

In decanter, vapors are condensed, separating into furfural layer and water layer

Decanter

Low-boiling-heads column

Dehydrating column

Low-boiling solvent

In dehydrating column, water passes overhead and is led to decanter; dry furfural is withdrawn from base

Water layer is led to low-boiling-heads column

Furfural layer is led to dehydrating column

Cooler

Filter

Furfural is cooled and filtered

Finished furfural is used in making synthetic rubber, solvents, dyes, lubricants, plastics and many other products

of potential chemical raw materials, such as straws, stalks and culls (rejected parts of the crop), for which there were no commercial markets. These agricultural materials would be especially attractive to industry because they would be inexhaustible. It is a sad fact that nature either does not replace minerals that are consumed, or else replaces them at a rate that can not begin to keep pace with our consumption. Plant raw materials, on the contrary, are always renewable. With many crops the entire production of a year may be consumed, except for seed, and then renewed by another year's growth.

The idea of applying chemurgy on a wide scale gradually took root. In the spring of 1935, a number of men prominent in agriculture, industry and science were invited to attend a conference at Dearborn, Michigan, in order to discuss the possibilities of chemurgy. As a result of this conference, the National Farm Chemurgic Council, Inc., was established. In 1955, its name was changed to the Council for Agricultural and Chemurgic Research, or the Chemurgic Council, for short. Its headquarters are now in New York City. This Council serves as a clearing house of information. It brings the subject of chemurgy to the attention of large numbers of people and it promotes support for public and private research.

Congress brought about a long forward step in chemurgic research in 1938. As an amendment to the agricultural act of that year, it provided for the establishment of four regional laboratories to be devoted entirely to finding new uses for farm products. These laboratories were set up at or near Philadelphia, New Orleans, Peoria and San Francisco. To each one was assigned a program of research in the crop generally grown in the nearest area. The laboratory

At the left, in the photograph, are shown several rolls of cigarette paper. At the lower right, we see the tow (that is, processed flax straw) from which the paper was produced.

Peter J. Schweitzer Division, Kimberly-Clark Corporation

buildings, costing $2,000,000 each, were completed and opened in 1940. Each laboratory was provided with space for erecting pilot-plant equipment, where new processes could be tested on a semicommercial scale.

A number of state experiment stations and universities are now taking part in chemurgic research. So are many industrial laboratories. Nearly every large food-processing company has also become a chemical manufacturer in order to make the maximum use of the materials at hand. As for chemical companies, they have naturally made an intensive search for the raw materials supplied by agriculture. Regional research institutions in several areas have also turned their attention to the development of new industrial uses of agricultural resources. Among organizations of this kind are the Midwest Research Institution, at Kansas City, the Southern Research Institute, at Birmingham, the Texas Institute of Technology and Plant Industry, at Southern Methodist University, Dallas, and the Southwest Research Institute, at San Antonio.

Chemurgy arose in the United States and has received by far its greatest development in that country. But there has been wide interest in the chemurgic idea elsewhere. The Chemurgic Council includes members from more than thirty nations; and farm chemurgic councils have been set up in other lands.

The goals
of chemurgy

Chemurgy attacks the problem of finding new industrial uses for plants in various ways. (1) It seeks new uses for the plants now commonly grown as crops. (2) It tries to find profitable uses for the wastes and residues of agriculture. (3) It develops new strains of established crops. (4) It seeks new and more profitable crops.

New uses
for old crops

Chemurgic research has made old crops more valuable by providing new industrial uses for them. Milo maize has long been grown in Kansas, Oklahoma and Texas. It is a species of corn whose grain develops at the top of the plant where the corn carries its tassel. It has long been valued as a feed grain, but until recently it had no industrial market.

Chemurgic researchers in the corn-products industries developed processes for making the grain into starch. A multi-million-dollar plant for this purpose started operations at Corpus Christi, Texas, in 1949. The plant supplies a market for thousands of acres of Texas milo maize; it

At the top is an ear of ordinary corn, which contains something like 25 per cent amylose. The center ear has from 50 to 60 per cent amylose; the bottom ear, from 70 to 75 per cent.

ARS—USDA

produces starch, dextrins, syrups and other products. Dextrose produced from milo maize is used in large quantities for the processing of cold rubber for tires.

Profitable uses for
agricultural wastes

The development of furfural is one illustration of how a waste farm material has been turned to profitable account. Here is another case in point.

Until the 1930's, nearly all the cigarette paper used in the United States was made by factories in France from old rags of linen (a flax product) collected in Central Europe. It then occurred to several manufacturers in the United States that the cigarette paper might be made from seed flax straw. The crop had been harvested, not for the fiber to be made into linen, but for the seed, from which was processed linseed oil, an important ingredient in paints. The flax straw was considered worthless. In some areas it was considered even worse than useless; it had to be raked up and burned since it did not decompose readily on the ground.

In the late 1930's, after a substantial research and development effort, the Ecusta Paper Company of Pisgah Forest, North Carolina, the Peter J. Schweitzer Company of Elizabeth, New Jersey, and the Smith Paper Company of Lee, Massachusetts, began to manufacture cigarette paper from seed flax straw. Today, Ecusta and Schweitzer * supply practically all the domestic requirements for cigarette paper (small quantities are imported). A certain amount of the processed fiber also serves for Bible paper, airmail stationery and currency paper. Most of the flax straw used for these operations is grown in the North Central States, primarily Minnesota and the Dakotas; some of it is produced in Texas, Montana, and California. The Howard Smith Paper Company of Beauharnois, Canada, is manufacturing cigarette paper from Canadian-grown seed flax.

Many other important industrial materials have been created from agricultural wastes. For example, bagasse, the fibrous substance that is left over from sugar cane after it has been pressed in the mill was once considered useful only as an inefficient but cheap fuel in the boiler houses of sugar mills. It is now being made into sturdy fiber board and other useful products. A syrup prepared from cull apples has been used instead of glycerin in order to keep the tobacco in cigarettes moist and even-burning.

* Ecusta is now a division of Olin Mathieson Chemical Company. Schweitzer, having previously absorbed the Smith Paper Company, is now a division of Kimberly Clark.

39

High-amylose corn — a new strain of an old crop

Amylose corn is one of the fascinating new chemurgic developments. Ordinary starch, which is a principal constituent of corn, is made up of globe-shaped molecules, and because of its starch corn has hundreds of industrial uses. However, corn also has some long, straight-chained molecules, known as amylose. Chemists believed that if a strain could be developed that had, say, three-fourths or more of amylose molecules, the starch would find many more profitable markets. So plant breeders, led by a young Illinois farmer, set out to produce a high-amylose corn. Their efforts, after a few years, resulted in strains with more than sixty per cent of amylose molecules and also individual plants that ran much higher.

Meanwhile the chemists, with enough of the new starch for experimentation, made it into a transparent food wrapping which can be eaten along with the frankfurter, candy or whatever it may contain. They demonstrated that the long molecules could greatly improve some kinds of paper. Many other uses are probable.

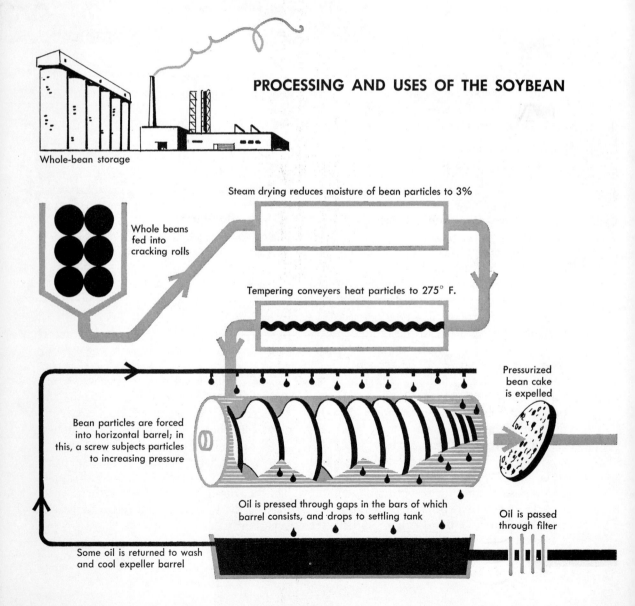

PROCESSING AND USES OF THE SOYBEAN

Whole-bean storage

Steam drying reduces moisture of bean particles to 3%

Whole beans fed into cracking rolls

Tempering conveyers heat particles to 275° F.

Bean particles are forced into horizontal barrel; in this, a screw subjects particles to increasing pressure

Pressurized bean cake is expelled

Oil is pressed through gaps in the bars of which barrel consists, and drops to settling tank

Oil is passed through filter

Some oil is returned to wash and cool expeller barrel

The search
for new crops

The search for useful new crops is an important part of chemurgic research. In some cases these new crops replace old ones that, for one reason or another, are no longer profitable.

In the years preceding the great depression of the 1930's, American farmers grew so much wheat and cotton that profitable markets could not be found. Because of the low prices, farmers could not prosper, and they had to receive substantial help in the form of large government subsidies. Chemurgists believed that in many areas it would be well to introduce other crops, which would command a higher price than wheat or cotton and which would supply materials that the nation lacked, instead of adding to surpluses. They attacked the problem with all the resources of organic chemistry, plant-breeding and agricultural engineering. The result was the development of several alternate crops, which had hitherto not been particularly important in the United States or had never before been introduced.

The photo (left) shows a soybean plant. It has been a food crop in China for thousands of years; in the United States it was grown mostly for hay and hog pasture until the 1920's. Now many products come from soybeans.

W. Atlee Burpee Co.

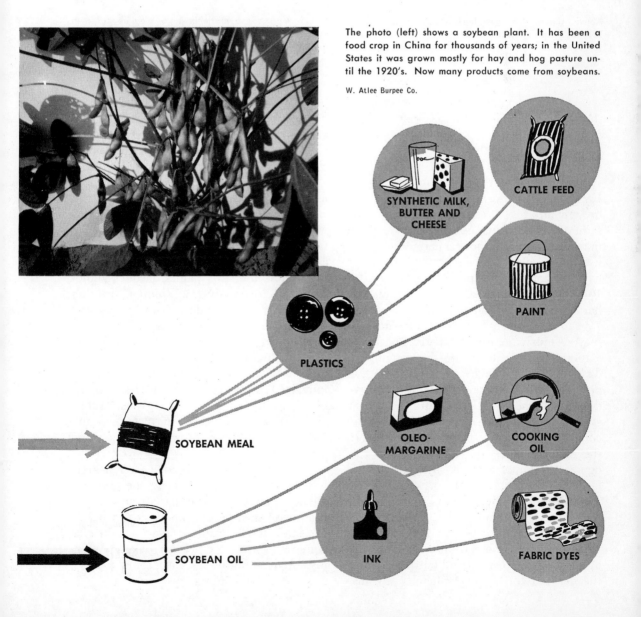

SYNTHETIC MILK, BUTTER AND CHEESE

CATTLE FEED

PLASTICS

PAINT

SOYBEAN MEAL

OLEO-MARGARINE

COOKING OIL

SOYBEAN OIL

INK

FABRIC DYES

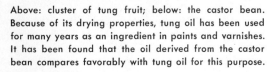

Above: cluster of tung fruit; below: the castor bean. Because of its drying properties, tung oil has been used for many years as an ingredient in paints and varnishes. It has been found that the oil derived from the castor bean compares favorably with tung oil for this purpose.

Safflower (above), a thistlelike herb, may be grown in semiarid regions or in irrigated land. The oil of this plant, refined by conventional methods, is used for cosmetics, drugs, food preparations and paints; the oil-free cake is ground into meal, mostly for livestock feed.

The most important of these crops is the soybean. This plant has been a staple food crop in China, particularly in Manchuria, for several thousand years; and as a matter of fact it was not unknown to America before the rise of chemurgy. Early in the nineteenth century, it had been brought to the Western Hemisphere by sailing-ship captains, and it had flourished in a small way. But no one had found much use for the plant; it was grown mostly for hay and for hog pasture. Not until after 1920 was as much as 1,000,000 bushels harvested in a year in the United States.

In the last year of the nineteenth century, American agricultural experiment stations began to look into the possible industrial uses of soybeans, and these experiments were continued in the first two decades of the twentieth century. It was shown that oil pressed from the soybean could be used for a variety of products. In the early twenties, oil mills began to recognize the importance of soybean oil, and farmers planted more and more of the beans.

Chemurgists became fascinated by the soybean; they found it the most versatile crop known to science. An almost infinite variety of industrial products can be produced from soybean oil and meal. To meet the demand for these invaluable raw materials, more and more acres have been planted to soybeans; annual production has grown until it exceeds 600,000,000 bushels a year. In Canada, too, the soybean has become an important crop.

The oil makes up one-fifth of the soybean, and more than twenty thousand tank carloads of it are separated every year. More than half the oil goes into cooking fats, about one-fifth into margarine and about one-twelfth into salad oils. The remainder is used for paints, lacquers, linoleum, oilcloth, window shades, printing inks, putty, insecticides, disinfectants and soap. The meal that remains after the oil is removed has become a foremost ingredient in livestock and poultry feeds. The meal is also used in the manufacture of industrial products like adhesives, textile fibers, waterproofing for textiles, waterproof glue, spreaders for insect sprays, finishing waxes, paper-sizing materials and plastics.

The soybean has become increasingly important as a food. Among the products we now get wholly or in part from the soybean are baked beans, canned green beans, breakfast foods, candies, chocolate drinks, coffee substitutes, crackers, flakes, flavorings, flour, infant foods, macaroni products, milk substitutes and soups.

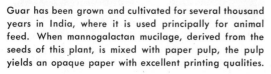

Stein, Hall and Co., Inc.

Guar has been grown and cultivated for several thousand years in India, where it is used principally for animal feed. When mannogalactan mucilage, derived from the seeds of this plant, is mixed with paper pulp, the pulp yields an opaque paper with excellent printing qualities.

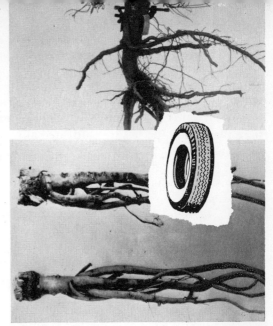

Lower photo, USDA

Above: guayule plant. Below: roots of the kok-sagyz, a Russian species of dandelion. The latex from which rubber is made has been obtained from both these plants, though not in commercially significant quantities. It is believed that kok-sagyz may yet be a profitable crop.

Thus, largely through the work of the chemurgist, the soybean has become a major crop in the United States and its cultivation has been steadily increasing. It is produced principally in the Midwestern states. The fields in which it grows would otherwise have been planted to corn, wheat and other crops which in normal years might produce surpluses.

Within the past few years safflower, a type of thistle, has become a successful crop in California, western Nebraska and eastern Colorado. Safflower seed yields up to 40 per cent of oil. The oil, which the ancient Egyptians used in their lamps, is sold for food preparations, drugs and cosmetics, as well as for paint. Because of its chemical structure, it is believed to prevent cholesterol deposits in the arteries (see Index, under Cholesterol), and much of it is sold for this and other pharmaceutical uses. More than 200,000 acres are planted each year, and the demand has been growing.

Tung oil is a valuable ingredient in paints and varnishes because of its drying properties. The United States had to import all its tung oil from China and Japan until the twentieth century, since the tung tree, from which the oil is derived, was unknown in the Western Hemisphere. In 1905, a few nuts were sent to America from China, and were planted at Brookville, Florida. Now, in Florida, Mississippi, Texas and other states along the Gulf Coast, tung orchards flourish on thousands of acres. Yet not nearly enough is grown to meet the demand. Because supplies of tung oil were insufficient, chemists cast about for another substance that would do some of its work. Finding no natural oil that was satisfactory, they tried to see what they could do by changing the composition of known oils. They found that by eliminating water (H_2O) from the molecules of castor oil they could produce a substance that was nearly as good as tung oil for many purposes. The castor beans that yield the oil in question are now imported chiefly from Brazil and India. But tests have shown that the plant grows readily in much of the United States, especially in the area between the cotton and corn belts. Commercial production of the castor bean is now well established in the Southwest.

Thus far we have dealt with the introduction into a country of crops that are well established elsewhere. In some cases, the "new" crops were new in the sense that they had not yet been cultivated anywhere on a commercial scale. As a matter of fact, we have hardly begun to tap the immense plant resources of the world. Well over

300,000 plant species have been identified by botanists. Yet man uses extensively not more than 2,000 species, and probably not more than 1,000 species are cultivated anywhere as crops.

The legume called guar (*Cyamopsis tetragonalaba*) was recently introduced to cultivation as a result of chemurgic research. Not long ago, paper manufacturers decided that they could make better paper if a material called mannogalactan mucilage were mixed in the pulp. The paper would be opaque and it would have better printing qualities. The only known source of the material was the carob bean, the product of a locust tree common in the Mediterranean regions. The total available supply, however, was much less than the industry required. Experts launched a search for some other plant that would yield the same product. It was found that guar, which is native to India, produces an abundant yield of a small seed from which the mannogalactan mucilage can be extracted. Guar is now being grown and processed in Texas.

In the effort to create a domestic rubber supply during World War II, experiments were made with two plants that had never been widely cultivated. One of these was the guayule, a shrub growing in the desert areas of southwestern United States and northern Mexico. It was found that the guayule can be cultivated and processed to yield natural rubber. However, since guayule rubber costs more than that obtained from the rubber tree (*Hevea brasiliensis*), the cultivation of the guayule shrub has not made extensive progress.

Another substitute for the rubber tree is a Russian species of dandelion, known as kok-sagyz. It is the roots of this plant that yield the latex from which rubber is made. Seeds were obtained from Russia and test plantings were made at various American state experiment stations. There were substantial yields of kok-sagyz in some of the northern stations; and extraction of the latex proved to be neither difficult nor costly. These tests were abandoned under wartime stress. The experimenters learned enough, however, to indicate that kok-sagyz might some day become a profitable crop.

The effects of chemurgy

There is no doubt that chemurgy has taken its place as a powerful industrial factor. What will its effects be?

For one thing, it will tend to decentralize industry. There is a growing tendency to erect processing factories near the farms where their raw materials grow. As more and more of these materials are made available for industry, processing plants will become more widely distributed. Since manufacturing costs are lower in rural communities, this development will benefit industry. It will benefit the workers, since they will be able to live amid more pleasant surroundings than in the crowded and smoky areas of big cities. It will also be a boon to the communities in which plant-processing enterprises are located; it will make more wealth available for health departments, schools and roads.

From the viewpoint of national defense, the development of new industrial products from agricultural crops will be helpful to the United States and Canada and other nations which have to import certain necessary materials from distant lands. If a war should interrupt production abroad or suspend shipping, the supply of imported materials would soon be exhausted and a serious crisis would develop. When the Japanese overran Malaya and the East Indies in 1942, they cut off the chief source of rubber supply for the Allies. Fortunately, the United States carried through a successful synthetic-rubber program; otherwise, the rubber shortage might well have undermined the entire war effort.

Chemurgy, then, will serve as a shield in time of war. It will also actively promote peace by raising the standards of have-not countries. It will enable farmers to make more effective use of the crops that now flourish in their fields, and to add new, profitable crops. It will make possible the variety of production that is the best guarantee of prosperity. Thereby it will help to do away with the poverty and discontent that breed wars.

See also Vol. 10, p. 282: "Chemurgy."

VARIABLE STARS

Eclipsing Binaries, Novae, Supernovae and Others

EVEN to the unaided eye, many of the stars display beautiful tinges of color. These colors are enormously increased in splendor and beauty by the telescope, which also reveals the delicate tints of many stars that without its aid appear white. Antares is a wonderful ruby red. Somewhat less deep and rich in tone is another red star, Betelgeuse. Aldebaran and Arcturus are red-orange. The sun is, of course, yellow; Procyon is yellowish white; Sirius is white with almost a bluish cast; Mirzam is definitely bluish. These illustrate the color range of all the stars in the sky, red to blue-white; but there are a hundred delicate tints in between — lilacs, amethysts and greens, forming a beautifully colored pattern in the night skies.

Homer, Cicero and some other ancient writers described Sirius, the Dog Star, as red, though it is today very clearly white. Can stars, then, change color? They can and do. It has been determined that the color is an index of temperature, red stars being comparatively cool (around 2,000°–2,600° centigrade) and blue-white stars being hottest (up to 50,000°). These are surface temperatures. Inside the stars, temperatures go up to tens of millions of degrees.

Astronomers believed formerly that all blue-white stars were "young" stars and that red ones were "old," with yellow and orange ones in between. Nowadays the accepted theory is that stars progress from red toward white.

This is what probably happens: a "young" star is a diffuse mass of elements in gaseous form. Its color is reddish, which means it is comparatively cool. (If you heat an iron red-hot, it is far cooler than if you heat it white-hot.) But it is,

nevertheless, hot enough so that many atoms are excited; they lose electrons. This means, of course, that they lose mass. The star contracts. Contraction creates pressure, and greater heat and nuclear reactions take place, in which some mass is lost but great energy is created. The star grows progressively hotter. Its color changes from red to orange to yellow to white to blue. For some reason that is not understood, in its hottest blue-white state a reversal sets in. The star starts to cool off, changing color again, growing white, yellow, orange, red and, finally, losing its luminosity, becoming a cold, dead mass. According to this theory of a star's life history, Sirius might possibly have been red "when Homer smote 'is bloomin' lyre," though it is white now and at some future date will be red again.

Sirius also changes periodically in brightness, or apparent magnitude. Thousands of stars do this. They are called variables. The variables have been classified according to the reason for their changing brightness. Sirius belongs to the class of eclipsing variables. That means that Sirius has a companion star, and the two revolve around a common central point. At times the smaller companion, or Sirius B, gets between Sirius A and the earth, cancelling out some of the luminosity of Sirius A.

Two-star companionships of this sort are called binaries, and very beautiful and impressive some of them are when viewed through a telescope. Stars that to unaided vision seem but a single point of more or less doubtful color reveal themselves in the telescope as pairs, displaying amazingly lovely harmonies of color, either contrasted or graded in pleasing nuances.

Algol, perhaps the best-known of the eclipsing binaries, was the first to be discovered by astronomers. Perhaps the ancients suspected that it was out of the ordinary; at any rate, the name "Algol," which means "the demon" in Arabic, may have referred to the bizarre quality of the supposed star's light, alternately becoming brighter and then dimming.

It was an English deaf-mute, John Goodricke, who in the year 1783 first discovered the nature of Algol's fluctuations. He advanced the theory that what appeared to be a single star was really made up of two: a bright star and a fainter companion that eclipsed the bright star as it revolved around it. He noted that the light of Algol becomes dimmer at intervals of about two days and twenty-two hours. Goodricke's theory was later fully confirmed.

Much has been learned since his day about the two stars that make up the eclipsing binary Algol. The brighter of the two — the primary star — has a diameter three times that of the sun. Its companion is somewhat larger, but is fainter by three magnitudes. The centers of the two stars are about 13,000,000 miles apart. Their orbits are almost edgewise to the earth; they are inclined from the edgewise position by only 8° or so.

Another interesting eclipsing binary is Zeta Aurigae. It is made up of a blue star, whose diameter is about seven times that of the sun, and a red star, with a diameter about fourteen times that of its companion. The two stars have approximately the same apparent magnitude. They revolve about each other once every 172 days, eclipsing each other at intervals as they move in their orbits. First we see a partial eclipse, lasting thirty-two hours, then a total eclipse of thirty-seven days and finally another thirty-two-day partial eclipse.

The components of certain eclipsing binaries are so close together that they appear as a single star even when viewed through the most powerful telescope. They have been discovered and studied, however, with the spectroscope. The two stars of a typical binary of this kind approach the earth and then recede from it in their orbits, unless these are at right angles to the line of sight of an observer on the earth. The motions of the stars can be analyzed with the help of the spectroscope because, in accordance with the Doppler effect (see Index), the lines in their spectra shift to the violet as they approach the earth and to the red as they recede from it. Binary systems that are analyzed by means of the spectroscope are called spectroscopic binaries.

The variable stars that we have considered hitherto in this chapter are called extrinsic variables. The changes of brightness that we note in them are brought about by external factors, such as eclipses. We shall now take up the intrinsic variables. These are the stars that vary in brightness because of various physical changes that take place in them.*

The novae,
or new stars

A fascinating class of intrinsic variables are the exploding stars known as novae, or new stars. They were given that name because to the older astronomers they appeared to be newly created. We realize now, however, that a nova is not "new"; it is a pre-existing star upon which an explosion has taken place.

A faint star, perhaps too faint to be seen by any except the most powerful telescopes, begins to be more brilliant. Within the space of a few days it becomes thousands of times brighter than it was before. After some days of maximum brilliance, the luminosity begins to fade. Ultimately, the star is about as faint as it was before. This would seem to indicate that the outburst is far more superficial than it would appear to be. The explosion of novae occurs without warning. On the average, they increase about 60,000 times in brightness, corresponding to thirteen magnitudes. Always the growth of brilliance in a nova is extremely rapid and its fading is gradual.

* Certain variables are both extrinsic and intrinsic. They are eclipsing binaries that also undergo internal changes.

A nova is designated by the word "Nova," followed by the genitive (possessive form) of the constellation in which it occurs and the year in which the outburst took place. For example, when we refer to Nova Persei 1901, we have in mind the nova that appeared in the constellation Perseus in the year 1901.

The brightest nova of which we have definite record appeared in the constellation Cassiopeia in November 1572; it was observed by the great Danish astronomer Tycho Brahe. Nova Cassiopeiae 1572, to give it its technical name, became so bright that it could be seen in daylight. It faded very gradually and finally disappeared from view in the spring of 1574. There were no telescopes in those days; otherwise, it would have been visible some time longer. Another particularly brilliant nova was observed by the German astronomer Johannes Kepler in 1604, in the constellation Ophiuchus; it remained visible to the naked eye for about eighteen months. This nova became known as "Kepler's star."

Astronomers speak of the "explosion" of a nova; but the term is quite misleading. It is not to be compared, say, to the detonation of a charge of dynamite. It occurs, rather, in the form of a series of powerful jets or sprays, ejecting vast quantities of material from the star. Its effect has been described as suggesting a Roman candle.

A gaseous envelope develops around a nova. In certain cases, this envelope is spherical. Sometimes it is elliptical; in such instances, it is probably composed of several "shells," each corresponding to a different explosion. Occasionally an envelope attains fantastic size. The one that formed around Nova Aquilae 1918 was something like 1,000,000,000,000 miles in diameter after about twenty years. The Crab Nebula in the constellation Taurus shows an envelope that has been expanding for hundreds of years. It seems to have been formed by a nova that appeared in the year 1054, according to Chinese and Japanese records. (Apparently, it was not observed in the West.) Actually, the outburst must have taken place four thousand years or so before 1054, since it would take that length of time for its light to reach us.

A number of novae explode only once. In others, which are called recurrent novae, two or more outbursts take place. These novae rise to maximum brilliance more slowly than those that explode only once, and they take less time to return to their former faintness. They also have smaller ranges of magnitude than most novae. Among the recurrent novae are Y Coronae Borealis, RS Ophiuchi and U Scorpii.

We can only conjecture about the cause or causes of the explosions that result in the formation of novae. It was formerly held that they were due to collisions between stars, or between a star and a nebula. Few take this theory seriously now. For one thing, how could we explain recurrent novae on the basis of collisions? We would have to assume that a single star collides again and again with some celestial object in the course of a century. This would be beyond the range of probability.

It is more likely that a nova arises as the result of the sudden release of internal energy. Perhaps this energy release is brought about by a gradual increase in temperature, which finally triggers a nuclear explosion. At the present time, all this is pure speculation.

The spectacular stars called supernovae

In some cases, the explosions accompanying the rise of novae are particularly spectacular. Such out-of-the-ordinary novae are known as supernovae. The first one recorded by astronomers appeared in 1885 in the central region of the galaxy called the great spiral in Andromeda. It was of the seventh magnitude; it was a tenth as bright as the spiral itself and ten times as bright as the average nova. At first, astronomers did not consider it as differing appreciably from the other novae known at this time, since the great spiral in Andromeda was thought to be a nebula and much nearer to us than it actually is. Once it was realized how far away from the earth the so-called spiral nebula is, it became obvious that this particular "new star" was an extraordinary one.

We pointed out that the explosion that causes the average nova to become so brilliant does not have far-reaching effects — that is, comparatively speaking. It is estimated that a nova throws off only a thousandth part, or even less, of the matter that it contains. The supernova is affected much more by an explosion; as far as we can judge, it throws off the greater portion of its mass. Furthermore, it does not return to its former state after the explosion takes place. If we adopt the internal-energy theory to account for the rise of supernovae, the energy release must be far greater than in the case of novae.

The pulsating
variable stars

The pulsating variables make up another important group of intrinsic variable stars. They have received the name "pulsating" because they grow alternately brighter and fainter according to a more or less definite pulse, or rhythm. The pulsating variables may be divided, rather arbitrarily, into two groups: long-period and short-period. In the first group, the period from one peak of brightness to the next ranges from a hundred days to a thousand; in the second group, from less than one day to about fifty.

Long-period
variable stars

The stars with the longest period are the large red giants, with low density and low surface temperature. Variations in brightness are not altogether regular. These stars are all fairly bright when they reach their maximum brilliance and have conspicuous variations; hence they can be studied by the amateur astronomer with a small telescope or even a pair of binoculars.

The long-period star called Mira, in the constellation Cetus, is the prototype of the long-period variable; as a matter of fact, the name "Mira stars" is sometimes given to them. Mira was the first long-period variable to be discovered. It was first observed by David Fabricius, a German theologian and astronomer, in 1596; he thought he had found a nova. It was seen again in the year 1638. The astronomers of that time also assumed that it was a nova, at first, but they soon became aware of its true character as a periodic variable. The star aroused the admiration of the older astronomers. They hailed it as *stella mira* ("wonderful star," in Latin); that is how the name "Mira" was derived.

Mira has a period of about 332 days. Neither the maximum nor the minimum brilliance in a given cycle can be determined beforehand with any precision. Either may be hastened or delayed by a week or two and occasionally by as much as forty days. There are also considerable variations in intensity of light. The maximum brightness of the star has been known to reach almost the first magnitude. In 1779, Sir William Herschel noted that Mira was almost equal in brilliance to Aldebaran, a first-magnitude star. At other times, however, Mira fails to attain the fifth magnitude at the peak of its brightness.

Its minimum brilliance is more uniform, on the whole, but it, too, is subject to irregularity. It usually falls between the eighth and tenth magnitudes, but it has been known to be considerably fainter than this limit. In 1783, for example, Sir William Herschel could find no trace of the star in a telescope in which all stars down to the tenth magnitude were visible.

If we consider the whole range within which Mira has been known to vary, we find that the light it emits at certain periods is about ten thousand times as great as that emitted at other times. In its usual range of brilliance, however, it is from twelve to fifteen hundred times brighter at maximum than at minimum.

In spite of its variations, we can mark a certain pattern — an average of waning and waxing. At intervals of about eleven months, the star begins to brighten. The progression from minimum to maximum brightness takes something like a hundred days. Mira retains its brilliance for some weeks; then it subsides to its former level of brightness. It takes the star almost twice as long to pass from maximum to

In the above photo, the arrow points to the star now called Nova Aquilae 1918 before the outburst that made it a nova.   Below, we see how much brighter the star (indicated here too by an arrow) became.

Yerkes Observatory

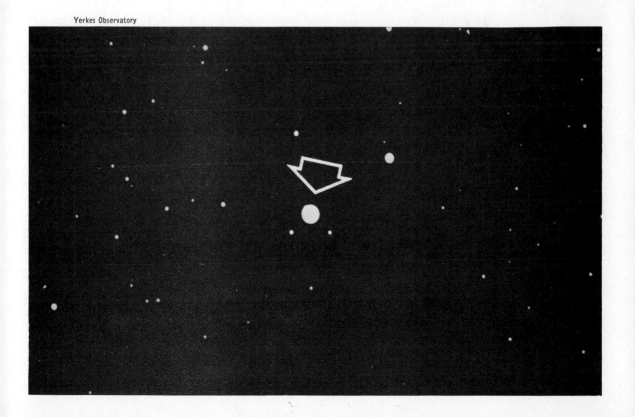

minimum brightness as from minimum to maximum. Considering the average of its performance, it remains at a high level of brilliance for about two months and at a low one for about four. It spends the rest of the time changing from one level to the other.

Another famous long-period variable is Chi Cygni. This striking star, which is of a beautiful scarlet color, is more than six thousand times as bright when it is at maximum brilliance than when it is faintest. It waxes and wanes in periods of about four hundred days through a range of eight magnitudes.

Mira and Chi Cygni show exceptionally wide ranges of fluctuations. The average amount of variation in seventy-five long-period variables that were kept under close observation at Harvard University was found to be five magnitudes. This means that the average long-period variable gives out a hundred times more light at maximum brightness than it does when at minimum.

Spectroscopic examination has shown that the increase in the brilliance of long-period stars is caused by periodical outbursts of incandescent gases, chiefly hydrogen. It is still not clear why these outbursts take place.

The short-period variable stars

The intrinsic variables with the shortest periods are called the RR Lyrae stars, because the first of the group to be studied was RR Lyrae (the star named RR in the constellation Lyra). They have been discovered in considerable numbers in the globular clusters; for this reason they are sometimes known as cluster-type variables. Their periods range from about an hour and a half to a little over a day. They remain at their peak of brightness for a very short time; they are at minimum brightness for a comparatively prolonged period. It is extremely difficult to observe these stars; often they can be studied only by photographic means.

Even at maximum brightness, they are so faint that we have identified only about fifteen hundred out of an estimated total of a hundred thousand in our own galaxy. As for the RR Lyrae stars in other galaxies, we shall probably never be able to discover more than an insignificant fraction. So far, astronomers have identified only a few in the neighboring galaxies.

Far better known are the Cepheid variables. Their name is derived from Delta Cephei, a star that can be made out with the naked eye. The total period of Delta Cephei from minimum to minimum is nearly five and a half days, and its light varies from a magnitude of 3.7 to 4.4. It takes only a day and a half for the star to reach maximum brilliance, while the descent to minimum requires four days. About fifteen hours after the downward progression begins, the star maintains the same degree of brilliance for a short time before resuming its declining course.

The average period of the Cepheid variables in our own galaxy is about seven days. Changes in magnitude vary from less than half a magnitude to somewhat more than a magnitude — a comparatively slight range.

Following are the periods and the variations in magnitude of a few representative members of the Cepheid variable group:

Name	Period (days)	Magnitude
SU Cassiopeiae	2.0	6.1-6.4
Delta Cephei	5.4	3.7-4.4
Zeta Geminorum	10.2	3.7-4.1
1 Carinae	35.5	3.6-4.8
SV Vulpeculae	45.1	8.4-9.4

We show the variations in brightness of a star by the so-called light curve, in which we plot brightness against time. Generally we cannot make observations of a variable star throughout the cycle of its variation, if the cycle is a day or more in length, because of the alternation of day and night. Hence we have to establish the light curve of a variable star from the data we obtain on different nights. It is a tribute to the perseverance and ingenuity of astronomers that they have been able to plot so many of the light curves.

These curves show considerable variation in the Cepheid variables. In some, the curve is smooth; in others, a steep ascent is followed by a gentle downward slope, which means, of course, that they brighten more rapidly than they grow faint.

Almost all the changes in brightness in Cepheid variables are due to the changes in temperature at the surface of the stars. The surface is hottest when the star is the brightest and coolest when the star is the faintest. The Cepheid variables that have the longest periods are the coolest.

Early in this century, a new and valuable method for sounding the depths of space was discovered by Henrietta S. Leavitt, an astronomer on the staff of the Harvard College Observatory, as a result of a study made by her of the Lesser Magellanic Cloud, in the southern heavens. This cloud was rich in Cepheid variables. Presumably all of these stars were at about the same distance from the earth. When Miss Leavitt had suitably arranged these Cepheids according to their period and brightness, she found that a definite relationship existed. The brighter the star, the greater the time that was required to go through the period of change from one time of maximum brightness to the next. If the distance and actual luminosity of any one of the Cepheid variables could be found, it would be possible to determine the luminosities and distances of other Cepheids from their periods. This conclusion was born out by Harlow Shapley's investigations of variable stars in globular clusters. Astronomers now determined the distances of several of the brighter Cepheids found in the heavens, and by using these results were able to calculate the distances of various other Cepheids on the basis of the period-luminosity relationship.

The Cepheids with like periods are assumed by the period-luminosity relationship to have the same intrinsic brightness. Consequently, the difference in observed brightness shows a difference in distance from us. We know that the light that we receive from a given source at a given distance will be four times as great if we halve the distance and only one-fourth as great if we double the distance. Let us suppose, by way of example, that a lighted candle is held at a distance of ten feet from us. If the candle is then set five feet from us, we receive four times as much light; if, on the other hand, it is moved to a distance of twenty feet from us, we receive only a fourth as much light. Utilizing this law of inverse relationship, we can tell that if a Cepheid variable appears to us to be sixteen times as bright as one with the same period (and therefore the same intrinsic brightness), it must be four times nearer to us than the other Cepheid variable in question.

Will our sun ever become a nova?

Astronomers are sometimes asked whether there is any danger that the sun might explode and become a nova or supernova. First, let us point out that the radiations of stars (including the sun) — their light, heat and other forms of electromagnetic energy — are the effects of nuclear reactions involving hydrogen. (We describe these reactions in the article The Sun, Our Star, in Volume 3.) It is when a star is reaching the end of its hydrogen supply that stellar explosions begin.

Suppose, now, that the supply of hydrogen in the sun should run low. The nuclear-reaction cycle would be affected. Other reactions in the sun's gases might then produce the periodic explosions we have observed in novae; or there might be a gigantic supernova explosion in which the sun would collapse and flare up and then cool down. In either case, the effect on earth would be disastrous. The oceans would start to boil and the lands would melt.

However, stellar explosions seem to occur in a star's old age; and the sun is quite young. Another point to note is that huge and very bright stars use up their hydrogen much more rapidly than smaller, dimmer ones; their span of life may be only a few hundred million years. The sun is not very massive and not very bright. As far as we know, it has enough hydrogen fuel to last for over 10,000,000,000 years.

See also Vol. 10, p. 269: "Stars and Constellations."

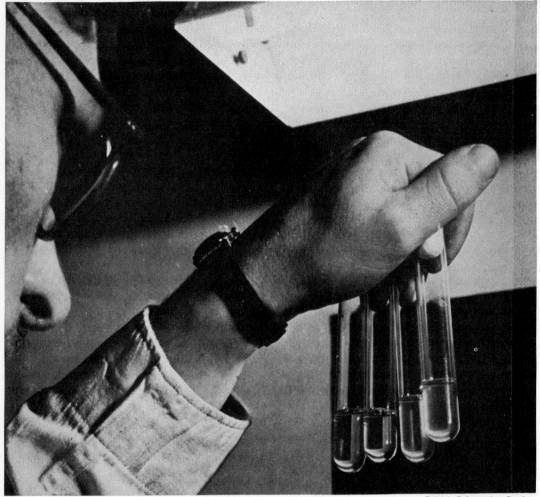

British Information Services

Tubes of clear diluted penicillin, to which bacteria called staphylococci have been added, are kept overnight at human body heat. If the penicillin is strong enough to kill the staphylococci, the tubes will remain clear. If the tubes become cloudy, more penicillin will be required in the solution.

PENICILLIN AND STREPTOMYCIN

Two of the most effective germ-killers that have been developed in recent years are penicillin and streptomycin. Penicillin is an acid produced by the mold *Penicillium notatum*. The acid was discovered by Sir Alexander Fleming in 1928, but it was not put to practical use until some ten years later. The mold is grown in vats or flasks in a liquid; the liquid is then treated chemically and evaporated. Penicillin is effective in treating acute infection of the blood, heart, eyes, ears and bones; it also serves to prevent infection in open wounds.

Streptomycin is sometimes successful in fighting diseases that resist penicillin. It is an extract of a bacterium that grows in the soil — *Streptomyces griseus*. Streptomycin was discovered in 1939 by a team of scientists at Rutgers University, headed by Dr. Selman A. Waksman. Streptomycin is invaluable in the treatment of pneumonia, typhoid fever, dysentery, undulant fever and gas gangrene.

Chas. Pfizer and Co.

Air compressor supplying air required in the fermentation step. In this, the solution in which the mold has been grown undergoes a series of chemical changes. The preparation of penicillin is now on a mass-production basis; machinery has taken the place of laboratory technicians in many operations.

The mold *Penicillium notatum* growing in a flask. Today molds are grown mostly in huge vats.

Three lower photos, E. R. Squibb and Sons

How the mold looks in its natural state. The drops of clear liquid on the surface of the mold contain a good deal of the germ-killing penicillin.

Penicillin is obtained by decanting (pouring off gently) the liquid matter contained in the flask or vat. The solid mold matter is then discarded.

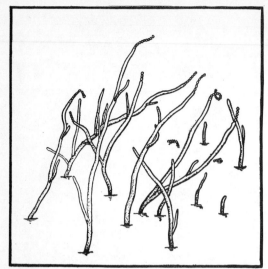

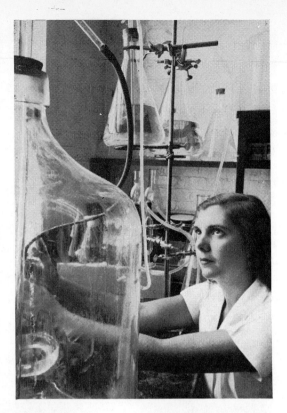

The grass-like growths in this drawing represent a microscopic view of *Streptomyces griseus,* the organism from which streptomycin is derived.

Right: a technician is preparing solutions that are to be used in testing freshly prepared streptomycin. The tests are given in order to find out how potent and stable the streptomycin is.

The final stage in the preparation of streptomycin. The machine in the picture is crimping on aluminum seals over the rubber stoppers of the vials, thus providing double protection for their contents.

INLAND WATER RESERVES

The Formation of the Earth's
Lakes and Marshes

CONSTANT transformations in the earth's crust have brought into being vast numbers of lakes — bodies of water lying in basinlike depressions. Lakes range in size from insignificant ponds a few yards across to magnificent expanses of water like Lake Superior in the New World and the Caspian Sea in the Old. Some lakes are isolated bodies of water; others are strung out like beads along the courses of great rivers. Most lakes are above sea level; a very few, like the famous Dead Sea, in Asia Minor, lie below sea level.

The beds of lakes have been created in various ways. Some have been formed by glaciers; some have been gouged out by the erosive force of running water; some have resulted from movements of the earth's crust; some have been due to the action of the wind, others to the action of the sea.

Perhaps a majority of lakes have been created as a result of glaciation — the action of glaciers. As these massive bodies of ice slowly made their way over the land, the rock fragments they carried dug out the underlying bedrock. Depressions formed in the rock; after the passing of the glaciers, water collected in these hollows and a lake formed. In other cases, glaciers deposited vast quantities of rock debris on the land; depressions in this mass of debris became the basins of lakes.

Most of the lakes formed by glaciation are quite small; a certain number of them, however, are of considerable size. Glaciation accounts, partly at least, for most of the great lakes in the North American continent. Many broad lakes of the Alps, including lakes Constance, Maggiore, Lugano and Como, were formed when debris carried by glaciers caused deep valleys to be dammed up.

Lake beds are quite often dug out by running water. Wide temporary channels may be formed by raging rivers in time of flood. When the flood waters recede, these channels remain; as water collects in them, they form the beds of shallow lakes. Sometimes tributaries, carrying tons of silt, may build up deposits across the valleys of the large streams into which they empty. The valley is dammed up in this way, and a lake is created. For instance, Lake Pepin, lying southeast of Minneapolis, was formed when a twenty-mile stretch of the Mississippi River was dammed up by the debris brought by its tributary, the Chippewa. The swift current of the Chippewa enabled it to transport material that was altogether too heavy for the larger and more sluggish river to remove.

Earthquakes lower some parts of the earth's crust and raise other parts: the hollows that are formed in this way often become the beds of lakes. For example, in 1811–12, the lower Mississippi Valley was shaken by a series of severe earthquakes. As a result of this shifting of the earth's crust, an area of five thousand square miles in Missouri was lowered and was converted into lakes and marshes.

Landslides sometimes raise a barrier across a valley, and the valley then becomes a lake bed. In 1893, a landslide in a deep valley near the headwaters of the Ganges River filled up about two miles of the valley with rock debris; the river, dammed up, became a lake. Later, the waters of the newly formed lake overflowed and cut a deep channel in the dam that held them back; a terrible flood resulted.

Many lake basins have resulted from movements of the earth's crust. During the formation of mountains, rocks are squeezed out of their horizontal positions as the earth's crust is forced upward. They fold or fracture as they yield to the immense stresses. Rain water gradually widens the fracture, which in the course of time becomes the location of a lake basin.

The earth crack called the Great Rift Valley cuts through the Middle East and eastern Africa from Syria to Mozambique. It covers a distance equal to one sixth of the earth's circumference. The lakes that fill its recesses are the Dead Sea, Albert, Tanganyika and Nyasa. The Dead Sea is the lowest-lying lake in the valley; its bottom is 2,600 feet below sea level. The valley also holds a series of brackish lakes, such as the Zawi, Abaya and Stefanie. Soda deposits are found in Lake Magadi, in Kenya, East Africa.

Huge masses of rocks may overturn so that older rocks lie on top of later formations. The hard and soft rocks upon the surface of the ground do not resist water erosion to the same degree. The weaker rocks are broken down much more rapidly, and they develop into the basins of lakes. The harder rocks stand as ridges that border the lakes.

Rocks that tilt upward may divide a body of water and form two lakes where before there was only one. Lake Ladoga in the Union of Soviet Socialist Republics was separated from the Baltic Sea in this way. This is the largest lake in Europe; it extends 130 miles in length and 80 miles in width.

Land distortions may shut off great areas from the sea, forming interior drainage basins such as the slightly saline lakes of Balkhash and Issyk-Kul in the Union of Soviet Socialist Republics. Lake basins develop where rising land masses sur-

The Märjelensee is a Swiss Alpine lake, formed by the Aletsch Glacier that bars its flow.

Swiss National Travel Office

round a plateau among still higher ranges. Lake Titicaca, between Peru and Bolivia in South America, was formed in this way. It is the highest navigable lake in the world; it lies at an altitude of 12,644 feet.

Lake basins formed
by the action of glaciers

Glaciers may gouge out lake basins or they may increase the size and depth of hollows that already exist in valley floors. Glaciers carry stones, sand and mud from the mountains that they drain; this stony material scratches and grinds the surfaces of the bare rocks over which it passes. Farther down the valley the ice gradually thins; it becomes less mobile on comparatively level ground.

When the glacier melts, it heaps a mound of stones called a moraine upon the valley floor. Moraines form barriers to the drainage of valleys; the water pent up behind them forms lakes. The Finger Lakes in west-central New York were formed by the action of glaciers long ago. Glacier National Park in northwest Montana contains more than sixty glaciers and nearly two hundred glacier-carved lakes.

Glaciers themselves may dam river valleys and cause lakes to appear. The Märjelensee, in Switzerland, was formed by the Aletsch Glacier; Lake Castain, in Alaska, by the Malaspina Glacier.

Hollows that later become the beds of lakes may be carved out by the action of the wind. The Seistan is an inland lake depression in eastern Iran and southwestern Afghanistan that has been deepened if not entirely formed by the wind. It is filled by a lagoon called Hamun-i-Helmand, which is something like seventy miles long during the rainy season.

The role of wind
in forming lakes

The action of wind on loose sand may also give rise to lakes, by damming up rivers where they flow into the ocean. During the flood tide, the ocean heaps up great quantities of sand on its shores; when the tide is low, the sand dries out and is caught up by the wind. If enough sand is blown into the river, its mouth may be closed and the water confined. Though the new body of water will at first be a salt-water lake, it may eventually contain fresh water. Lakes formed in this way may be temporary; the river, prevented from dropping its load of debris into the sea, may choke the lake with silt.

Lakes are sometimes created through the action of volcanoes. Lava floods may build lake basins, either by damming rivers or by creating depressions. Snow or rain may fill the extinct core of a volcano. Crater Lake in southwest Oregon lies in an enormous pit formed by the destruction of the crest of an extinct volcano now called Mount Mazama. The circular lake lies at an altitude of 6,164 feet; it is 6 miles across and 1,983 feet deep. Cliffs from 500 to 2,000 feet high line the shore. Several crater lakes are found on Mount Taal, a volcano in the Philippines.

In the tundra, or treeless plains of the northern arctic regions, a great number of frozen lakes occur. These lakes are called organic lakes; their basins are formed in areas where snowdrifts linger and discourage the growth of vegetation. When the drifts melt, such areas have a thinner growth of vegetation than the surrounding regions. Gradually these snow-covered areas sink deeper and deeper, and thus lakes are formed.

Movements and
temperatures of lake waters

When winds ruffle the surface of a lake, they heap water at the end toward which they are blowing. The degree of change in level depends upon the advancing waves. Another kind of temporary disturbance of level, called a seiche, occurs in many lakes in which the whole mass of water in the lake swings from shore to shore. These rocking movements are usually in the direction of the length of the lake. Each swing may last from a few minutes to several hours. Lake Geneva in Switzerland has experienced seiches of from several inches to several feet. In Swiss lakes they seem more common by night than by day, and in spring and summer than in winter;

Fishing in Lake Garda, the largest in Italy. It is over 1,000 feet deep in many places.

Beautiful Lake Tenaya, in California, lies between Lake Mono and the Yosemite Valley.

Around Lake Como, the third largest lake in Italy, are many fine villas and resorts.

NYSPIX — Commerce

Lake Placid, as seen from the summit of Whiteface Mountain. The lake is in Essex County, in the northeastern part of New York State. It is five miles long and has a maximum width of one and a half miles.

they frequently occur when the sun "suddenly begins to shine from amid heavy clouds." In Lake George, on the other hand, seiches are frequently associated with thunderstorms.

Seiches have been carefully studied at the Lake of Geneva; much of the information that we have about them comes from this source. The cause of seiches must be traced to some agency that heaps up the water at one end of the lake and then lets it fall back, thus producing the oscillations. Suspected agencies are minute earthquakes, or microseisms (see Index), prolonged winds blowing steadily in one direction and then stopping suddenly, and variations in barometric pressure. Seiches observed in the Baltic Sea (in northern Europe) were definitely related to changes in barometric pressure.

Investigations of lake temperatures have revealed that at the bottom of most lakes there is a mass of cold water whose temperature varies little from season to season. In lakes of 500 feet depth or more, a constant temperature of 42° Fahrenheit has been found at the 500-foot level and below, regardless of changes in temperature at the lake's surface. Measurements made at Loch Lomond, Seneca Lake, Loch Ness, Loch Oich, Loch Morar, the Lake of Geneva, Lake Maggiore, Lake Lugano and many other lakes seem to substantiate this particular finding.

Lakes perform a number of important geological and climatological functions. Like the ocean, they tend to temper extremes of air temperatures winter and summer, day and night. They also filter the water of the rivers that flow into them; river-borne silt settles to the lake bottom. The most important function of lakes, however, is that of regulating the amount of water on the surrounding plains. In periods of flood, lakes may be able to prevent rivers from overflowing their banks by taking up the excess water. In time of drought, lakes can supply the extra water needed by the rivers. Hence they are really storehouses of water.

Lakes are able to regulate the flow of water in rivers because they can hold so much more than rivers can. A large amount of water can be added to or removed from a lake without causing much change in its surface level. The levels of rivers, on the other hand, would be greatly affected by the same quantity of water. Suppose a river one-tenth of a mile wide flows into a hundred-square-mile lake and then continues for ten miles to the sea. If a quantity of water sufficient to raise the last ten miles of river twenty-five feet were added to the lake instead, it would raise the lake's surface level only three inches.

The regulating action of the Lake of Geneva on the Rhone River was measured and described by the nineteenth-century French geographer Jean-J.-E. Reclus. "The gauges used at Geneva," he wrote, "establish the fact that the discharge of the Rhone as it issues from the lake is, at its maximum, 753 cubic yards. Now, as the various tributary streams of the lake supply more than 1,400 cubic yards during their highest floods, it is evident that the Lake of Geneva acts as a complete regulator. It keeps back at least half of the inundation water, which it subsequently empties down gradually when its tributaries have retired to their usual level. It is certain that, owing to the regulating action exercised over the discharge of the river, the plains on the bank of the middle course of the Rhone, from Geneva to Lyons, are comparatively protected from floods."

There are both fresh and salt water lakes. Some salt lakes, such as the Caspian Sea, were formed by an uplift of the land that cut off an arm of the sea. Others, like the Dead Sea, derive their salt from the waters brought by their tributaries. As a rule, lakes with an outlet contain fresh water; those without one are saline.

Lake Superior is the largest fresh water lake; it is 350 miles long, 160 miles wide and covers an area of 31,820 square miles. Second in size, with an area of 26,640 square miles, is Lake Victoria, or Victoria Nyanza, in British East Africa. The Aral Sea in the USSR is third in size with an area of 24,600 square miles; it has no out-

Lake freighter, on Lake Superior, being pulled into dock. Lake Superior is the largest of the Great Lakes and is also the largest fresh-water lake in the world. It is 350 miles long and 160 miles wide.

Standard Oil Co. (N. J.)

let and is slightly saline. The five North American Great Lakes — Superior, Huron, Michigan, Erie and Ontario — cover a total area of 94,680 square miles.

The Great Lakes, large as they are, were dwarfed by a much larger lake that once existed in the area now occupied by the states of North Dakota, Minnesota and Michigan. Geologists have named this great body of water Lake Agassiz after the eminent nineteenth-century Swiss naturalist Louis Agassiz. It was a shallow lake, 700 miles long and more than 250 miles

from the rivers that drain into them. Salt lakes of marine origin have been cut off from the sea by movements of the earth's crust. They are not very numerous and tend to dry up.

The Caspian Sea is the largest salt lake and largest lake in the world. It covers an area of 152,540 square miles and has a maximum depth of 3,104 feet. It is believed that at one time an arm of the Mediterranean Sea reached northward from Asia Minor across Russia to the Arctic Ocean. The belief that it stretched all the

Michigan Tourist Council

Waves dashing against the shores of Drummond Island, on Lake Huron. This lake is the second largest of the Great Lakes. It is bounded on the north and east by Ontario; on the south and west by Michigan.

wide; its waters were derived from the glaciers that had invaded the land. When the great ice sheet receded, the waters of the lake drained away into Hudson Bay. All that remains of it are thousands of tiny lakes and ponds scattered over its former area. They give Minnesota its nickname "Land of a thousand lakes."

Salt lakes, as we have said, may be divided into two groups, on the basis of origin — lakes that have derived their salt from the sea and those that receive salt

way to the Arctic is supported by the fact that an Arctic species of seal is present in great numbers in the Caspian Sea, where a profitable seal industry is still flourishing.

The connection between the Mediterranean and the Caspian Sea is apparent. Between them lie the Black Sea and a chain of small salt lakes and marshes. Great deposits of sea shells stretch directly from the Mediterranean to the Caspian. Stretching northward from the Caspian to the Arctic are salt plains and more salt lakes that seem to trace out the former connection to the Arctic.

The surface of the Caspian Sea is now about 92 feet below sea level. Although this is partly due to a lowering of the land, evaporation is also a factor. At present, the influx of water to the sea from rivers just about balances the rate of evaporation so that the level has been stabilized. This is not true, however, in the case of the Aral Sea, which was once connected to the Caspian. It is slowly evaporating away and its level is becoming lower and lower each year.

Since rivers constantly add salt to the Caspian and since evaporation prevents an increase in the quantity of water, one would expect the Caspian to be excessively salty. It is rather surprising, then, to find that on the average it is not so salty as the oceans or even the Mediterranean. The reason for this apparent anomaly is that the shelving shores, and particularly the wide shallow inlet of Kara-Baghas, act as natural salt-pans, evaporating the thin layer of water covering them and causing a deposit of crystalline salt. The salt is thus being gradually withdrawn from solution, while evaporation is made good by a continual supply of fresh water from the rivers.

In the Kara-Baghas itself, which is an offshoot of the Caspian Sea, the waters are much saltier. It has been estimated that every day 350,000 tons of salt are swept into it through its narrow channel. So salty has it become that the seals have left it, and no vegetation will grow on its banks. Still saltier is the arm of the sea called Karasu, where the salinity rises to 5.7 per cent. All around the Caspian Sea there are shallow salt pools and lagoons where salt is concentrated and deposited. "One basin," says Reclus, "still occasionally receives water from the sea, and has deposited on its banks only a very thin layer of salt. A second, likewise full of water, has its bottom

Off the Iranian shores of the Caspian Sea, a fleet of small fishing boats is engaged in harvesting a rich sturgeon catch. The Caspian Sea is the largest lake in the world; its area is 152,540 square miles.

Romanoff Caviar Co.

hidden by a thick crust of rose-colored crystals like a pavement of marble. Others have dwindled to small pools of water or dried up completely."

The Dead Sea in Palestine is noted not only for its salinity but also for its depth below sea level. It is about five times as salty as the Mediterranean; its surface lies 1,300 feet below that sea's level. Fed by the Jordan River, the Dead Sea covers an area of about 360 square miles. The high salt concentration, about 26 per cent, prevents any sort of plant or animal life from existing in its waters; hence the name. It is not the concentration of salt alone that causes this complete absence of any form of life. The salt that makes up the largest proportion of the dissolved salts in the Dead Sea is magnesium chloride, rather than the normal sodium chloride of other salt-water bodies. This high percentage of magnesium chloride, about 15.9 out of the 26 per cent, is particularly destructive to life. Although the Dead Sea was at one time somewhat larger than it is today, it was never connected to the Red Sea. As evidence of this, it may be pointed out that iodine and certain fossilized foraminifera have been found in the Dead Sea, but not in the Red.

To the west of the Caspian lies Lake Elton, a very saline lake, whose bed consists of layers of salt constantly thickening from year to year. In summer, when evaporation is rapid and the lake diminishes in size, its shores look like immense, glistening snow fields.

In the northwestern part of Utah lies the Great Salt Lake, with an area of 2,000 square miles and an average depth of 20 feet. The water is extremely salty. Resembling ocean water in composition but nearly six times as dense, it contains 18 per cent solids. It has been estimated that the lake holds about 400,000,000 tons of sodium chloride and 30,000,000 tons of sodium sulfate. So dense is the water that it is buoyant enough to float in, but difficult to swim in. The only living things in the lake are some small worms, seaweed and a few shrubs that can tolerate the saline soil.

At present rainfall and evaporation about balance each other and the lake preserves a more or less constant level. It is a remnant of a greater lake, an inland sea as large as Lake Michigan, covering 20,000 square miles and over 1,500 feet deep, to which the name Lake Bonneville has been given. This was a fresh-water lake, which drained into the Pacific by the Snake and Columbia rivers; its beaches and shore terraces can still be traced along the slopes of the surrounding mountains. To the west of this great lake, there was a second fresh-water lake almost as large, now called by geologists Lake Lahontan, also draining into the Pacific Ocean. As meteorological and geological conditions changed, these immense lakes dried up and became more saline. The deserts of the Great Basin are largely due to the precipitation of salts from the retreating lakes.

Some forms of sedimentary rocks (see Index) are formed by the precipitation of salts by lakes and oceans. Evaporation of lake water having a composition similar to that of sea water would produce a precipitation of gypsum (calcium sulfate) followed by a precipitation of rock salt. This order of formation is found in all saliferous deposits. In cases like the Dead Sea and Lake Elton where the highest concentration is now of magnesium chloride, the less soluble sodium salts have already precipitated resulting in formations of gypsum and salt at the bottoms of these lakes. At one time these salts were dissolved in the waters of the lakes.

As we have already said, most lakes tend to be filled up. In the course of such filling up they are converted into a marshy condition. Many marshes are simply half-dried lakes. Not infrequently a lake full of water in the rainy season may dry up partially and become a marsh in the dry season. But marshes also originate in other ways. Along the course of rivers subject to overflowing, which traverse low-lying land, there are always swamps and marshes. Even apart from overflowing rivers, any great stretches of level land become marshes if there is sufficient rainfall.

The great plains of Brazil, crossed by the Paraguay and its tributaries, show a

Israel Office of Information

One of the many plants built along the shores of the Dead Sea to recover valuable potash from its waters. The potash, used for fertilizer, is exported from Israel to various parts of the world.

An air view of the Negev mountains, with the Dead Sea beyond, shows the region almost a desert. The Dead Sea, one of the world's saltiest lakes, is the source of salt and many valuable minerals.

series of immense marshes. Lake Chad, in Africa, is so surrounded by bogs that its true dimensions cannot be defined. Where the ground is rich in decayed vegetation, the marshes are known as peat bogs; these are especially interesting as the "factories" where coal is being prepared for future generations. There are enormous peat-bog formations in Russia, Ireland, the United States and Canada; there are also hundreds of thousands of acres of bogs in Germany, Sweden, Norway, England, Scotland, Holland, France and Denmark.

This is how peat is formed. The dead mosses and other marsh plants that are to be found in these bogs sink to the bottom as they are decomposing. As time goes on, the deposits become thicker and thicker; the lower layers, under the pressure of the water and of fresh deposits of decaying plants, become compressed and carbonized and yield peat. In the process of formation the peat contains tannin, organic acids and various salts, which make the substance antiseptic. Animals entombed in peat bogs are remarkably well preserved.

The reclamation of bogs for farming purposes involves draining and plowing up the soil to permit necessary oxidation processes to take place. Once reclaimed, the soil is very fertile; it is particularly well adapted for the growth of vegetables like celery, onions and the various roots.

The salt-water marshes along the shores of the sea are among the largest marshes in the world. A belt of marshland five to twenty miles wide extends along the coast of Florida for many miles. The marshes of this area are often known as mangrove swamps, because they are covered with thick jungles of mangrove trees, which have bird nests on their branches and crabs and barnacles among their roots. Along the shores of the Gulf of Mexico, there are millions of acres of marshes covered with cypresses; they are known as cypress swamps.

For centuries it was believed that swamps and marshes gave off poisonous vapors that caused malaria. It is now known, however, that malaria is not due to poisonous vapors but to a protozoan carried by mosquitoes that have been infected.

See also Vol. 10, p. 269: "General Works."

Tennessee Conservation Dept.

Cypress trees in Reelfoot Lake, a big, swampy tract in the northwest corner of Tennessee.

JET-PROPULSION PROGRESS

A New Era in the History of Aviation

BY JAMES J. HAGGERTY, JR.

ON AUGUST 27, 1939, the first flight of a jet-propelled airplane took place, and a new era in the history of aviation was ushered in. It has been an era of exciting progress. Planes powered by jet engines (also called reaction engines) are not subject to the power limitations imposed on piston-driven planes; as a result they have attained speeds and altitudes that even the most visionary scientist would hardly have predicted previously.

The principle of jet propulsion is quite simple. All jet-propelled craft are operated, wholly or in part, by jets of hot gases spurting from a tube — the jet nozzle — at the rear of the plane. At one time it was thought that these gases struck the air at the rear of the plane and then bounced back, causing the plane to move forward. What actually happens is this.

The thrust of gases out of the jet nozzle produces an equal thrust against the front part of the engine. This is an example of Newton's third law of motion: that for every action, there is an equal and opposite reaction. It is this "equal and opposite reaction" that accounts for the recoil of a gun that has just been fired. In the case of the jet plane, as hot gases spurt out of the jet nozzle, there is a "recoil" — a push against the front part of the engine. Since there is no thrust against the rear part of the engine, which is open to the outer air, the forward push propels the plane.

Practical work on jet propulsion applied to aircraft started long before 1939. The first to explore the field was A. A. Griffith, whose name has been all but forgotten. Griffith, a scientist with the British Royal Aircraft Establishment, conceived

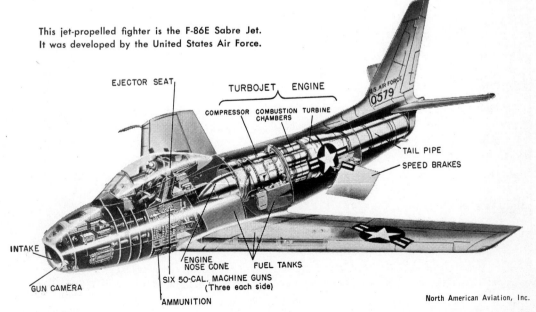

This jet-propelled fighter is the F-86E Sabre Jet. It was developed by the United States Air Force.

EJECTOR SEAT

TURBOJET ENGINE

COMPRESSOR COMBUSTION TURBINE
CHAMBERS

TAIL PIPE

SPEED BRAKES

INTAKE

GUN CAMERA

ENGINE
NOSE CONE

FUEL TANKS

SIX 50-CAL. MACHINE GUNS
(Three each side)

AMMUNITION

North American Aviation, Inc.

the idea of employing a gas turbine engine to propel an airplane; the turbine was to drive a propeller. He conducted some preliminary tests in the late 1920's.

In the meantime a young flyer of the Royal Air Force, Frank Whittle, had also been studying the application of the gas turbine to aircraft propulsion. His ideas differed from Griffith's in one very important respect. He proposed to eliminate the propeller; the plane would be driven entirely by the engine exhaust. The Air Ministry felt that Whittle's proposed engine had no military application and therefore it refused to finance the construction of an experimental engine. For several years, Whittle was unable to go forward with his project. In 1935, however, he obtained financial backing and in the following year he founded Power Jets, Ltd., a company devoted to jet research.

Power Jets, Ltd., constructed a prototype engine that was bench-tested in the spring of 1937. By current standards it

an airframe, which was to be powered by the first Whittle engine designed expressly for flight — the W-1. In May 1941, the airframe-engine combination was successfully flown for the first time. The engine provided a meager 850 pounds of thrust. (Thrust is the term used in the measurement of jet power; at 30,000 feet and at a speed of 375 miles per hour, one pound of thrust is the equivalent of one horsepower.)

Although Whittle is commonly held to be the father of jet propulsion, his jet engine was not the first to power an airplane in flight. In Germany, Hans von Ohain had been working on the jet principle since 1935, independently of Whittle and apparently unaware of Whittle's research. By 1937 Von Ohain had successfully bench-tested a small jet engine. On August 27, 1939, Von Ohain's Heinkel 178, powered by a jet engine called the Heinkel S3B, made the first successful jet-powered flight in the history of aviation. The en-

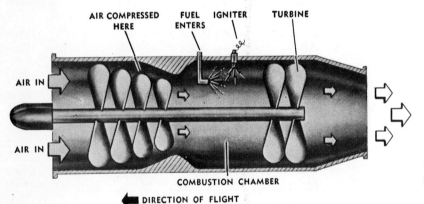

AIR COMPRESSED HERE   FUEL ENTERS   IGNITER   TURBINE

AIR IN

AIR IN

COMBUSTION CHAMBER

Diagram showing turbojet parts and how they work.

← DIRECTION OF FLIGHT

Curtiss-Wright Corp.

was not much of an engine, and it fell considerably short of Whittle's expectations; but it proved that jet propulsion was feasible. The company built a second and then a third engine. By the middle of 1939 the Air Ministry had changed its mind concerning jet propulsion. It decided to sponsor the construction of an engine designed for actual flight, and at the same time contracted for a new airplane in which the revolutionary engine could be used. The Gloster Aircraft Company received the contract for the construction of

gine developed slightly more than 1,000 pounds of thrust. In Italy, a Caproni-Campini plane made a jet-propelled flight on August 27, 1940.

Jet-propelled planes played no part in the early years of World War II. Later the Germans developed a pilotless jet plane, known as the V-1, or flying bomb or buzz bomb, which was launched from sites on the western European coast; the first one was fired on June 12, 1944. The V-1 flew until its fuel was exhausted; then it dived to earth and exploded. The Germans also

Boeing Airplane Co.

The Stratotanker-Stratoliner, a tanker transport plane which is driven by four jet engines.

built a huge rocket, called the V-2, which was first fired on September 8, 1944; it crashed into target areas in England at a speed much greater than that of sound.

Piloted jet planes appeared toward the end of the war. The British introduced the combat plane known as the Gloster Meteor I; it was powered by the Rolls Royce Welland engine, based on the original Whittle design. The Germans countered with the Messerschmidt 262, a jet fighter with a Junkers 004 engine. Neither the Gloster Meteor I nor the Messerschmidt 262 had much of a war record.

In the United States, jet-propulsion progress had lagged from lack of official interest, although some American engineers experimented with the principle in the late 1930's. In the early 1940's both the General Electric Company and the Westinghouse Electric Corporation began to work on jet-propulsion engines. In October 1942, the Bell XP-59, with a GE I-16 engine, took to the air. June 1944 marked

the highly successful debut of the Lockheed XP-80, which was powered by a GE turbojet known as the I-40 and later as the J-33. The P-80 became America's first operational fighter.

Jet propulsion made tremendous progress in the years immediately following the end of World War II, as engine efficiency increased. The cold war with Russia, the Korean "police action" and the continuing international tension lent new impetus to jet development, particularly as applied to fighters and bombers. Today engines of 10,000 pounds of thrust or even more are common.

There are four basic engine types that use the jet-propulsion principle: the rocket, the turbojet, the turboprop and the ramjet. There are also several variations and combinations of these types.

The rocket principle goes back many hundreds of years, but it has only recently been applied to aircraft. Germany put two rocket-powered aircraft into the air during

This is an experimental ramjet model shown in flight.

World War II, and one of them, the Messerschmidt 163, saw brief but ineffective service as a combat craft. Since the war, rocket engines have been used extensively as power plants for guided missiles and unmanned research planes.

A rocket engine is really only a combustion chamber in which fuel, either solid or liquid, is burned. The resulting gases expand enormously under intense heat and are forced out from the rear of the rocket tube at high velocity, creating thrust. Unlike other types of jet engines, the rocket does not draw air from the atmosphere; it carries its own oxygen with it, in the form of liquid oxygen or a compound, such as hydrogen peroxide.

The great disadvantage of rocket power is the extremely high rate of fuel consumption; most rocket engines can fly for only a few minutes. The chief advantage of the rocket lies in its ability to generate very high thrust in a comparatively small package. Rockets called jato units have been attached to jet planes or propellered planes to provide extra power for take-offs. (The word "jato" comes from "jet-assisted take-off.") Planes powered exclusively by rockets are still in the experimental stage. Because the rocket does not require air, it is the only known power plant that could be used in spaceships flying beyond the earth's atmosphere.

The United States has been actively engaged in developing rocket planes. The first one, the MX-324, took to the air in 1944. The SX-1, or X-1, was the first aircraft to exceed the velocity of sound (in 1947). It was powered with four alcohol and liquid-oxygen rockets, developing a total thrust of 6,000 pounds. In 1953 and 1954, the X-1A attained an altitude of more than 90,000 feet and a speed well above 1,600 miles an hour. The X-2 bettered these marks in 1955, reaching an altitude of 127,000 feet and flying at 2,200 miles an hour. Its rocket engine had a thrust of 16,000 pounds and could be controlled more effectively than the ones in earlier experimental craft. A number of other rocket planes were designed to solve different research problems. Generally, they were somewhat modified versions of the machines we have mentioned above.

The X-15 is a particularly powerful rocket vehicle. Equipped with a rocket engine fueled with liquid oxygen and ammonia and developing a thrust of 50,000 pounds, the X-15 can fly at more than 3,600 miles an hour 100 miles above the earth. As the first manned spaceship, it will answer many questions about travel in the uppermost reaches of our atmosphere.

The rocket power plants that are used in missiles are of two basic types. Some have only a short range and require only a brief duration of power. Power duration is longer in long-range missiles; after the fuel is used up, the rocket continues its flight on the momentum provided in this way. We give more details on rockets in another article. (See Index, under Rockets.)

The jet-type engine best known to the layman, perhaps, is the turbojet (turbine-jet). In this form of propulsion, air is

Fuel is injected and burned in the afterburner. This causes further expansion of the gases escaping from the engine, adding to the basic thrust and increasing the velocity.

In the early days of jet development, the turbojet engine used a great amount of fuel; continued research, however, has made its operation much more economical. Fuel consumption in this type of engine is measured by the pounds of fuel required to create one pound of thrust for an hour. At first it took well over one pound of fuel to create one hour's thrust-pound; in some modern engines, however, the consumption rate is as low as three fifths of a pound of fuel for an hour's thrust-pound. This is not much higher than the comparable rating of a piston engine. It is believed that the jet engine will use even less fuel than the piston engine in the not too distant future.

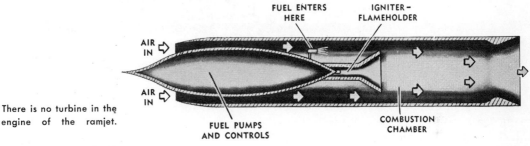

FUEL ENTERS HERE   IGNITER – FLAMEHOLDER

AIR IN

AIR IN

There is no turbine in the engine of the ramjet.

FUEL PUMPS AND CONTROLS

COMBUSTION CHAMBER

Curtiss-Wright Corp.

◀ DIRECTION OF FLIGHT

drawn from the atmosphere through intakes, usually located either in the nose or the wing roots of the plane. The air is compressed, led into a combustion chamber, mixed with injected fuel and ignited. The resulting gases are forced rearward at high velocity. They drive a turbine connected to the main engine shaft on which the compressor is mounted. Then they escape through the nozzle; the reaction from the spurt of gases provides thrust. The escaping gases therefore not only propel the plane but also serve to compress the air drawn in through the intakes.

Some turbojet engines are equipped with a device called an afterburner, or reheat tube. This is simply a tail pipe attached to the rear of the engine proper.

Speed is the major advantage offered by the turbojet type of engine, although up to the present time such speed has been obtained at the expense of range. Today, there are several types of jet-propelled craft capable of exceeding the speed of sound (about 660 miles per hour at high altitude) in level flight.

A variation of the turbojet principle is the turboprop (turbine-propeller) engine. The intake, compressor, combustion and turbine units are like those of the turbojet. The difference is that most of the energy released by the escaping gases is used to drive a propeller; some of it, however, provides thrust as the gases escape. Thus propeller drive and exhaust thrust are provided by a single engine.

The turboprop represents a compromise between the jet and the piston engine. It is slower than the jet but uses less fuel; it is faster and freer of vibration than the piston motor. It burns more fuel than the latter; but it can use low-grade fuel, costing less than aviation gasoline.

The ramjet, or "flying stovepipe," is the simplest of the jets that draw in air, since it has neither compressor nor turbine. Essentially it is a long tube containing a fuel injection system and a spark plug. To work effectively the ramjet must be moving at high speed; to operate most effi-

less interceptor" developed by the United States Air Force. The Bomarc is fired from the ground, using a rocket engine for the take-off. By the time the rocket fuel is exhausted, the Bomarc has attained a speed at which its two ramjets can operate efficiently. Ramjets are capable of far greater power than anything yet attained by turbojets; but they also have a very high rate of fuel consumption.

In all ramjet applications to date, the engine has been used in combination with the rocket or the turbojet. It is conceivable that it may be developed into an auxil-

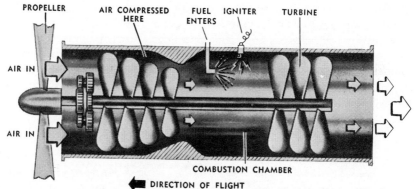

PROPELLER   AIR COMPRESSED HERE   FUEL ENTERS   IGNITER   TURBINE

AIR IN

AIR IN

Diagram of turboprop engine driving a propeller.

Curtiss-Wright Corp.

COMBUSTION CHAMBER

DIRECTION OF FLIGHT

ciently it must travel at speeds in excess of Mach 2. (Mach, or the Mach number, named after the Austrian physicist Ernst Mach, is the ratio of the speed of a body to the speed of sound in the surrounding atmosphere; therefore Mach 2 is twice the speed of sound.) Compression is provided by the so-called ram effect — that is, the sudden compression of the air as it enters the tube at high velocity; no mechanical compression is required. The compressed air passes through the combustion chamber; here the burning fuel creates the working gas, which is expelled rearward as thrust.

Because of the high speed required for ram compression, the ramjet has as yet found no application in manned aircraft. For one thing, having no mechanical compressor, it is incapable of taking off from the ground, that is, at zero speed. It does, however, have an application in guided missiles which operate within the atmosphere, such as the Boeing Bomarc, a "pilot-

iary power plant for combat fighters, which are now approaching the speeds at which a ramjet can operate. In such applications it would probably be used only for combat emergency power, as the afterburner is used today. It would not necessarily replace the afterburner, which could be used on take-off and in climbing.

Jet power is being applied at a rapidly increasing rate to commercial aviation. Today, giant turboprop and turbojet airliners, measuring well over a hundred feet in length and in wing span, carry up to two hundred or more passengers each, on sustained flights of thousands of miles and, in some cases, at velocities near the speed of sound.

Great Britain was the pioneer in the turboprop field. In 1950, Vickers-Armstrongs brought out the Viscount, a four-engine turboprop transport that has since become highly successful in Europe and in North America. The Viscount model 840 seats up to 75 passengers; with its four

2,080-horsepower Rolls-Royce engines, it cruises at 400 miles an hour; the maximum range may be 1,650 miles.

The Vickers Vanguard 952 is a much larger turboprop machine than the Viscount. Its four Rolls-Royce engines generate over 5,000 horsepower apiece; it can transport as many as 139 passengers and cruise at 424 miles per hour. The Vanguard's extreme range is 3,400 miles; its cruising ceiling, as much as 24,000 feet.

The Bristol Britannia 320 is another large English turboprop transport, with four Proteus engines of 4,445 horsepower each. Its maximum passenger capacity is 139; its cruising speed, about 400 miles an hour; its maximum range (at 357 miles per hour), nearly 4,700 miles. A number of other passenger and freight turboprop planes are manufactured in Great Britain.

Great Britain has also led the world in civilian turbojet flying. In 1951, the British Overseas Airways Corporation began long-range jet service between London and Johannesburg, South Africa, with the de Havilland Comet 1. After some serious accidents the Comet was improved, until today it is one of the most widely used jet liners. The Comet 4C is powered by four Rolls-Royce engines of 10,500 pounds thrust each, cruises at 509 miles per hour at 33,000 feet, carries up to 99 passengers and has a range of 2,450 miles.

Other countries have followed Britain's lead in the jet race. In the United States, the Boeing Company flew the first American jet-powered commercial transport — the 707 prototype — in 1954. Several different models of this four-engine plane are now being used by a number of airlines the world over. These craft are among the largest jet passenger planes in existence; they are powered with Pratt and Whitney or Rolls-Royce engines of different types, each providing a thrust of 12,500 to 18,000 pounds. An intercontinental version, the Boeing 707-320, has a wing span of 142½ feet and a total length of nearly 153 feet. It can carry 189 persons, cruise at better than 600 miles per

This plane, shown here in flight, is the British Vickers Viscount turboprop.

hour and fly over 4,700 miles at heights of up to 40,000 feet.

Convair is bringing out two jet transports, the 600 and the 880. The Convair 600, with four General Electric engines, can fly 635 miles an hour at 35,000 feet; its maximum range is 4,400 miles and its maximum passenger capacity, 121 persons. Each jet motor of the 600 is equipped with an additional turbine and fans at the rear; these devices compress the air flowing through the engine and exhaust it through special openings, thus increasing the operating efficiency. The slightly smaller Convair 880 has four General Electric jet motors, each with 11,200 pounds of thrust, and is able to haul as many as 110 persons. Its extreme range is over 4,300 miles; its highest cruising speed, 615 miles an hour at 25,000 feet. The 880 can reach 40,000 feet in altitude.

The Douglas Aircraft Company has also entered into commercial jet aviation with its DC-8 airliner. Four Pratt and Whitney or Rolls-Royce jet motors push the intercontinental version of this craft through the air at a height of 30,000 feet and at speeds approaching 600 miles an hour, for a range of well over 5,000 miles. At 550 miles per hour, the range is nearly 7,000 miles. Seating capacity for up to 176 persons is provided.

The Lockheed Electra 188 is a turboprop craft of medium capabilities. Its four Allison engines, each rated at 3,750 horsepower, can maintain a speed of more than 400 miles an hour; the general range is from somewhat over 2,700 miles to more than 3,000 miles. Passenger capacity is 66 to 98 persons.

The U.S.S.R. (Russia) has come out with an amazing variety of regularly used jet and turboprop transport planes in a relatively short time. Its Tu-104 went into routine service in 1956. This is a twin-jet craft, capable of cruising at better than 500 miles an hour and carrying 50 persons for almost 2,000 miles at nearly 33,000 feet. Its engines develop a combined thrust of close to 30,000 pounds. Modified versions of the 104 have maximum speeds up to 600 miles per hour

or more and have greater passenger capacity (but with reduced range and ceiling). The Tu-110 has four jet motors; it can cruise up to 36,000 feet at nearly 500 miles an hour for about 2,000 miles nonstop. Its highest speed is 620 miles per hour; its passenger capacity, 100 persons.

The Soviets have built one of the largest passenger planes in the world, the Tu-114, a turboprop ship that rivals jet transports in speed and range and exceeds them in carrying capacity. Capable of seating as many as 220 passengers, this giant, powered with four 12,000-horsepower engines, reaches a cruising speed of 565 miles per hour at nearly 30,000 feet of altitude. Its normal range is something like 6,200 miles. The Russians claim that the extreme range of the Tu-114 turboprop is 9,000 miles or even more.

Jet and turboprop airplanes are used for different purposes. The large turbojet serves particularly for longe-range and intercontinental flights or for shorter hauls with large passenger and freight loads. The principal application of the turboprop and the small turbojet is in short-range and medium-range flight, where high speed is not so necessary. Turboprops use less fuel than turbojets and generally they do not require very long runways. In recent years, aeronautical engineers have narrowed the differences in operation and capabilities between the jet and the turboprop airplane to a certain extent.

The future of the jet plane

As the world moves into the era of supersonic and spatial flight, jet and rocket engines will undoubtedly be used to a greater extent, wherever very high speeds are necessary and wherever it is impossible to use propeller-driven craft.

The practical use of atomic power in jet and turbojet machines has still not been realized, but experiments go on. A small nuclear reactor will probably heat air or water to operate the engines, so that a plane will be able to fly around the world without stopping to refuel.

See also Vol. 10, p. 285: "Aviation."

THE DEFENSES OF PLANTS

How Plants Ward off Their Animal Enemies

IN THE constant struggle for life, plants develop a great variety of defenses to insure the survival of their species. Certain plants find safety in numbers. For example, the grasses, which are otherwise quite defenseless, grow so quickly and flourish so abundantly that there is little danger of their being completely annihilated by browsing animals, such as cattle and sheep. Other plants manage to survive because they have efficient defensive weapons. The mountain laurel and rhododendron, for instance, have a strong poison in their leaves; other plants are effectively protected by barbs or spines or disagreeable odors. Certain plants have developed ways of attracting insect friends that either aid in pollination or destroy other insects injurious to these plants. For example, flowers of many plants contain nectar, which attracts bees and moths; these insects distribute pollen in their trips from flower to flower.

Unfortunately for most plants that are in danger of being eaten, it is the green shoots and the leaves that are consumed by herbivorous, or vegetation-eating, animals. The green parts of plants contain chlorophyll, and photosynthesis (see Index) takes place in these areas. If they are destroyed, the plant loses its food-making machinery. Therefore any adaptations that protect these vital centers make it more likely that the plant will survive.

An adaptation is a characteristic feature that allows a plant to successfully live under certain environmental conditions. The environment includes not only the conditions of the soil, the amount of moisture, the temperature and so on, but it also involves the other plants and animals that live in the same general area. An adaptation is the result of a long series of evolutionary changes. These changes are due to variations in the plant's heredity. If the variations are favorable and if they are passed on to the offspring, the species as a whole will have a particularly good chance to survive. Plants developing favorable traits of this kind are said to be naturally selected.

The statement (often heard) that a single plant or a whole plant species adapts itself to a certain environment is not quite accurate. It implies that plants are able to know beforehand the problems that arise from a change in the environment. Furthermore, it suggests that plants can change their structures or manner of functioning in order to meet such problems. This has not been scientifically demonstrated. With our present knowledge of plants and their responses to the environment, it is better to say that they become adapted by the natural selection of hereditary variations.

A herbivorous animal, desiring food, does not reason that by eating the green parts of a plant it will destroy the plant's capacity to produce more food. Men act somewhat differently because they usually save enough of an individual plant or a sufficient number of plants to replenish what they have taken for themselves. But this protection given to plants through man's foresight is only extended to a very few species — namely, those that he requires for his food, clothing or other products of his civilization. If plants were dependent for survival upon man's foresight, they would be hard pressed indeed, for the plants that he seeks to protect because they are useful to him are few in number. Plants must, therefore, depend upon their own defenses for survival.

There are a great many different kinds of plant defenses. Certain types protect the plant from all kinds of dangers, while others are effective only against certain enemies. There are defensive devices that serve only to repel enemies; others may do considerable harm to any animals that may attempt to attack the plant.

Plant poisons are efficient defenses

Among the most efficient plant defenses are the plant poisons, which take effect when the plant is touched or eaten. It is interesting to note that not all parts of the plant are equally toxic, or poisonous. Most of the poisons are concentrated in the plant parts that are most susceptible to attack: the leaves, in which food is manufactured, and the fruits and flowers, which serve for reproduction.

Different poisons produce varying effects on different animals. Poison ivy (*Rhus toxicodendron*), for example, brings about a very unpleasant skin rash in human beings, but it serves as the main source of food for certain insects which devour the leaves of the plant.

The many kinds of plant poisons

The plant poisons are made up of complex organic compounds — alkaloids, glucosides, resinoids and organic acids. They produce a variety of symptoms in the victims. Plants that are poisonous by contact, such as poison ivy, secrete a number of poisons belonging to the alkaloid group. The main effect, as we pointed out, is a rash — the condition known as contact dermatitis.

Plants such as sorghum, arrowgrass and certain species of wild cherries secrete a glucoside-type poison called prussic acid.* This poison kills within minutes after animals feed on a plant containing it. Another active ingredient in many plant poisons is the compound known as tremetol,

---

* Prussic acid (HCN) is more properly known as hydrocyanic acid; in its dilute form, it is called Scheele's acid. Its gaseous form, hydrogen cyanide, is one of the most deadly and fastest-acting poisons known. Hydrogen cyanide gas is used in gas chambers for the execution of criminals.

which is found in the white snakeroot and the rayless goldenrod. Tremetol causes the disease called "trembles" in grazing animals; it is characterized by muscular weakness and shaking. This form of animal poisoning is of particular concern to mankind, because the poison is very soluble in cow's milk and can be transmitted to man through that medium. In the human form, the disease is commonly known as "milk sickness."

Plants that secrete photosensitizing poisons

Some plants secrete what are known as photosensitizing poisons. These substances make an animal highly sensitive to light. If the animal remains in the shade, it will be unaffected, but if it steps into strong sunlight, it will develop a wide variety of symptoms; death may result. Photosensitizing poisons are secreted by such plants as the buckwheat and St.-John's-wort.

Some poisonous plants do not manufacture their poisons at all, but take them up directly from the soil. These plants are often referred to as seleniferous because their poison is so commonly the element selenium. They can absorb this element through their roots from certain types of shale. The locoweed, wild aster and prince's-plume are plants of this type. They cause the conditions known as alkali disease and "blind staggers."

Substances that make plants undesirable as food

Other plants take up from the soil certain substances which, though they are not toxic, render the plants undesirable to animals looking for food. One of the most common substances of this type is silica, the oxide of silicon. In the stems of horsetails, the leaves of rhododendrons and the needles of certain evergreens, one finds large deposits of silica. As a result, the plants become unappetizing and quite indigestible for animals. The same is true of certain kinds of vegetation found in the Australian bush.

Plants have found other ways of making themselves undesirable as food for animals.

PLANTS DEFENDED
BY THEIR POISON

The leaves, flowers or fruits of the plants shown on this page are avoided by animals because they are poisonous.

Barberry (*Berberis thunbergii*).

Hellebore (*Helleborus niger*).

Nightshade (genus *Solanum*).

Some rhododendron blossoms.

Left: monkshood (*Aconitum napellus*).

In order for an animal to like a food, it must have a pleasant taste. Certain plants are either very bitter or very sour. Most animals sampling vegetation of this kind will leave it severely alone thereafter. The sense of taste is very closely allied to that of smell. Plant defenses may be in the nature of more or less offensive smells, which cause animals to pass by what might otherwise be considered as a palatable food. The dog fennel and the hound's-tongue are among the plants whose odor is evidently very disagreeable to the ordinary grazing animals. In the case of poisonous plants with unpleasant odors, such as the water hemlock and the tobacco plant, an odor of this kind will protect not only the plant itself, but also the grazing animal, which will seek its food elsewhere. In fact, many of the poisonous plants that give off foul odors are also disagreeable to the taste — an effective double protection for the plants and for the animals that otherwise might be tempted to feast upon them.

Evil-tasting substances in fruits, flowers and seeds

In some plants, such as the peppers, mustard and horse radish, the leaves are not the worst-tasting parts. It is the reproductive structures — the fruits, flowers and seeds — that contain particularly unpleasant, though not poisonous, substances. These substances are chemically related to the poisons found in some plants. For example, the taste of mustard comes from a glucoside that is transformed by myrosin, an enzyme, into an oil (C_3H_5NCS). (You will recall that many of the plant poisons are glucoside- and alkaloid-type compounds.) Pepper gets its flavor from an alkaloid called piperine. Horse radish contains a volatile oil, butyl sulphocyanide, which is related to one of the deadliest poisons known to man.

Drops of water may protect plants

Most grazing animals do not care for plants on whose leaves drops of water or dew are present. Some plants have developed a special capacity for retaining drops of water, derived from dew, much longer than other plants do. Among the varieties using this defensive device are the teasel and the compass plant.

How animals are able to avoid poisonous plants

Although the species of poisonous plants are comparatively few in number, the number of individuals is large. With so many poisonous plants occurring over a wide range, one might wonder how animal life has survived so well. The answer is that animals seem to be able, in some way that we do not understand, to avoid these dangerous forms of vegetation.

It is an old belief that in the jungle, if a man eats only what he sees monkeys eat, he will run no risk from poisonous vegetation. Certainly, animals do seem to know instinctively which plants are safe for them to eat. As we have seen, they shun some plants because they give off an unpleasant odor. Even if poisonous plants do not give off any odor — or at least an odor that is perceptible to human beings — animals seem able to avoid them. Certain grazing animals also keep away from various mosses and ferns which, as far as we know, are not poisonous. We can only guess that such plants are harmful to the animals in some way.

Sharp weapons that defend plants

Some plants protect themselves by warding off their enemies with sharp, pointed weapons such as spines, prickles and thorns. Spines are modified leaves. The spines of the cactus plant, for example, are formed when most of the leaf dries up and falls off, leaving only its central structure. The spine-type leaf helps the plant conserve water by presenting much less surface for evaporation. Thorns are modified twigs; they serve exclusively for protection. Prickles are derived from epithelial, or surface, tissue. They also serve only for protection.

As in the case of plant poisons, the spines, thorns and prickles are distributed strategically on the plant so as to pro-

N. Y. Botanical Garden

The sharp spines of the cactus.

The thistle is a prickly plant.

L. W. Brownell

PLANTS THAT WARD OFF THEIR FOES

J. Horace McFarland

The leaves of the eryngo plant, above, have several sharp spines along the edges.

The leaves of the aloe plant (left) have a very sharp apex and their margin is spiny.

Standard Oil Co. (N. J.)

79

tect its important parts — the leaves, fruits and flowers. They are arranged in such a way as to come in contact with any animal attempting to feed on the plant. The distribution and size of the protective structures depend upon the kind of animal most likely to attack the plant and the probable point of attack.

The huge floating leaves of *Victoria regia* are protected only on the under surface and at the margins, because only at these points is it vulnerable to attack. Other plants are armored only at the lower portions of their stems. Above a certain height, they are quite safe from the attacks of grazing animals.

Defensive structures
that protect young plants

In some cases, a plant has defensive structures only during a part of its life. A young plant that is weak and tender may be armed with thorns until it grows into a more sturdy individual that is much less likely to be attacked. Plants that are armored only on their lower parts are of this particular type.

Earlier in this article, we mentioned that the leaves of certain plants are undesirable as animal food because they contain silica deposits. Other leaves are not suitable for grazing animals because they are formed in such a way as to cut the mouths and internal organs of animals that try to eat them. The pine needle is a case in point. It is interesting to note that not only are such leaves very similar to needles in appearance, but that in certain parts of the world they are actually used by the natives as sewing needles.

Defensive weapons
determined by geography

Plants with needlelike leaves are very common in dry, arid regions; the leaves protect the plants and help conserve water. Protection against animals is especially important in these areas because of the scarcity of other types of plants on which the animals can feed.

Similar in shape to the needlelike leaves are the long, narrow sturdy ones of the South African aloe plant. This type of leaf is not only long but also pointed; an animal would have to be something of a sword swallower to eat it. The margins of the leaf are protected by small spines.

Spines that are like
the teeth of a saw

A number of other plants have spines along the margins of their leaves for defensive purposes. The spines are like the teeth of a saw; they cut the lips and mouths of any grazing animals foolish enough to try to eat them. To reinforce these cutting edges, the leaves are often impregnated with silica deposits. Some plants, such as cut-grass, contain so much silica that the leaves do not even require spiny margins. They cut like knives and are capable of inflicting serious wounds.

The hypodermic needle
of the stinging nettle

Certain plants have combined the needle defense with poison to produce even more effective protection. The stinging nettle is of this type. Its barbs act in much the same way as hypodermic needles, piercing an attacking animal and injecting poison at the same time.

When one brings one's hand in contact with certain leaves, such as those of the stinging nettle, there will be unfortunate effects. Stinging hairs, each composed of one large cell, will break off from the leaves and will fasten securely upon the hand. They will inject an irritating substance into the skin — a substance related to formic acid ($HCOOH$), which is the "venom" of certain red ants and some other insects. If a number of hairs, concentrated in a small area, have penetrated the skin, the injected substance may bring about considerable inflammation.

The defenses of the stinging nettle are effective only against attack from above. The hairs, which are quite elastic, are pointed upward; they are also restricted to the leaves, the lower parts of the plant being bare. If one passes one's hand upward so as to press down upon the hairs on the leaf, the hand will glide over them without mishap; the hairs will simply be pressed against the leaves. But if one reverses the direction of motion, the slightest touch will break off a number of hairs, releasing their poison.

Plants, therefore, have a great variety of natural defenses. Essential plant parts, such as leaves, fruits and flowers, may be disagreeable in odor and in taste or even poisonous to animals that might want to feed upon them. Some plants are provided with spines, prickles or thorns that discourage browsing animals. Others are protected by pointed, barbed and stinging hairs.

Plants that protect themselves by mimicry

Plant defense, however, is not entirely a matter of nature's having provided actual defensive weapons. The plant has still other resources. It may turn to subterfuge, for example, to protect itself. Thus we have very curious instances of mimicry, where a striking resemblance to another plant or object is unmistakable. Perhaps the best-known example of this kind of mimicry is the common dead nettle. It takes its name from the fact that in general appearance it looks extremely like the common stinging nettle. Actually, however, it has none of that plant's unpleasant properties. Some plants in South Africa look very much like pebbles, while others greatly resemble in coloration the ground in which they grow.

Nature's ingenuity is seldom shown more clearly than in the relationships that come under the general heading of symbiosis (from the Greek word that means "a living together"). This term describes a relationship formed for the mutual benefit of two dissimilar organisms. These may be two different kinds of plants or animals, or a plant and an animal. In many cases, the two organisms involved can live independently, although their association brings benefit to them both.

The lichen provides an example of symbiosis

The lichen is perhaps the classical example of the working of symbiosis, for it is a plant that has developed entirely from the interaction of two other plants upon each other. These two plants are an alga and a fungus, each of which contributes, by its own characteristics, to the formation of the lichen. The alga is a colony of green cells embedded in the fungus. In the alga the process of photosynthesis goes on, while the fungus threads absorb necessary moisture and extract needed minerals from the air and water.

Certain plants, or trees, known as ant plants, form an almost symbiotic type of relationship with ants; they provide the ants with food or protection, or both. In return, the ants protect the plant from the invasion of other insects. The actual protection here may be carried out by the

The leaves of the aquatic Victoria lily are often five feet across, with a rim several inches high.

William Lavarre, from Gendreau

ants' hurling their excretion of formic acid against the unwelcome visitors. Sometimes ants have a kind of symbiotic relationship with plant lice. Their attempts to maintain too large a population of lice on a given plant may destroy that plant. The corn-root louse, for example, is cared for by an ant that lays bare the roots of the corn plant and then places the lice on the exposed root surfaces and thus greatly damages them.

Surrounding water
may protect plants

A gardener may surround a plant with water, to protect it from some of its small enemies that can neither fly nor swim. He may stand the potted plant in another pot containing water. Many water plants, such as the water lilies, are protected by nature in a similar way. Bees and other flying insects that visit the plant in search of nectar and pollen are not kept from the plant, while the injurious snails and insects, such as ants, are effectively warded off. In some cases, a slight reservoir of water habitually accumulates at the base of the plant leaves, as in the compass plant and the teasel, and this water pocket serves much the same purpose, though on a considerably smaller scale.

Sticky secretions
may provide protection

Even more common is another kind of effective defense. Extremely sticky secretions from the plant tissues lie upon the surfaces of some plants, to trap insects that crawl over them. This excretion may come from the epidermis (the external skin or coating) or from special glands or hairs. It is most commonly found on the flower stalk or on the principal stem. The protection derived from such sticky secretions sometimes extends not merely to the flower but to the whole foliage, as in some of the primulas and saxifrages. On the leaves of these plants the dead bodies of insects are often found, caught in the sticky protective substance.

A waxlike covering, present in so many flowering parts, is still another means of protecting the flower. It is no defense against snails, which can pass over it with no difficulty, but it does protect the plant against the many varieties of hard-bodied little insects that so often launch such extremely damaging attacks on both leaf and flower.

Plant defenses against winged insects, usually in the form of hairlike structures inside the flower, are constructed in various ways. In addition, the actual arrangement of the flower structure itself, especially as regards the hiding and protection of the nectar, is often of the greatest ingenuity.

How undesirable insects
are kept away from nectar

In many plants, nature has provided mechanical arrangements to protect the flower nectar against insects that could creep up the stems from the ground. One of the most interesting of these is found in some of the balsams. Special glands developed from the stipules at the leaf bases secrete a sweet discharge that attracts insects climbing up the stems. They come in contact with a drop of this sweet substance at the base of each leaf and find it as satisfying as they would the flower nectar. It also has the advantage of being closer at hand. Ants in search of nectar go so far, therefore, and no farther, and the nectar is safe. In the absence of such sweet stuff at the leaf bases they would, of course, go on to search out the nectar within the flower itself and thus interfere with the visits of the winged-insect pollen-carriers, which are so essential for the successful development of the plant.

Plant defenses may
attract as well as repel

Plant defense involves not only the repelling of enemies but the attracting of friends. Under this heading we include nature's many devices to attract insects to different parts of the plants where they will find food; the arrangements of petals to aid in cross-pollination; the provisions for the dispersion of seeds and fruits and their protection.

Ewing Galloway

A deadly enemy of insects, this open-mouthed fly-catching goose plant is here pried open with a stick. When an insect touches it, the lips snap shut and the tiny victim is smothered and then digested at leisure by the plant, which was imported from the West Indies to Ceylon.

American Museum of Natural History

An adult walrus taking his ease. The walrus is related to the seals; it is a heavily built animal, which reaches a length of thirteen feet. Its formidable tusks are sometimes almost three feet long.

FIN-FOOTED CARNIVORES

Seals, Walruses and Sea Elephants

A PERSON unfamiliar with sea lions may wonder, when he first sees these comic and entertaining creatures at the zoo or circus, just what kind of animals they are. They have coats of hair like mammals, but their flippers seem to have nothing in common with the feet of mammals. Are these flippers like the fins of fish, or are they like the paddles of the whales and dolphins? The truth is that sea lions are flesh-eating mammals that long ago returned to the sea to live an aquatic life. The limbs of the sea lions and their kin — the walruses and seals — have bones that correspond to those found in the limbs of the land mammals. Moreover, the teeth and skull bones of these aquatic creatures show they are related to the carnivorous, or flesh-eating, mammals of the land.

Scientifically, the seals and walruses are classified in the suborder Pinnipedia, which means fin-footed. The pinnipeds are large, bulky mammals with rudimentary tails. Their external ears are very small or absent altogether. The body hair of the different groups varies from the finest of fur to coarse bristles. The pinnipeds prey upon fish, squids and other mollusks, crustaceans and, occasionally, sea birds. We can trace the pinnipeds back as far as the Miocene period (about 25,000,000 years ago). Some zoologists think that the finfoots were derived from the earliest and most primitive carnivores called the creodonts. Others believe that they evolved from a later carnivore stock that produced the dogs, bears, weasels and otters. The fossil remains of a mammal with seallike characteristics were found in a Siberian geological deposit that was about 10,000,-000 years old. This mammal had a long, slim tail and limbs that indicate it could walk on land as well as swim. It also had features somewhat similar to those of the otter. This ancient carnivore itself could not possibly have been an ancestor of the pinnipeds; however, it may have been derived from the ancestral group that gave rise to the modern seals and walruses.

The limbs of seals and walruses have undergone a strange transformation. The upper and lower leg bones have been much shortened; the front and hind feet have been lengthened. There has been no increase in the number of toes. A web of skin grows between the toes, thus forming the flipper, or fin. The fin-foots, unlike all other aquatic vertebrates, swim entirely by means of their limbs. The two hind legs are directed backward to serve, in swimming, as a substitute for a tail. The front legs are well suited for steering. The eared seals, which include the sea lions and fur seals, can bring their hind legs around in a forward position when they are traveling on the land. This is also true of the walruses. The earless seals, of which the elephant seal is an example, cannot turn their hind limbs forward at all. When on the beaches these animals are barely able to drag themselves along.

Why should seals and walruses retain leglike limbs for land travel when in other respects they are so remarkably adapted for life in the sea? The reason is that all the members of the pinnipeds must leave the water to breed and raise their young and, consequently, need supports on land for their large bodies. When the ancestors of the fin-foots first took to the water, they were probably as agile on the land as most other carnivores. But in time they became adapted to an aquatic environment and came to the land less frequently.

There are three families of pinnipeds. Included in the first (family Otariidae) are the fur seals and sea lions. These animals have long necks, small external ears and front limbs nearly equal to the hind limbs in length. The northern fur seals (genus *Callorhinus*) of the north Pacific and the southern fur seals (*Arctocephalus*) of the Southern Hemisphere possess a luxuriously soft and dense underfur. The sea lions, which lack this underfur, include the California sea lion (*Zalophus*), the northern sea lion (*Eumetopias*) and the southern sea lion (*Otaria*). The California sea lion has a high forehead and glistens black when wet. It gives honking barks. It lives along the coast of southern California and through the central Pacific to New Zealand, Australia and Japan. The northern sea lion, which is larger and has a low forehead, ranges chiefly in the North Pacific, though it will go as far south as southern California.

The walrus is a
large pinniped with tusks

The walrus (*Odobenus*) of the Arctic is confined to a family (Odobenidae) of its own. Its great size and large tusks distinguish it from other aquatic mammals.

The earless, or hair, seals (family Phocidae) lack external ears, have a very short neck and possess forelimbs much shorter than the hind ones. Common seals (*Phoca*) of the northern seas include the small and spotted harbor seal; the ribbon seal, which has yellow bands around its neck, rump and front flippers; the ringed seal, with whitish rings on its sides; and the harp, or Greenland, seal, which has along its sides dark bands that join across the shoulders and over the rump. The large bearded seal (*Erignathus*) of the Arctic is distinguished by bristles on the sides of its muzzle. Another large seal of Arctic waters is the gray seal (*Halichoerus*), a uniformly grayish animal.

The male hooded seal (*Cystophora*) of the North Atlantic differs from other seals by having a pouch atop its forehead and snout. This bag is inflated when the animal becomes annoyed. The closely related sea elephants, or elephant seals (*Mirounga*), are giant species living in the Antarctic, South Pacific and Indian Ocean and off the coast of Lower California.

Other seals of the Antarctic and South Pacific are the crab-eating seal (*Lobodon*), which varies in color from brown to yellowish white; the sea leopard (*Hydrurga*), the largest Antarctic species aside from the elephant seal; Ross's seal (*Ommatophoca*), a small blackish seal; and Weddell's seal (*Leptonychotes*) of the south polar seas, blackish, peppered with silver and gray. The monk seal (*Monachus*) is a large animal, black above and whitish below, inhabiting the waters of the Mediterranean, Caribbean and central Pacific.

Pinnipeds are specialized
for swimming and diving

Pinnipeds are graceful and powerful swimmers. Sea lions, when they are swimming slowly, use their front flippers as paddles. When they go faster, they also work the hind limbs. The elephant seal moves only its rear flippers to swim. Weddell's seal presses together the bottom surfaces of its hind feet and waves this apparatus from side to side. Pinnipeds often dive to considerable depths and remain submerged for some time. The maximum time spent under water has been recorded for several species; for instance, the elephant seal can stay submerged twelve minutes, the harbor seal, fifteen and the harp seal, twenty minutes. How is this possible? First of all, these mammals, unlike land species, almost completely renew the air in their lungs when they breathe. The respiratory center of their nervous system does not respond quickly to an increase of carbon dioxide in the blood and so does not demand more rapid breathing. Finally, pinnipeds, when under water, efficiently distribute breathed-in oxygen so that it goes only to those parts of the body involved in essential activity.

Almost all pinnipeds are gregarious, or social, animals, though the sea leopard, Ross's and the bearded seals seem to lead solitary lives. Of the earless seals, only the male gray seals and elephant seals

Northern fur seal (*Callorhinus ursinus cynocephalus*).

Hawaiian monk seal (*Monachus schauinslandi*).

Galapagos fur seal (*Arctocephalus philippii*).

SEALS AND SEA LIONS

Two fine specimens of the Peruvian "snub-nosed" sea lion (*Otaria flavescens*).

Photos, Zool. Soc. of San Diego

gather harems of females in the mating season. Some earless seals migrate to their breeding ground, but other species rear their young upon shores or ice floes wherever they happen to be. Sea lions and fur seals, on the other hand, make long migrations to their breeding territories, where the males gather a large number of females together into harems.

Description of a
fur-seal rookery

The bulls arrive at the breeding grounds about the end of May or the beginning of June and take up positions along the shore, awaiting the coming of the cows a few weeks later. The cows are heavy with young when they arrive. Each male now seeks to obtain as many cows for his own harem as possible. He emits loud roars designed to frighten away his competitors. Violent fights ensue; the bulls inflict serious wounds with their sharp, powerful teeth. As the bulls seek new cows, the ones they have already secured may be stolen by other seals farther away from the water's edge.

The stronger the bull, the more mates he obtains. The harems of the large bulls average about thirty cows; some bulls may manage to gather as many as a hundred cows. The competition is so strong and so continuous that the bulls will not leave the beach to feed at sea. As a result, they may go for several months of the year, fighting and mating, without eating at all. In the competition for mates, the larger bulls obviously have the advantage. Apparently, the seals that produce the largest male offspring are favored in the process of natural selection. At any rate, the male seal often weighs five or six times as much as the female.

Shortly after the cow arrives at the breeding ground, she gives birth to the single pup conceived the previous season; about a week later she mates for the next year. Four to six weeks after birth, the seal pup learns to swim. During the summer, the cow frequently swims out to sea to feed, returning at gradually lengthening intervals to nurse her pup. The senses of the seal are apparently so acute that she has no trouble in finding her own offspring among the many thousands she encounters when she returns to the swarming beach. About four months after it is born, the young seal is weaned.

With the approach of winter, the seals move out to sea until the following summer. By this time, the seal pups are far enough along in their development so that they are able to accompany their mothers on the long swim to warmer latitudes. On their migration to and from the summer nesting grounds, seals are subject to attacks by many sea predators, particularly the killer whale, which may swallow as many as twenty or more seals at a single feeding. It is estimated that not more than half of the seal pups ever reach maturity.

Because of its habits, the seal is an easy prey. When the Russians claimed the Pribilof Islands in the eighteenth century, Russian seal hunters wrought havoc among the huge hordes at the rookeries. The seals, quite helpless on land, were clubbed to death by the thousands. They were skinned on the beach and the carcasses simply left to rot. The animals were hunted so avidly that one species, the *Hydrodomalis stelleri,* became extinct within twenty years of its discovery. Several other species became too rare to be hunted profitably. In time the Russian government realized the seriousness of the situation and put a stop to this destruction of valuable resources. Then, in 1867, the islands were purchased from the czars by the United States; the sealing rights were leased to private companies and the slaughter was resumed. As the demand for furs grew, pelagic sealing — that is, catching seals at sea — became common. Since the herds migrated in huge groups at fairly regular times, they were easily located. In the late nineteenth century, over one hundred ships were engaged in sealing. They claimed many cows as their victims. Unfortunately, when a cow was killed, the herd lost not only that female but the pup it was carrying. If the female was killed in the late summer or early fall, it also meant the death of the pup she was nursing.

But the fur seal, happily, has been saved from extinction, at least for the present. The nations most interested in the sealing industry and therefore concerned with the rapid depletion of the herds finally agreed upon measures safeguarding the fur seals. These nations were Great Britain, Russia, Japan and the United States. They drew up a pact called the fur-seal treaty of 1911. Under its terms, sealers of any of these nations could not kill seals at sea in the waters of the North Pacific and adjacent seas north of the 30th parallel. Exceptions were made for the native people of the North Pacific who depend largely upon seals for their subsistence and use primitive methods in hunting them.

The two most important fur-seal rookeries are those located at the Pribilof Islands, Alaska, and the Commander Islands, in the Bering Sea. The Pribilofs belong to the United States and the Commander Islands to Russia. According to the treaty, the management and killing of the fur seals were left to these two countries. Provision was made for Japan and Canada to each receive 15 per cent of the skins taken by the United States and Russia. This fur-seal treaty was terminated in 1941.

A new Alaska fur-seal law, agreed upon by Canada and the United States, still prohibits the killing of seals at sea. Also, the law provides that Canada is to receive 20 per cent of the skins taken on the Pribilof Islands. The remainder of the skins are to be kept by the United States. Sealing operations on the Pribilofs are carried on by natives of the islands who are under the direct supervision of the United States Fish and Wildlife Service.

Since one hardy old male, or bull, seal may maintain up to one hundred cows in a

U. S. Fish and Wildlife Service

A fur-seal rookery at one of the Pribilof Islands. A large bull is in the foreground.

Baby walruses are nursed by their mother almost two years before their tusks grow enough to help them dig for food.

Zool. Soc. of San Diego

harem, and since an equal number of male and female seals are born, there is an excess of bulls in every rookery. These surplus bulls, which are called bachelors, are either too young or too weak to fight for a harem; they accordingly separate themselves from the rest of the herd. These are the seals that are killed for their pelts. A certain number of bachelors are allowed to escape, however, so that they may become harem masters in the future.

The bachelors are usually killed and skinned, or pelted, in July. The pelts undergo preliminary preparations, such as salting, at the islands. Then they are shipped to a government-supervised factory in St. Louis where they are dressed, dyed and sold at public auction.

The walrus is another fin-foot that may visit the Pribilofs, though it usually keeps to the Arctic seas. It is a good swimmer but cannot endure a long-distance trek such as is made by the fur seal. Often it will climb aboard ice floes and let the ocean currents carry it along. Walruses begin their northward migrations in May, and during the journey young are born and mating occurs. When the Arctic waters freeze over, walruses may go as far south as Labrador in the Atlantic and to the Bering Sea in the Pacific.

The walrus is a ponderous seallike mammal, weighing as much as 3,000 pounds. It has a massive neck, which supports a proportionally small head. Its muzzle is broad and blunt and is studded with coarse, stiff bristles. Short, reddish hairs scantily cover its thick and wrinkled hide. Like the eared seals, the walrus turns its hind feet under its body to assist it when walking on land or ice.

The characteristic mark of adult walruses is the possession of a very fine pair of ivory tusks, which are really elongated canine teeth of the upper jaw. The tusks of the male are quite straight and may grow to a length of thirty-nine inches. Those of the female are shorter, more slender and somewhat curved. The tusks make formidable weapons in defense and in fighting for a mate. They are also efficient tools for digging in the sediment of the ocean floor for mollusks, such as clams, snails and oysters. Besides mollusks, walruses relish starfish and shrimp. When the young, tuskless walrus is still nursing, it will often cling by its front flippers to the mother's neck as she swims and dives in search of food.

Sociable by nature, walruses keep together in small herds, spending their time on shores or on ice floes. Though usually mild in manner, they become aggressive during the mating season, or when attacked or when their young are threatened. Man is their greatest enemy; they also fall victims to polar bears and killer whales.

The prosperity of some Eskimo tribes of the far north depend to a great extent upon the herds of walruses. These Eskimos use every bit of a walrus carcass. The flesh is eaten, the hide is made into leather, the intestines serve as a substitute

and hooded seals are somewhat similar. These animals often migrate together. Early in February they head northward for the Strait of Belle Isle or the Gulf of St. Lawrence, where the young are born, or whelped, sometime in March. Until the young harps are able to swim, they are left on sheets of thin, drifting ice. The old seals leave and approach their babies through holes made in this ice. The hooded seal young are whelped on ridges

American Museum of Natural History

Sea elephants are giants among seals. The male has a long, flexible snout that can be inflated.

for window glass, the tusks are carved and the oil is burned to provide light.

Of the true, or earless, seals, the harp and hooded seals are the ones chiefly hunted by commercial sealers for their hides and blubber. Like other true seals, they do not move well on land or ice. Not only are the hind limbs of little use, but their fore limbs are quite small and their necks are so short that they can barely lift their heads. Their pelts have long, coarse hair and lack the characteristic underfur of the fur seals. The habits of the harp

of ice, which are approached directly from the sea. Hooded seals scatter in small groups away from the herds of harp seals.

Other true seals are found along rocky coasts or about islands and ice floes of most seas. Certain species even live in inland waters. The Caspian seal dwells in the Caspian Sea; the Baikal seal flourishes in Lake Baikal, which is a large freshwater lake of Eurasia. Since these seals are awkward on land, their presence here suggests that these waters had a former connection with the open sea.

But seals can travel on land, if necessary, — a short way in a long time. It is recorded that a gray seal traversed fully thirty miles of snow-covered land in Norway, the time taken, it is believed, being about a week. Whether they would have endurance enough to find their way from some other watercourse to the inland seas mentioned is, of course, another matter.

Many seals are destroyed on land by polar bears, and many in the waters by killer whales. But the shark is also an enemy of this animal. The common seal along our coasts is called the "harbor" or "leopard" seal. It is one of the small-

Because of lack of space we cannot mention here all the various genera of the true seals; but the sea elephant, or elephant seal, must be noted. This is indeed a prodigious beast, with a length of from to 20 to 22 feet, and a girth of from 15 to 16 feet, these dimensions relating to males only, for the females are considerably smaller. The enormous coating of blubber by which the animal is enveloped is a perfect protection against cold, for it has been found that the body of a sea elephant that has lain for twelve hours in the icy water of the Arctic fully retains its internal heat. Formerly to be found in

New York Zoological Society

A California sea lion sunning itself. The animal's coat glistens black when it is wet.

est of the true seals, rarely exceeding a length of five or six feet, and is whitish in color, heavily spotted with gray. It sometimes finds its way into the Great Lakes or other inland bodies of water having a connection with the sea, but it is nowhere regarded with friendly interest by fishermen whose nets it frequently robs or even destroys in its pursuit of fish. It is found in both the Atlantic and Pacific as far south as North Carolina and Lower California though not abundant south of Maine and British Columbia. It is not as gregarious as other seals and, therefore, in no present danger of extermination.

enormous herds, sea elephants are now relatively few in number.

One species, the California sea elephant, was nearly exterminated in the early years of the twentieth century. Conservation practices in those days were decidedly sketchy, to say the least; the animals had been considered fair prey for years by whalers and sealers. Naturally there had been an indiscriminate slaughter of the helpless beasts; the few that remained made their final bid for life on the island of Guadelupe, off the coast of Lower California. Since there was no law, written or unwritten, to protect them, it seemed only a matter of time before

Herbert J. Pontine

A Weddell seal about to dive for its prey.

they would all be slaughtered by the rapacious hunters of whales and seals.

But this was not the only threat to their existence. A rich English collector, who had become interested in these huge animals, was appalled to learn that there was not a single adult specimen in any museum in the world. He determined, therefore, to send out an expedition to Guadelupe in order to obtain a number of specimens for museums before the animals should be utterly exterminated. He was hotly denounced by many naturalists and by animal lovers generally. They felt that it was all very well to provide specimens for our museums, but that it was not right to seek specimens of animals that

Australia News and Information Bureau

A fur seal that is found in Australian waters.

were so near to extinction. But the expedition was sent to Guadelupe nevertheless.

The members of the expedition, with true scientific zeal, photographed the animals as they lay on land, then killed them and sent the carcasses home to be preserved. At the same time a horde of whalers and sealers descended upon the island and began their own slaughter of the innocents. The sea elephants on Guadelupe, menaced by commercial killers and scientific killers, seemed doomed to destruction. Fortunately, however, some of the animals either managed to escape detection or else were not

U.S. Fish and Wildlife Service

The harbor seal frequents harbors and river mouths.

on the island when the slaughter took place. When Guadelupe was visited by an exploring party in 1919, a small herd of the enormous beasts was still holding its own.

These seals breed from February to June; during this period they are often to be found on the shore. Here they are sluggish and ungainly. They appear to be quite unaware that they have a deadly potential enemy in man; they show no signs of fear when humans approach them. When they are attacked, their attempts to escape are pathetically futile, since they move about with difficulty on land and quickly become exhausted. They are more truly in their element when swimming about in the sea; here their chief enemy is the killer whale.

Another species of sea elephant makes its home in the vicinity of various islands in the Antarctic region; the big animals live on friendly terms with their neighbors, the penguins. They have a better chance to survive than their northern cousins.

See also Vol. 10, p. 275: "Mammals."

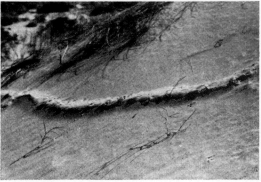

Mounds made by *Scapanus townsendi*, a species of mole found in Pacific coastal areas in the United States.

The two upper pictures show *Scapanus townsendi* and its trademark — the surface pushed up by its burrowing.

U. S. Fish and Wildlife Service

A series of typical ridges and mounds made by moles. The sectional view in the foreground shows part of a somewhat deeper runway that connects with the different mounds and also with a hunting path that lies farther underground.

THE INSECT EATERS

Humble Animals That Maintain the Balance of Nature

THE group of small mammals that specialize in eating insects includes the hedgehogs, shrews, moles, tenrecs and solenodons. These insectivores, or insect eaters, as they are called, are widely distributed in the temperate and tropical portions of the Old and New Worlds. Hedgehogs are found in Europe, Asia and Africa; moles and shrews, in those three continents and also in North America. The tailless tenrec dwells in Madagascar; the ratlike solenodon, in Haiti and Cuba. Australia and the oceanic islands have no members of this group. South America has practically none; the only representatives of the insect eaters in that large continent are a small number of shrews.

The insectivores are not seen so frequently as other mammals. They are shy and retiring animals; they are mostly active at night, frequenting places where abundant growths of vegetation serve to attract numerous insects. For the most part they are small-bodied creatures with elongated and mobile snouts and small eyes and ears. They possess musk, or scent, glands, from which a strong-smelling substance is secreted. Their toes are armed with very strong claws; certain species of insectivores use these claws as tools for burrowing into the earth.

The burrowing types of insect eaters, such as the moles, excavate a vast network of tunnels in the ground. The shrews and the hedgehogs, on the other hand, live above ground, though shrews sometimes use the tunnels made by mice. Tree shrews dwell almost entirely in trees; the African otter shrews, the water shrews and the water moles are particularly well fitted for an aquatic type of life. Most of the insectivores supplement their staple diet of insects with worms, snails, lizards and small birds. Shrews often devour mice; otter shrews feed upon fish; hedgehogs and tree shrews consume fruit.

Though the insectivores make up only a small part of today's mammalian population, they are particularly interesting because they show us what some of the first mammals were like. Between 160,000,000 and 230,000,000 years ago, certain reptiles gradually evolved into the mammalian type of animal. The first mammals were quite small; they ate either insects or flesh and they lived in trees. Their small size and tree-dwelling habits undoubtedly helped them escape the attention of the formidable flesh eaters belonging to the dinosaur group. From this early mammalian stock several types of mammals evolved. One branch developed into the egg-laying mammals, such as the duck-billed platypus; another, into the pouch-bearing animals; still another, into the true insectivores. The bats, which still feast upon insects, were an early offshoot from these insectivores.

The insect eaters of today seem to have departed but little from their ancestors. It is true that the hedgehogs now have a coat of bristling spines and that the moles have gone underground to live; but their skeletal structure, teeth and small brains show typically primitive characteristics. Perhaps the shrews have changed least of all; they are still unspecialized, mouselike creatures. The tree shrews, which have larger brains than their relatives, are generally considered to represent a link between the insectivores and the primates, which include the lemurs, the monkeys and the apes.

The tree shrews, also known as tupaias, are now included in the order of Primates, to which man belongs. The name "insectivores" was first applied to the group by the great comparative anatomist Baron Georges Cuvier (see Index).

The characteristics
of the true insectivores

Though many animals, such as birds, fish, bats, frogs and snakes, eat insects, they are not all insectivores. The true insectivore is a mammal distinguished by the peculiar character of its teeth. These have the shearlike action of the typical carnivore, as opposed to the grinding action found in most other animals. In addition, the insectivore's teeth are furnished with sharp projections enabling the animal to seize and pierce the body of insects. Nearly all members of the order possess a long, narrow, mobile muzzle. The brain shows several primitive characteristics; large olfactory bulbs (extensions of the brain concerned with the sense of smell), few folded surfaces and an underdeveloped cerebrum. The cerebrum is well developed in the higher mammals, such as the apes. In man this portion of the brain is concerned with the thinking process.

It has been extremely difficult for zoologists to classify the insectivores. The taxonomist (one who classifies living things according to their natural order) looks for similarity of anatomical features when he tries to fit animals into a logical classification. When he deals with the insectivores, he finds that some types have retained many ancient, specialized features, while others have been profoundly modified through the ages.

According to the modern classification, the eight surviving families, arranged according to increasing specialization, are the Solenodontidae (including the solenodon of the West Indies); the Tenrecidae (great tenrec, aquatic tenrec, rice tenrec); the Potamogalidae (otter shrew); the Chrysochloridae (golden mole of South Africa); the Erinaceidae (hedgehog, moonrat, gymnura); the Macroscelididae (elephant shrew); the Soricidae (common shrew, pygmy shrew, short-tailed shrew or "mole" shrew, musk shrew, web-footed shrew); the Talpidae (desman, "shrew" mole, eared shrew mole, star-nosed mole, all the American moles).

The ancestor of the insectivores is believed to have been a hedgehog type. Its teeth and bones have been found in fossil deposits that date back to the Upper Cretaceous period, some 60 million years ago. At this time the climate was changing from warm and humid to temperate and dry. The great reptiles disappeared, and the small, shy, warm-blooded creatures called mammals began to increase and multiply. Certain mammals (the bats) became aerial; some (the whales) became aquatic; others (the hoofed mammals and rodents) became plant eaters; still others (the cats, civets and wolves) became meat eaters. The insectivores, however, followed the ancestral mode of life and became even more specialized in it. Considering the fact that they survived for such a long period of time, the insectivores must have been astonishingly successful in adapting themselves to changes in climate and to the predacious (preying) habits of the carnivorous animals, their greatest enemies.

Adaptations to an
underground environment

This phenomenon of adaptation is very important for the modern theory of evolution. An adaptation may be defined as a characteristic that is advantageous to an animal in the particular surroundings in which its life is spent. Sight is a distinct advantage for all animals living on the earth or in the air. Genetic factors (see Index) can cause hereditary poor sight in animals. The blind animal would have to develop other senses such as smell and touch to survive. In an underground environment it would even have the advantage over a sighted animal, which needs vision to orient itself. It is therefore not surprising that the insectivore known as the mole, an animal with small, weak eyes, has successfully survived. Regardless of its visual shortcomings it has evaded its enemies and found food.

In adapting to its environment, this animal chose to lead the life of an underground burrower. The form of its body shows the adjustments made for this mode of living. The mole's head is conical in shape; there is no distinct neck, and the body is in the form of a compact cylinder. The eyes are minute with the power to discriminate only between light and darkness. Some moles even have the eyes covered with a thin layer of skin. External ears are lacking. The soft, thick fur is like velvet, lying smoothly in whatever way it is pressed; thus it keeps friction at a minimum as the mole plows through the soil. The powerful front feet are its digging tools. These feet are very broad and are equipped with stout, strong claws. The

monly occurs in low meadows, gardens and lawns in the eastern half of the United States. The Pacific mole and related species live along the Pacific coast from Lower California to northern Washington. They are similar to the eastern mole but are larger and have a slightly hairy tail. The hairy-tailed mole is a small animal, whose tail is decidedly hairy; it is confined to the northeastern states and southeastern Canada. The most unique American mole is the star-nosed mole, which has a disk of short, fleshy tentacles at the end of its nose. It lives in the eastern states and southeastern Canada. The common European mole represents the mole family in the Old World; it occurs from England and France to the islands of Japan.

A. R. Thompson

The common shrew of Europe and central Asia is one of the smallest and most pugnacious of all mammals.

front legs of some species are set so that the soles of the feet face outward. The feet, fleshy nose and hairy or naked tail are sensitive organs of touch. The mole's sense of smell is keener than its vision. The sense of touch is best developed and is used to locate prey.

Moles inhabit much of the world's north temperate region, but are absent in the Southern Hemisphere. North America has several species. The eastern mole is a moderate-sized animal, five or six inches long, with a short, naked tail. It com-

Moles require a fairly moist, loamy soil, which is easily worked. Their burrows must be humid, for these creatures cannot live long in a dry atmosphere. They prefer open terrain, such as lawns, low meadows and gently sloping hillsides. The star-nosed mole, however, chooses the damp soil and grassy swamps near streams or lakes, while the hairy-tailed mole likes upland wooded areas.

Moles excavate two types of tunnels. There are temporary channels that lie close to the surface, forcing the soil into ridges; there are permanent galleries that run through the soil about a foot or so below the ground. In the tunnels near the surface, the moles hunt for food. The digger literally swims through the soil at an amazing speed. The soil is not removed but merely pressed aside by the mole's body as it forces its way through.

Deeper galleries are more carefully excavated

The deeper galleries are more carefully constructed and serve as passageways to the hunting ground and to the mole's nesting chamber. Using first one front foot and then the other, the mole loosens a bit of soil and passes it back along the side of the body, kicking it to the rear with its hind feet. When a quantity of loosened soil has accumulated, the mole then pushes it along the tunnel to an outlet at the surface; this makes the characteristic molehill. The little insectivore is careful to keep the outlet plugged with dirt when it is not in use. The mole has a keen sense of direction; it digs at about the same level below the surface whether the ground forms a ridge or a hollow. The eastern mole digs tunnels at the rate of about twelve feet an hour. During a mole's lifetime, an extensive network of these tunnels is excavated. Some tunnels may be used in common by dozens of moles living in the same area.

The mole's nesting chamber, or fortress, is excavated about a foot below the surface, under a good-sized mound. If the animal lives close to water, the chamber is located above the water level. Several galleries lead from the upper part of the chamber and form a network of channels around it. It is believed that these may aid in giving proper aeration to the chamber. At the base of the fortress another gallery is excavated. This serves to drain the nesting site of any excess moisture and gives the mole an emergency exit in case of danger. The nest itself is composed of leaves and grass. During the breeding season, female moles construct a similar but smaller nesting chamber in which to have their young. Moles are usually solitary creatures; they defend their fortress and its approaches and their hunting territory against intruders.

Mating occurs during the early spring and occasionally in the fall. The female carries the young in her womb from four to six weeks, depending upon the species. From two to six offspring are produced. The young develop quickly and are ready to take care of themselves when but six or eight weeks old. The eastern mole may live about three years, barring accidents.

Since moles are extremely active creatures, they require a great deal of food to keep themselves alive. It has been estimated that a mole will eat its weight in food every day. Earthworms, beetle larvae, cutworms, adult beetles and snails form the principal fare. Apparently, when there is an abundance of earthworms, moles store them to be eaten later. The mole bites the worm at the front end; this immobilizes the prey but does not kill it.

Moles spend the greater part of their lives underground. Occasionally they come to the surface in order to hunt for prey, to find water and to obtain grass and leaves for the nest. They can swim, sometimes crossing small rivers. The star-nosed mole is an accomplished swimmer and diver. It has even been observed swimming under ice during the winter.

Water moles, or desmans, are semiaquatic insectivores

The active swimmers known as water moles, or desmans, which live in Spain, northern Europe and Russia, are related to the true moles. These animals have webbed hind feet, a vertically flattened tail and long, oily guard hairs that protect the skin and thick fur from water. The elongated, tubular snout forms a proboscis. Desmans construct their nesting chambers in river banks. They prey on small aquatic animals of all kinds, including fish.

The tiny shrew mole is another insectivore related to the moles; it has features of both the shrews and the moles. For instance, its face is long like that of a shrew, but its naked muzzle is molelike. Its moderately broad, stoutly clawed front feet are larger than those of a shrew but not so well developed as those of a mole. Shrew moles live in the humid coastal

The long-eared hedgehog, at the right, is a native of India. Its sharp spines serve as a defense against its enemies.

areas of the northwestern United States. A similar species, the eared shrew mole, occurs in Japan. These insectivores live partly underground, partly on the surface.

Hedgehogs and their kin comprise a family living entirely in the Old World. The common hedgehog has a small rounded body, short legs and tail and a sharply tapering muzzle. Its back and sides are covered with hairs modified into sharp but unbarbed spines, or quills. Coarse hairs clothe the legs, under parts and face. It inhabits Europe and Asia from the British Isles to Korea. Closely related species live in northern and eastern Africa. Southern Asia is the home of two rather curious relatives, the moonrat and the gymnura. These are longer-bodied animals with long pointed muzzles and moderately long tails. They lack spines.

When approached by a predator, the hedgehog rolls into a ball, tucking its head and feet into the center. The erect spines form a formidable defense against attack; this is not a completely efficient defense, however, for weasels, owls, vultures and foxes sometimes made a meal of the hedgehog. The moonrat has an odor peculiarly like that of onions. This may be offensive to certain predators.

The hedgehog is, for the most part, a nocturnal roamer, spending the day in any place that affords warmth and dryness, such as a hollow log, a burrow in an embankment or a den among rocks. It feeds on insects, worms, slugs, birds' eggs, frogs and snakes. In hunting prey, the hedgehog will sometimes take to water, swimming fairly well. It will cross small rivers when pursued. During the winter, it hibernates in a snug retreat; at this time its body temperature may fall to 43° F.

Usually hedgehogs breed twice a year. The gestation period lasts from five to seven weeks, and the number of offspring, which are covered with soft, white spines, average from four to six. A nest of leaves in the shelter of a tree's roots or some hidden burrow serves as a nursery. Here the young remain until a month old; then they follow the mother in her wanderings.

The author has become familiar with the doings of a hedgehog family that sleeps beneath the floor of a summerhouse; his observation of the family has given him an insight into the lives of these insectivores. The hedgehogs feast on the insects of the garden and find an additional food supply in the thousands of frogs living within the boundaries of their domain.

All that they have to do is to eat and drink and be merry. There are only two pitfalls in the garden, yet both have almost proved the undoing of the hedgehogs. One is a miniature lake, which is more easily entered than left. Though there is plenty of water elsewhere for drinking purposes in the garden, a hedgehog once sought to quench its thirst in the water of the lake, blundered in and was almost at its last gasp when it finally was rescued.

The second pitfall is supplied by a steep drop in the ground at one place in the vicinity of the fence that encircles the garden. This results in a penned-in area about a foot wide and many yards in length. Hedgehogs, some of them more than half

N. Y. Zool. Soc.

The elephant shrew, or jumping shrew, is an inhabitant of Africa. Note the trunklike extension of the snout.

grown, blunder into this "trap" and cannot get out. On one occasion a little mound of stones was built in order to enable two little trapped animals to climb up to the higher ground. A couple of days later, the two were still imprisoned. In spite of their exceedingly sharp claws they had apparently been unable to scale the "wall" that enclosed them; nor had they availed themselved of the bridge provided by the mound of stones.

The nearest relative of the hedgehog is a rat-like animal called the gymnura, or rat shrew. The latter name describes the external characteristics of this animal; but its teeth and its internal structure show that it is closely allied to the hedgehog. The

gymnura is found in Burma, the Malay region, the Philippines and parts of China. There are several species, some as big as good-sized rats, others as small as mice. All are purely insectivorous. The discovery of fossil remains of an extinct genus of gymnura in France has been cited as evidence of the similarity between the fauna of Europe in prehistoric times and that of the present-day Malay Archipelago.

The shrews are widely distributed

The most numerous and widely distributed of all the insectivores are the shrews. The common shrew, or shrew mouse, found in Europe, Asia and North America, is about the size of a mouse and somewhat resembles that animal, particularly in the shape of its body, feet and tail; but it is distinguished from the mouse by its muzzle, which projects far beyond the lip. The common shrew feeds upon insects and their larvae, and inhabits dry places, making a nest of various kinds of leaves and grasses. The young, numbering from five to seven, are born in the spring. Common shrews are very voracious; they sometimes kill and devour one another.

The pygmy shrew is even smaller than the common shrew. Rarely more than three inches in length, this tiny animal is the smallest mammal of the North American continent.

The short-tailed, or mole, shrew is the best-known species in the eastern United States. It is about four inches long, with a hairy tail measuring one inch. It remains quite active in the winter; it is sometimes found burrowing in the snow. At this season of the year it feeds largely on beechnuts as well as chrysalids and larvae.

The tupaia, or tree shrew, is widely distributed in the countries of the Orient. The head with its elongated muzzle is typical of the insectivores and the character of its teeth is unmistakable. For a long time this animal was classed in the order Insectivora, but close inspection of the brain and eyes revealed the animal to be more closely related to the lemurs, belonging to the order Primates.

The African water shrew or otter shrew, although a true insectivore, is an excellent swimmer and fine fisher.

The great tenrec of Madagascar can inflict a severe bite. Note the long muzzle and insectivorous teeth.

The European water shrew can dive and swim with great facility. It feeds mainly on larvae and young fish.

The jumping shrews, also called elephant shrews because of the trunklike extension of the snout, are the African representatives of the tupaias. They are ground animals and move by leaps, after the fashion of the kangaroo. The jumping shrews are entirely restricted to Africa. There are many species of these interesting animals. Some haunt the thick undergrowth; others dwell in stony wastes, where clefts and crannies in the rocks offer a secure haven during the dreaded hours of daylight.

China has an aquatic shrew that is more highly specialized than any of its relatives — the web-footed shrew. The soles of its feet are provided with disclike pads; these enable the animal to grip securely the surface of smooth stones and rock as it seeks its food in a river bed.

The musk shrew—
the bane of insect pests

The musk shrew of India is remarkable for the strong, musky odor that emanates from glands situated on the sides of the body. Though it leaves an offensive musky trail, it is generally tolerated in the houses that it enters, because it feeds greedily on cockroaches and other insect pests.

The otter shrew of West Africa finds a place among the insectivores because of the insectivorous characteristic of its teeth. The muzzle is broad, not pointed, after the fashion of other insectivores; and this expert swimmer's habits — particularly its skill in catching fish — decidedly suggest the otter rather than the shrew.

The tenrecs and the
solenodons are closely related

Among the most fascinating of the insect-eaters are the tenrecs and the solenodons, which are closely related. The former are natives of Madagascar, the latter of the West Indies. The existence of these closely allied animals in such widely separated areas points to ancient land connections that have vanished under the seas.

The great tenrec is the most fearsome beast of the tenrec group. Tailless, it attains a length of sixteen inches and considerable girth; it possesses formidable

N. Y. Zool. Soc.

The little solenodon, above, is an inhabitant of Haiti.

teeth and it can inflict a serious bite. Its diet consists almost exclusively of earthworms; it hunts this prey only during the hours of night. To defend itself against its enemies, the tenrec has developed a formidable coat of armor, consisting of bristles and spines. This animal hibernates like the European hedgehog and it lives in burrows, which it excavates by means of its strong claws. Unlike the hedgehog, however, the great tenrec does not roll itself up into a ball-like form for defense.

One species of tenrec has become aquatic. Another, the rice tenrec, carries its search for insects beneath the roots of growing rice to such lengths as to become a serious pest; more so, even, than the insect pests upon which it feeds. Tenrecs rank among the most prolific of all mammals; as many as twenty-one at one birth have been recorded. The flesh of these animals, especially before they hibernate, is highly esteemed by the natives of Madagascar.

The solenodons, found only in Haiti and Cuba, have long snouts and tails and large ears. The body of the solenodon is hairy; by way of contrast, the tail is naked.

We shall end our account of the insectivores with the animal known as the golden mole. It is to be found in eastern, southern and central Africa. This animal is more nearly related in structure to the tenrec than to the mole; still, in its habits of burrowing and tunneling it is clearly molelike. The ears are tiny; the eyes are buried beneath the skin. The metallic luster of the coat gives the golden mole its name.

See also Vol. 10, p. 275: "The Mammals."

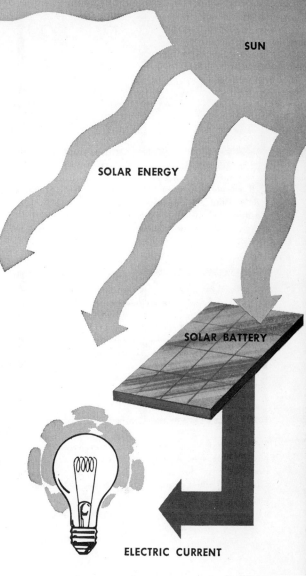

SUN

SOLAR ENERGY

SOLAR BATTERY

ELECTRIC CURRENT

THE SOLAR BATTERY*

How Sunlight is Used to Generate Electricity

OUR star, the sun, radiates energy at the enormous rate of 3.79×10^{33} ergs a second. (See Index, under Ergs.) The earth is so small, comparatively speaking, and so far away from the sun that it receives only about a two-billionth of this energy; yet even this represents a staggering quantity.

We would be able to put our slim quota of solar energy to even better use if we could harness it effectively. Within the last few decades, much progress has been made in this respect. Heat produced by the sun's rays has served to generate electricity indirectly,** to provide warmth for houses and even to cook food. In the useful device called the photoelectric cell, the light of the sun (as well as light from other sources) has been converted directly into electric energy. (See Index, under Photoelectric cells.)

The electric power produced by the photoelectric cell is very feeble. A far more effective device for generating electric power from sunlight is the solar battery, developed by the scientists of the Bell Telephone Laboratories. It is made up of a number of cells, each consisting of a wafer of pure silicon, to which certain impurities have been added.

To have some idea of the workings of a solar cell, let us recall that electrons — particles with a negative electrical charge —

* Article prepared with the assistance of Bell Telephone Laboratories, Inc.
** Solar energy is made to heat a liquid; the resulting vapor runs turbines which are connected to generators of electricity.

revolve around the nucleus of a typical atom in several concentric shells.* Atoms combine with one another through the agency of the electrons that make up the outermost shell. A silicon atom has four electrons in this shell. In a silicon crystal, each atom is normally joined to four of its neighbors by its four outer electrons.

If an electron is jarred loose from a silicon atom by some outer force such as light, heat or electrical energy, it will wander freely through the crystal, leaving a hole in the crystal lattice, or framework. This hole behaves like a positive charge of electricity. Every time an electron from an adjacent atom moves in to fill such gap, it

* Ordinary hydrogen has only one electron; the atoms of hydrogen and helium have only one shell.

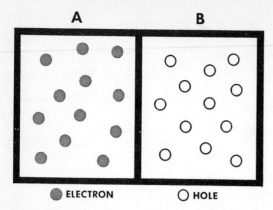

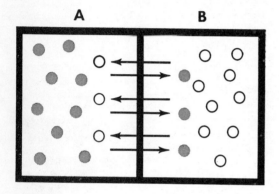

1. If we put arsenic in half a silicon wafer (A), there will be more electrons than holes; if we put boron in the other half (B), there will be more holes than electrons. "Electrons" and "holes" are explained in the text.

2. The negative electrons and positive holes cross the boundary between A and B. Side B will acquire a negative charge as the electrons pass into it; side A will acquire a positive charge as it receives holes from B.

3. Typical silicon solar cell, described in this article.

Bell Telephone Laboratories, Inc.

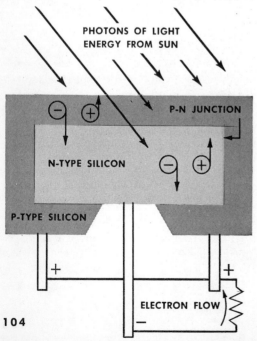

leaves a new gap behind it. In this way holes can travel from one part of the crystal to another, just as electrons can. When we apply an outside source of energy to a crystal, therefore, a certain number of freely wandering negative and positive charges are created within the crystal. They travel at random and balance one another.

Suppose that we introduce a trace of arsenic in a silicon crystal. The arsenic atom has five electrons in its outer shell. If it combines with four adjacent atoms of silicon (just as a silicon atom would combine with its neighbors), one of the arsenic electrons will be left over and will become a free electron. Even a trace of arsenic contains a very large number of atoms. Therefore we bring about an excess of freely movable electrons by introducing the arsenic. The crystal as a whole remains electrically neutral, since the extra free electrons are balanced by positively charged protons in the nuclei of the arsenic atoms.

There would be a different state of affairs if we put boron in the crystal. Each boron atom has only three electrons in its outer shell. If it combined with four silicon neighbors, the lack of one electron would leave a positive hole in the crystal lattice around the boron atom. As we would be introducing considerable numbers of boron atoms in the crystal, we would now have an excess of positively charged holes. Again, the neutrality of the total crystal would not be affected.

Suppose now that we put arsenic in one half of a silicon crystal (A in Figure 1) and boron in the other half (B in Figure 1). Each half of the crystal would be electrically neutral. Yet the concentration of free electrons would be greater in the arsenic half than in the boron half; and, similarly, there would be more wandering holes on the boron side than on the arsenic side. Since both kinds of charges move about in a haphazard way, they would tend to diffuse throughout the entire crystal, crossing the boundary between A and B. Side B would acquire a negative charge because excess electrons would be passing into it. Side A, on the other hand, would become positively charged, since it would receive an excess

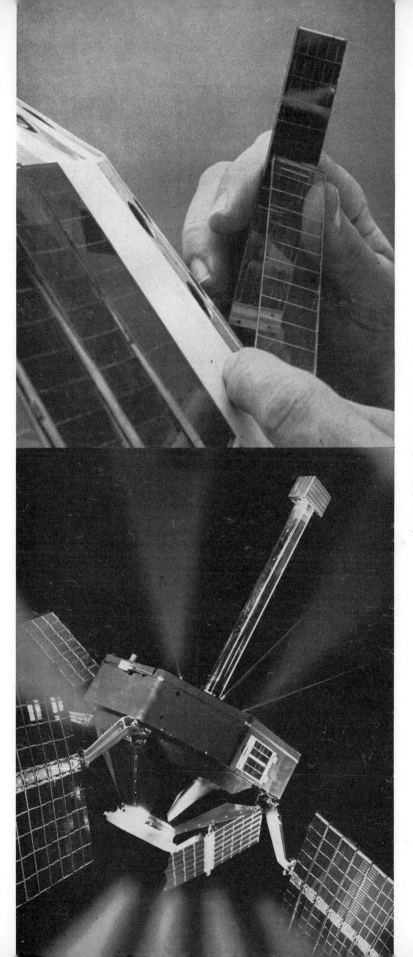

Upper photo: attaching solar cells and transparent sapphire slices to a developmental model of the *Telstar* communications satellite. Solar cells are covered with thin slices of sapphire, set in frames of platinum, in order to protect the cells from space radiation. Lower photo: an S-3 energetic-particles satellite, designed to analyze the energy of charged particles encountered in space. Solar cells are installed on the four paddles of this satellite.

Upper photo: Bell Telephone
Laboratories, Inc.
Lower photo: NASA

of holes. (See Figure 2.) These charges would build up as more electrons and holes passed through. In time they would become strong enough to prevent any more diffusion at the boundary. Negatively charged side *B* would repel electrons; positively charged side *A* would repel holes. Equilibrium would be established and we would now have a built-in electric field, as the inventors of the solar battery call it.

We have just such an arrangement in each of the cells that make up the solar battery. Each cell (Figure 3) is a thin wafer. Its body consists of silicon with a trace of arsenic. This is called n-type silicon ("n"="negative") because the addition of the arsenic provides an excess of negative charges. The surface of the wafer is made up of a thin layer of silicon with a trace of boron; this is called p-type silicon ("p"= "positive") because it has an excess of positively charged holes. The boundary between the silicon-arsenic body and the silicon-boron surface is called the p-n junction. The p-type silicon, corresponding to *B* in Figure 1, develops a negative charge; the n-type silicon, corresponding to *A*, develops a positive charge.

To use the solar battery, terminals are attached to the n-type and p-type material, as shown in Figure 3, and the terminals are connected by wiring. When sunlight strikes the surface of the wafer, it knocks out electrons from the crystal lattice in the vicinity of the junction and produces electron-hole pairs. In so doing it upsets the equilibrium established between the n-type and the p-type silicon. Electrons are pulled across the junction into the n-type silicon and holes are pulled into the p-type silicon. Electrons then stream from the negative terminal by way of the electric wiring to the positive terminal. Current will flow in this way as long as sunlight strikes the silicon wafer. Practical solar batteries consist of a number of such wafer-cells set side by side and electrically connected in series.

The solar battery can deliver a maximum voltage of a little more than half a volt in bright sunlight, and a maximum power of about a hundredth of a watt per square centimeter of exposed surface area.

It is estimated that it can convert up to 11 per cent of the sun's energy it receives directly into electric power.

The solar battery was put to work on a very modest scale at Americus, Georgia, in order to provide power for telephone amplifiers, or repeaters. These are required at intervals in a telephone line, in order to maintain the strength of the signal that is being transmitted. The power requirement for such repeaters is small; and it was found that a solar battery could readily meet it.*

The most promising use of solar cells at the present time is in the field of space exploration. They have already been used quite extensively in artificial earth satellites to provide electric power for various purposes. For example, the satellite *Explorer XII*, launched in August 1961 and used in radiation research, had 5,600 solar cells, capable of delivering 5 watts of power.

Solar cells play an important part in the Bell System's experimental *Telstar* communications satellite, which receives a radio signal beamed at it from the ground, amplifies the signal ten billion times and retransmits it on another frequency.** The satellite, roughly spherical in shape, has 72 flat faces, or facets. Solar cells are set on most of the facets; there are 3,600 of the cells in all. They are mounted on ceramic bases in platinum frames; coverings of clear man-made sapphire protect them from bombardment by charged particles and tiny meteorites. The *Telstar* solar cells convert sunlight into electricity and charge 19 nickel-cadmium cells. The latter in turn provide power for the electronic circuits in the satellite.

It is believed that as the solar battery is perfected, it will find many useful applications. It should offer some striking advantages. It has no moving parts or corrosive chemicals, such as we find in various other generators of electricity; hence it should have an unusually long life. Besides, the sunlight on which it operates is our most abundant resource.

* The Americus project was technically successful, but was later discontinued.
** *Telstar* is also designed to analyze its own performance and provide information about energetic particles in space.

THE LAW OF PARITY

Significant Discoveries in the World of Subatomic Particles

BY EDMUND GREKULINSKI

THE LAW of the conservation of parity has been considered very important in modern physics. This law states, in effect, that for the absolute universe of space and time, distinctions between left and right and up and down, between an object and its image and so on have no real significance. These directions mean something only to an observer or in spatial relations among objects in various locations.

It is true that in our familiar everyday world, as opposed to the universe at large, distinctions between right and left assume great importance. We find a difference between right and left arms, right and left legs, right and left shoes and gloves. The structures of various organisms tend to be oriented to the right or to the left, more toward one than the other in different groups (the coiling in the shells of snails, for example). Thus, one direction may be preferred over another; even the sun, planets and moons mainly rotate and revolve in a prevailing direction. And

yet, is it not possible to imagine a system like ours in all respects, but constructed on a pattern where directions are the reverse of ours? Except for this one basic difference, would it otherwise differ markedly from the one we now inhabit? Probably it would not.

Consider a pair of shoes. Except for the difference between rightward and leftward directed structures, no one doubts that both shoes are identical in style, size and function. Suppose you have a pair of shoes, *a* and *c*, as shown in Figure 1. If you hold shoe *a* to a mirror, the reflection, *b*, will be identical with shoe *c*, the other shoe of the pair. If we use symmetrical kidney beans instead of shoes in the demonstration (Figure 2), we see that bean *b*, the mirror image of *a*, is exactly the same as bean *c*. We also observe that bean *c* is exactly the same as bean *a* turned upside down. In this case, then, there is no basic difference between the object (bean *a*) and its image (bean *b*).

107

1. Shoe *b* is the reflection of shoe *a* in the mirror; *b* is identical with shoe *c*, the other shoe of the pair.

2. Bean *b* is the mirror image of bean *a*; *b* is also the same as bean *c*, which is simply *a* turned upside down.

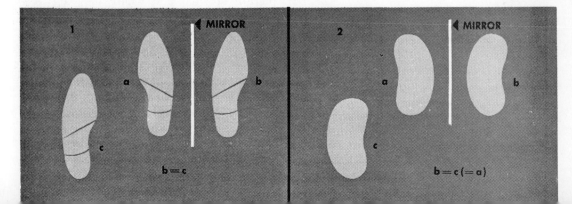

Parity in the subatomic world

If we could directly see the particles that compose the atoms of matter and the waves of energy, we would notice, in certain situations, that particles spinning or moving in one direction tend to be balanced

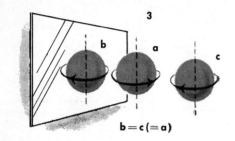

3

$$b = c (= a)$$

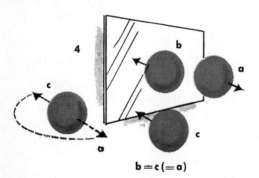

4

$$b = c (= a)$$

3. Ball b is the mirror image of ball a; b is identical with ball c, which rotates in the same direction.
4. Ball b is the mirror image of ball a; b is identical with ball c, which is a turned 180° to the left.
5. Three straight lines, x, y and z, meet at right angles at O and thus define a three-dimensional space.
6. When the same axes, x, y and z, are extended beyond the O point, the extended portions receive minus signs.

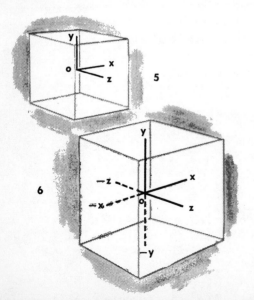

by identical particles spinning or moving the opposite way. To illustrate this aspect of parity, hold a plain ball, a, up to a mirror (Figure 3). Ball and image are absolutely identical and cannot be distinguished at all. Now spin the ball between thumb and forefinger along an axis parallel to the plane of the mirror. You will notice that the image of the ball in the mirror spins in the opposite direction. Reflection b is identical with a real ball, c, rotating in the same direction; a corresponds to c turned upside down. The introduction of motion into the concept of parity here is of the greatest importance, as we shall see shortly.

Another important factor to consider in balanced motion is symmetry. We illustrated mirror-image parity for the sphere by rotating it, in Figure 3. Suppose now that, as in Figure 4, we move the ball a in a straight line directly *away* from the mirror; its image, b, of course, will move in the opposite direction. If we move another ball, c, directly away from ball a (which moves in the opposite direction), ball c will correspond exactly to the mirror image b. These motions may be considered as symmetrical or balanced. In Figure 4, balls a and c may be said to constitute a pair. If we turn a 180° to the left so that it is "upside down" in relation to its former position, it corresponds perfectly to c, even including the arrows. Therefore, a, b and c are identical.

We can detect the nature of particles of subatomic size only indirectly, since they are too minute and move far too rapidly for us to see. We observe how they interact with other particles and waves, and we measure these interactions. But this system of measurement necessarily disturbs the particles, so that we cannot get an exact picture of what and where they are at a given moment. However, using statistics as a guide, physicists can "guess" about the probable positions of particles.

These bodies travel and may also spin, or rotate, on their axes as well. They are affected by electrical and magnetic fields. They react with each other, giving off or forming different particles, as in radioactivity. Some (called electrons) have nega-

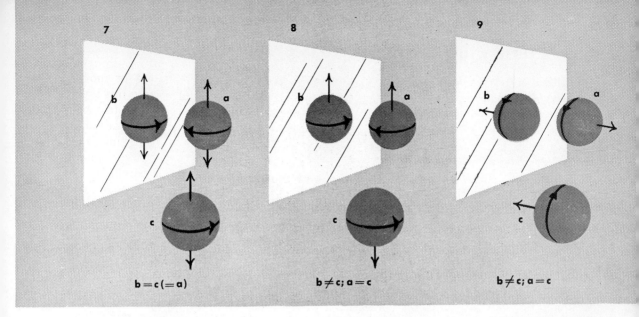

7 8 9

$b = c (= a)$ $b \neq c; a = c$ $b \neq c; a = c$

tive electrical charges; others are positive (positrons and the heavier protons) and still others are electrically neutral (neutrons and the far lighter neutrinos and antineutrinos). Certain particles move freely through space, while others (electrons, protons and neutrons) are also in atoms.

To locate such a particle in three-dimensional space, we construct three straight lines, called axes, at right angles to each other (Figure 5). The position of the particle is related to distances marked off on the three axes, or co-ordinates, which are given arbitrary symbols or letters (usually x, y and z). Suppose this system of axes is reflected in mirrors (Figure 6). We could call the reflections negative (labeling them $-x$, $-y$ and $-z$) to distinguish them from the positive axes, x, y and z, that are being reflected. Axes $-x$, $-y$ and $-z$ might also be called left-handed.

The actual picture in the world of subatomic particles is a bit more complicated than we have drawn it so far. Under certain conditions, groups of particles behave like waves in their motion. Physicists have made a careful study of these wave motions. They have determined that the probable positions of particles can be derived by a formula involving the square of what is called the wave function. (The wave function represents the amplitude, or height, of a wave.) If the axes which determine the position of a particle are reflected and the wave function becomes negative as the axes do, that particle is said to have *odd* parity.

7. Ball *b* is the mirror image of ball *a*; *b* is identical with ball *c*, which is simply ball *a* turned upside down.
8. Ball *b* is the mirror image of ball *a*; but *b* is not the same as ball *c*, which is ball *a* turned upside down.
9. Ball *b* is the mirror image of ball *a*; but *b* is not the same as ball *c*, which is just ball *a* turned about.

If the wave function still remains positive after reflection, however, the parity is called *even*. It is as if, despite the "extension" of the axes into the "negative" region, the particle's "image" is still "positive." In a system with a given number of particles interacting, parity is not supposed to change from the beginning to the end of the process. It is either even or odd, but not both; that is, parity is theoretically conserved.

Certain scientists began to express doubts about the "law" of parity during the 1950's. It was noted that certain particles, called mesons,* during their transformation, or "decay," into other kinds of mesons and various particles, had different parities, although they were absolutely alike in all other respects. One type of meson decays into two other mesons of different mass, each of the two having odd parity, so that the over-all parity is even. This is so because, when two negative quantities are multiplied together, the product is positive. In other words, two "odds" make an "even." Another kind of meson decays

---

* Mesons are subatomic particles with masses intermediate between those of electrons (which are light) and those of protons (which are heavy). Mesons may be electrically positive, negative or neutral; they occur commonly in cosmic rays or as the result of cosmic-ray activity. Some physicists believe mesons represent the binding force in atomic nuclei.

into three mesons, each of the latter identical with the products of the first-described process and each one also having odd parity; the over-all parity is odd. The reason is that when three negative numbers are multiplied together, the result is negative; or three "odds" make an "odd." The two original mesons (at the beginning of decay) are alike except for their parity — one odd, the other even. It is as if identical objects have different mirror reflections. Even if we assume the same parity for both, then somehow parity has changed in one of the two decay processes above. It would appear that the "principle" of parity has been violated, that certain reactions in the subatomic field are not symmetrical and that there is a difference between right and left and between particle and image.

To test the parity law for mesons, physicists at Columbia University in New York City and at the University of Chicago conducted a number of experiments. The Columbia group of M. Weinrich, L. Lederer and K. Garwin, for example, bombarded a target in a cyclotron (atom-smasher) with protons. As a result, the target emitted a stream of mesons, known as "pi" mesons. The latter then decayed into so-called "mu" mesons (somewhat different from pi mesons) and into extremely light particles with no positive or negative electrical charge — antineutrinos. This process was carried out in a magnetic field, which lined up the mu mesons' axes of rotation in one direction; it also made these mesons spin in a single direction. This was important, because now (unlike the situation in earlier experiments) the movements and positions of the reacting particles could be observed and measured. The mu mesons were then collected in a carbon block, where they in turn decayed, giving off electrons, antineutrinos and neutrinos — light particles with no positive or negative charges (unlike the electrons). It was noted that *more* electrons were emitted in one direction along a mu meson's spin axis than in the opposite one. According to the law of parity, the electrons should have come off equally in all directions or in equally opposed ones, as in Figure 7 (indicated by the arrows).

That is, they were supposed to come off each end of each rotation axis. Parity in this case would have been preserved; note that image *b* is identical with *c*; *a* and *c* constitute a pair, and *a* is *c* turned upside down, so that *a* = *b* = *c*. (Compare this with Figures 1-4.)

The reader should note that each sphere in Figure 7 is symmetrical in that its upper half, with its arrow, is a mirror image of the lower half, with its arrow. That is, imagine a mirror inserted through the equatorial region of each ball, perpendicular to the axis of spin. Figure 7 shows what the situation would be if the law of parity held true.

The actual experimental results are shown in Figures 8 and 9. The meson *a* emits electrons off one end of its spin axis. Its paired opposite, meson *c,* spinning in the *opposite* direction, emits electrons off the *opposite* end of its axis. We see that *a* is *c* turned upside down. Is *a*'s reflection, *b*, identical with meson *c*? It is not, of course. In Figure 8, *b* spins in the same direction as *c*, but its electrons stream off its axis in the opposite direction. In Figure 9, electrons move in the same direction in *b* and *c*, but the spin directions are opposed. It is as if *a* has two different "mirror" images: *b* and *c*.

Other experiments, conducted by Tsung D. Lee and Chien S. Wu of Columbia University, Chen N. Yang of the Institute of Advanced Study and Ernest Ambler of the National Bureau of Standards, indicated similar parity-law violations. This time they occurred in beta-ray emissions (of electrons and positrons) from the aligned spinning nuclei of cobalt atoms. Again one direction was favored over another.

So far, breakdown of the parity principle has been observed only in a few, so-called "weak interactions," involving low orders of energy. In other subatomic processes (strong interactions), parity seems to stand up. Future experiments will have to determine to what extent the law of the conservation of parity holds true. The findings thus far seem highly significant; they indicate that we may have to change our ideas about certain basic concepts.

PLANT SCOURGES

Weeds—a Standing Menace to Our Fields

by CLYDE M. CHRISTENSEN

A PLANT can be a thing of beauty and yet not be a joy forever. The bright trumpets of bindweed, a pasture that is sunny with buttercups, Queen Anne's lace in a meadow — here are spectacles to delight the passing motorist. But the farmer finds them less delightful, for the bindweed strangles his corn, the buttercups contain an irritating juice that cattle dislike, and Queen Anne's lace, which is a wild carrot, harbors a destructive weevil. These and other plants that spring up unbidden the farmer calls weeds. This name has nothing to do with plant classification. A weed is any plant growing where it is not wanted.

A dandelion in the front lawn is a weed; it is not a weed when cultivated for fresh greens. Sweet clover sown in the meadow is not a weed; when some of the ripe seeds drift over on the wind to the flower garden, they grow into weeds.

Weeds do enormous harm. In actual figures, many billions of dollars a year are lost, over the world, because of weeds. The sum is greater than the combined loss due to both plant diseases and insects. The loss is likely to be highest in regions where agriculture is most intensive and where the crops and land are of greatest value, but weeds in open-range country are also costly.

The cereal-grain crops offer a good example of the tax we pay on weeds. Seeds of weeds in the grain field are harvested along with the grain. On the average, wheat brought to the elevators contains one per cent of weed seeds. This means that in, say, a billion-bushel wheat crop, ten million bushels of weed seeds were harvested, threshed, stored, hauled to mills or terminals and there removed and dis-posed of. Even a small percentage of seeds of certain weeds, if blended with the grain, would give an off flavor to the flour. The grower receives a reduced price for his grain when it contains weed seeds.

Weeds that grow in hay or forage crops are of low feed value; many of them are unpalatable to stock, and some are actually poisonous and cause the loss of valuable animals. Others, when eaten by cows, give an unpleasant flavor to milk and butter. Weeds with pointed or spiny seeds or seed pods may get caught in the coats of animals, causing acute discomfort and also reducing the value of wool or hides.

Along railroad rights of way and highways, weeds are often fire and traffic hazards. They clog drainage and irrigation ditches and canals. For example, the water hyacinth, imported to Florida from the tropics, in a short time blocked many coastal and inland waterways. Constant and costly work is now required to keep these channels clear.

Some plant diseases have spread by means of weeds. The stem rust of wheat is caused by a fungus that spends part of its life on the barberry. To aid in controlling this disease, millions of ornamental barberry bushes have had to be uprooted and destroyed. The blister rust of white pines is caused by a fungus that spends part of its life on the leaves of wild and cultivated currant bushes. Some virus diseases that attack cultivated plants may be traced to weeds that not only serve as a source of the virus but also furnish a breeding place for leaf hoppers, aphids and other insects that carry the viruses. A number of insects that directly injure cultivated plants breed on or in weeds.

Hay-fever sufferers know that ragweed (as well as some other weeds and cultivated plants) causes heavy human misery. Poison ivy, poison sumach and other weeds are even worse pests than ragweed.

Probably the greatest harm done by weeds is in competing with crop plants. The weeds grow rapidly and send strong root systems far into the soil. In less than three weeks of growth, a vigorous mustard plant in a grain field will produce more than four hundred feet of roots. A dozen of these plants will produce nearly a mile of roots in three weeks. They rob the crop plants of soil minerals and water, and their leaves screen sunshine from the cultivated plants, further reducing the grower's yield.

Many weeds have remarkable powers of reproduction. An average-sized pigweed will produce from 100,000 to 200,000 seeds in one season, and a large pigweed may ripen 10,000,000 seeds. An annual crop of about 20,000 seeds per plant is a good average for each of 200 different kinds of annual and perennial weeds.

The seeds of many weeds are tenacious of life. In a famous test in Michigan, the seeds of twenty different kinds of weeds were buried in the soil (too deep for growth) in 1879. After forty years, some seeds of ten of the kinds were still able to germinate and, after sixty years, seeds of two kinds were still viable (able to grow). Nearly 70 per cent of the seeds of one weed, the moth mullein, germinated after having been buried in the soil for sixty years.

The seeds of some weeds can lie under water for months, or even years, and can still remain viable.

Quackgrass and certain other weeds can multiply rapidly by underground root stems or by root fragments. Perennial sow thistle and wild morning-glory, two of our most pestiferous weeds, not only produce seeds in abundance, but can also grow from severed roots. A small piece of morning-glory root carried into a field may in one year produce a root system fifteen feet across and extending down twenty feet into the soil. When these roots are cut by a cultivator, each piece will send up a new plant.

Weeds are famous travelers, and every continent today has a weed population made up of both native and immigrant pests.

The seeds are carried in almost every imaginable way. Those of the dandelion and some thistles have a buoyant parachute, called a pappus, on which they can ride for miles, drifting along with the wind. Many burrs, the beggar-ticks and others, have sharp or barbed spines by means of which they attach themselves to the fur of animals or the clothing of man. The puncture vine, a native of the Sahara Desert, has been spread by railroads and is now at home in many fertile lands. Motor cars and airplanes also carry it; the seeds cling to the tires and may be dropped off hundreds of miles from where they were picked up.

Animals and birds swallow seeds whole and later excrete them. When livestock are shipped long distances, seeds are transported in this manner. Farm machinery, especially harvesting machines and threshers, will carry seeds from one field or farm to another. The seeds of a few weeds can be carried by streams or by the water in drainage or irrigation ditches.

Man has only himself to blame for some of the weeds that plague him. Most of North America's worst weeds, for example, are aliens, brought from the other continents. The early colonists, when they set out from the Old World, brought seeds of plants to raise in the New. Almost no at-

Standard Oil Co. (N. J.)

Water hyacinths. These plants, imported to Florida, have become a pest, blocking waterways.

tempt was made to insure the purity of this seed stock, and so the common weed pests of these plants came along, too. Before 1700, dozens of different weeds common in Europe had a foothold in New England, and, as time passed, hundreds more were brought in. Not all of them prospered, but many did. Among them was the plantain. .Here are a few examples of weeds that thrive far from their native heaths.

Russian thistle — tumbleweed — was unknown in the United States before 1873. At that time a small amount of flax seed containing some thistle seed was imported and planted in South Dakota. Within ten years the Russian thistle had spread extensively in South Dakota and Missouri, and within another ten years it had infested sixteen states, from Missouri to the west coast.

The Canada thistle is not native to Canada but was brought from Europe early and became established around Montreal. When General Burgoyne's army invaded New York from Canada in 1777, hay brought for the cavalry horses contained thistle seeds. The Americans soon got rid of Gentleman Johnny Burgoyne, but the Canada thistle is now a costly pest in thirty-six states, from coast to coast.

Johnson grass, brought from Turkey, was tested as a forage grass in California. It proved of little value, but soon escaped and became a most detrimental weed along irrigation ditches.

The prickly pear, a kind of cactus, was introduced into Australia from North and South America for use as an ornamental plant. Within a short time it had taken over hundreds of thousands of acres of valuable agricultural land and amounted to almost a national calamity.

If a piece of land is stripped of vegetation and then left idle, weeds are the first plants to take over, and they take possession of the land in definite order. First, annual weeds form a dense cover. They not only hold the soil in place but also, as they rot, enrich the soil. Gradually the annual weeds are replaced and crowded out by biennials, which have the advantage of a two-year lease on life. The biennials are crowded out by perennials. If the climate

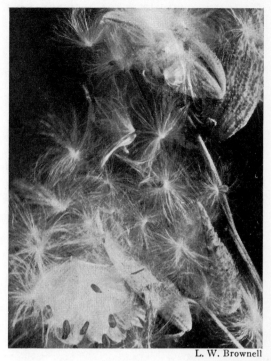

L. W. Brownell

The seeds of the milkweed, shown above, are carried far and wide on their buoyant "parachutes."

is suitable, these in turn may give way to forest trees. The trees will persist until fire, flood or other violence of man or nature again lays the ground bare, when the cycle begins anew. You can see the beginning of such a cycle, on a small scale, on newly filled-in lots or abandoned garden plots or even in untended yards around vacant houses in city or country.

Man fights the war of weeds with a variety of weapons, old and new. No one method of fighting will control all weeds, and certain weeds can be stamped out, or checked, only by the use of several methods at the same time. The attack to be chosen depends on the weed concerned, the extent of its infestation, the location and the purpose for which the land is used.

Governments long ago recognized the danger of importing alien weed seeds mixed with other seeds and the danger of introducing new cultivated plants that might become pests. Plant quarantine laws are helpful in this connection. Where there is such a national plant quarantine, seeds, plants or plant parts for propagation can be brought into a country only after rigid inspection at the port of entry; and new plants brought into the country are tested in isolated plots. The tests not only show whether the plants have a tendency to run wild as weeds, they also show whether the plants might be disease-carriers or hosts to harmful insects.

Within some countries, seed for sale is most carefully inspected, and very low limits are set on the kind and number of weed seeds that it may contain.

Cultivation is the time-honored method of keeping weeds in check. In small gardens, the hoe is still the most valuable weed-control tool we have. In large-scale growing, various kinds of cultivators are used. Regardless of the type or frequency of cultivation, it is practically impossible to get rid of all weeds, because seeds remain in the soil, and new seeds are carried in constantly. Cultivation remains, however, one of the most successful means of combating many weeds and the only method of controlling some of them.

Certain weeds are associated with particular crops and can be kept down by rotation of crops. By sowing a field to small grains one year, a cultivated crop, such as corn, the next year and a hay crop, such as sweet clover, the third year, a good share of the weeds associated with each of these crops can be kept in check. A combination of cultivation and crop rotation is the best general means of control.

Along railroad rights of way, highways and irrigation ditches, weeds are sometimes kept in check by burning or mowing, but this is likely to be only a temporary control.

In pineapple and sugar-cane fields in Hawaii, wide strips of heavy paper are laid and the plants are set out through holes in the paper. The weed seedlings, of course, cannot push through the paper, nor can vagrant seeds blown along by the wind find resting place in the soil. This good device is sometimes used in home and market gardens, but it is pretty costly.

Insects may sometimes be called on for aid. The prickly pear, which overran much valuable land in Australia, was controlled by introducing from South America an insect that tunneled through the plants and killed them. It was necessary, of course, to be sure that the insects used would not attack any valuable plants.

Within the last hundred years or so, there have been many attempts to control weeds by means of chemicals. Some of these have been very successful. However, until around 1940 there were almost no chemicals that would kill weeds without injuring other plants or poisoning animals or birds or doing other harm. Now there are chemicals that are highly selective — that is, they poison certain plants but do not injure other plants or animals. One, known as 2,4-D, has had a great deal of publicity. However, it does kill or injure some cultivated plants as well as weeds, and so it must be used with caution. Weed-killing chemicals will doubtless be more helpful in our fight against weeds, but the man with the hoe is likely to be needed for a long time to come.

Du Pont

A growth of poison ivy in a vacant lot in Wilmington, Delaware, is sprayed with ammonium sulfamate.

THE BUILDER'S MATERIALS

A Survey of Substances Used in Every Age

by JOHN H. CALLENDER

IN HIS extensive career as a builder, man has worked with mud, baked clay, stone, glass, copper, brass, iron, steel, felt, hide, logs, boards, leaves, grass and even snow. Some of these materials he found ready for use; others had to be most carefully prepared before they could serve for dwellings, religious temples and monuments.

All of the builder's materials, so abundant and so varied, may be divided into three general classes: primitive, traditional and modern. Primitive materials are readily available in nature and require very little preparation; they include logs, saplings, reeds, leaves, grass, skins, clay and snow. We list under traditional materials those that have been in use over a long period of time and that require more or less extensive preparation — stone, brick, tile, mortar, timber, plaster, metal, glass and the like. Modern materials include new or relatively new substances such as structural steel, reinforced concrete, vermiculite, fiberboard, plywood and plastics.

The names "primitive," "traditional" and "modern" suggest the order in which building materials were first employed by man. They do not stand for definite historical periods, for both primitive and traditional materials are still popular.

We think of a builder nowadays as a specialist who devotes all his time to his chosen lifework. But the history of building with primitive materials generally shows no such specialization.

Perhaps an example or two will make this clearer. American colonists frequently came to a land of almost unbroken forest. Before a man could settle anywhere and plant crops, he had to clear away the trees.

Many a colonist took the trees that he had cut to make a shelter for himself and his family. Selecting trunks that were straight and uniform in size, he laid one on top of another, notched and lapped the corners and built a log cabin. The spaces between the logs he chinked with clay to make the house watertight, and the log chimney was lined with clay to keep it from burning. The complete house was built by the colonist and his sons, with help from neighbors.

A small log cabin erected by American pioneers.

Of course, the American Indians built with primitive materials. The tribes of the western plains were hunters, especially of the bison that roamed the plains in enormous herds. Bison and deer meat were the food of these tribes; the skins of the animals were fashioned into clothing. It is not surprising to find that the Plains Indians made their houses of these hides, stretched over a pyramidal frame of saplings.

Primitive materials are generally used in their natural state, without processing; but this is not always true. Many Arabs live in tents of cloth woven from sheep's wool,

Lehnert and Landrock

The massive temple of Rameses II at Abu Simbel, Egypt. It was carved out from the living rock.

goat's wool or camel's hair. In Mongolia, where the herds are mostly short-haired animals, such as cattle and horses, the tents are covered with felted cloth.

Primitive man occasionally built homes of field stone, putting the rocks in place with a minimum of cutting. More often, stone was reserved for temples to the gods or monuments to heroes.

Traditional materials

Stone

For thousands of years, stone has been without a rival as the material for buildings of a monumental nature — that is, those that are intended to endure and to have dignity and authority. We find the earliest monuments of this sort in ancient Egypt. The Egyptians believed that the earthly body would be needed in the next world, and they invented methods of embalming to preserve the body almost indefinitely. They buried with the body various belongings of the dead person — food, clothing, money, jewelry, cosmetics, musical instruments, books, weapons and toys. The tomb in which the body and all of its possessions were to be buried was built to last "forever."

Nothing seemed more permanent than the stone cliffs that rimmed the Nile Valley; hence, the Egyptians built their tombs of stones cut from these cliffs, or else they carved out tombs in the very cliffs themselves. The tombs withstood the ravages of time and weather and the attacks of all

marauding animals except man. As a building material, the stone of Egyptian tombs has proved to be as durable, thus far, as the Egyptians hoped it would be.

Stone provided a material out of which the images of the gods could be carved; it also seemed appropriate for buildings dedicated to these supernatural beings. The kings, who were considered to be gods, came to live in temple-palaces of stone, as did the high priests. This set the style for the wealthy nobility, who also built stone palaces for themselves.

Greece is a rocky country with rich veins of marble; therefore the inhabitants turned naturally to stone for building after wood became scarce. The architecture of the Greeks, executed in white marble, has been an important influence in Western architecture down to the present day.

The Romans built first of brick, but as they came under the influence of Greek culture, they, too, found stone more beautiful. Instead of using solid blocks of stone, as the Greeks and Egyptians had done, the Romans often put up walls of brick, or rubble or concrete, and faced them with thin slabs of stone. For this purpose, they imported rare marbles and porphyries from distant parts of their vast Empire. These stones were selected for their beautiful colors and patterns, and were often cut and laid in such a way that the veining of adjacent stones formed a pattern. Floors were paved with mosaics, made up of small

Pan American World Airways

This ornate pavement of mosaic work was found in the remains of a Roman villa in Bignor, Sussex.

pieces of stone of various colors. These were usually laid so as to provide an abstract pattern; sometimes, however, they formed definite pictures.

One of the disadvantages of stone, from the viewpoint of Egyptian and Greek builders, was the very short distance that a single stone spanned between supports. This made it impossible for them to build a large open room. The interiors of Egyptian and Greek temples give the impression of a forest of columns. The only way in which the span can be increased in an all-masonry building is by the arch; and it is remarkable that the Egyptians and Greeks, master builders as they were, made no use of the arch. The Romans, however, used it extensively in its various forms, especially the vault and the dome, which permitted them to build vast open halls, uncluttered by columns.

In the Middle Ages, stone continued to be an important building material. As medieval civilization moved northward, a new factor was introduced — the desire for more light. The dimness of the interiors of most earlier buildings had been rather a pleasant relief from the brilliant Mediterranean sun. But in northern Europe, light was desirable and man wanted it in his buildings. By the thirteenth century, there had developed in northern France a method of building in which glass took up a good deal of the space formerly occupied by stone in the side walls. Yet far overhead was a solid stone roof. This construction is known as Gothic.

How was it brought about that walls with so much glass could be made to support roofs of heavy stone? The Gothic builder combined arches to form immensely strong vaults, which could be flung up to great heights. He made most liberal use of clerestories — upper stories that were provided with windows lighting up the interior of the building. He provided support for the building by side structures called buttresses.

Thus, he utilized the natural advantages of stone — its strength and its durability — as a building material and he overcame its disadvantages — its heaviness and its inability to span very far. The natural character of stone is massive, inert strength,

The choir of England's Salisbury Cathedral, a splendid example of medieval stone construction.

exemplified by the Egyptian pyramids. But in the late Gothic cathedrals, stone seems to stretch aloft, to soar, in complete disregard of its natural limitations.

A well-built, solid stone building is as durable a structure as man has ever devised. The forces of nature are rarely able to injure it seriously; masonry buildings have successfully withstood hurricanes, tidal waves, long submersion under water and burial by sand dunes and by volcanic ash. Only a severe earthquake or a major landslide, among natural forces, would be likely to destroy them.

But man can easily raze the sturdiest stone buildings. High explosives, in time of war, have destroyed in a few seconds structures that had stood for thousands of years. In time of peace, too, man has deliberately dismantled buildings, stone by stone, to make way for something else or to get the stone for another building. Most of the picturesque ruins of Rome, such as the Colosseum, are ruins because medieval builders made quarries of them.

In spite of today's wealth of new materials and new techniques of building, stone still seems to be considered the noblest material. Our principal public buildings — cathedrals, railroad stations, museums, libraries and other buildings that are intended to be impressive — are faced with stone, regardless of their construction.

Most desirable are the even-grained stones that are fairly easy to cut, such as sandstone, limestone and marble. If too soft, the stone will probably not be very durable. Some coral limestones can be cut with a knife, but, unless they are stuccoed, they weather very badly. Granite, which is stronger and more durable than limestone but harder to cut, is excellent for foundations and for whole buildings — also for curbstones and paving blocks. In rubble, or roughwork, masonry, as distinguished from ashlar (cut stone), practically any type of stone may be seen. Slate, which splits easily in one plane, makes good flagging and roofing.

Ashlar masonry is usually set in mortar, although the Greeks and the Incas and some other fine masons used none. The purpose of mortar is not to hold the stones together (metal dowels provide for that) but to give a small space between the stones, to allow for minor errors in cutting and to waterproof the joints.

The exposed face of stone masonry may be given various textures, ranging from highly polished to extremely rough, or rock face. Or it may be carved into various designs. The joints may be recessed in such a way as to form a series of channels in the outer surface of a wall. A rugged effect, called rustication, is produced.

Brick and Tile

Second only to stone as a permanent building material and second only to wood as a utilitarian material is brick. It is made of clay, molded into blocks while moist and then hardened in the sun or by fire. Standing midway between wood and stone in cost, brick has been used as a substitute for either or both of these materials in places where they were not available.

The first bricks were hardened by being exposed to the sun. At about the same

Photo by Horydozak

A marble masterpiece: the Thomas Jefferson Memorial, erected in Washington, D. C., in 1941.

time that the Egyptians were first learning to cut stone into blocks and build with it, the people in Mesopotamia discovered that if bricks were baked in an oven, instead of merely being dried in the sun, they would be much harder and more durable. Entire cities were then built of brick, from the most imposing temples and palaces down to the least important buildings.

Brick is perhaps not quite so enduring as stone, but it will last well over a thousand years, which should be durable enough for most purposes. Furthermore, it is fireproof, which wood is not; hence, it is replacing wood in many areas.

Bricks are always laid in mortar. The type of mortar, the width of the joint and the skill with which the work is done all have an important effect upon the appearance, the durability and the waterproof qualities of a brick wall.

A wall consists of at least two thicknesses of brick. In order to tie these two thicknesses together, some of the bricks are turned so that their length extends back into the wall, and only their ends are exposed. These headers, as they are called, may be arranged to form a pattern, which can be quite decorative. The Dutch and Flemish, who are very skillful builders in brick, have developed patterned brickwork to a high point.

In modern times, brick walls are sometimes built with a more or less continuous hollow space between the two thicknesses. This gives better insulation and prevents water from coming through. In this case,

Standard Oil Co. (N. J.) photo by Webb

The patterned brickwork of this small house at Lisse, Holland, produces a most pleasing effect.

the two brick walls are tied together, not by headers but by metal ties built into the brickwork.

Brick varies in color, according to the clay; pigments are rarely introduced. Although most bricks are brownish red, they range from deep maroon to light pink and are occasionally in such colors as yellow, buff, gray or dark brown. Face bricks of extra hardness and having special colors and surface textures are sometimes used for the outside of the wall, the remainder of the wall being common brick.

In order to span even a small opening in brick masonry, an arch must be built, or else a lintel (beam) of some other material must be used. The Babylonians, who had no other materials, used the arch extensively. In other countries, stone lintels were often used over small openings, such as windows and doors, and, in more recent times, iron and steel have served the same purpose. But, as in all masonry construction, long spans require the arch in some form.

The Babylonians discovered that, if certain minerals and pigments were added to the surface of clay, they would melt during the firing and form a hard, smooth, glazed surface. These glazed bricks were used decoratively to relieve the monotony of unbroken brick masonry. Glazed clay, generally in thin, flat shapes, called tiles, later became an important building material. Tile has been used for decoration by most people who have built extensively in brick — the north Italians and the Dutch, for ex-

ample. But it was the medieval Mohammedans who carried tile to its highest point of development, especially in Persia and in Spain. There, entire buildings, inside and out, were faced with tile in a thrilling variety of patterns and colors.

Burned clay (terra cotta) has served from ancient times and still serves for roofing tiles and drainage pipes. Hollow building blocks of terra cotta have been much used in modern times for fireproof partitions and walls that do not bear loads. Special firebrick is made for lining the fireboxes of boilers and for other industrial structures where very high temperatures prevail. An extremely hard type of brick is made especially for paving streets.

Wood

Wood is the world's great utilitarian building material. Probably there are more buildings of wood today than of any other substance that is used for building purposes; more space is enclosed by wood than by any other building material.

Wood is in some respects ideal for building purposes. It is still plentiful in many parts of the world and could be grown as

Ewing Galloway

Intricate tile patterns adorn the façade of the Great Mosque in the city of Córdoba, in Spain.

a crop in most other places. Unlike the masonry materials we have been discussing, it is light but very strong for its weight. It is easily worked, is a good insulator, is warm to the touch and springy to the step. It also has certain disadvantages. It is inflammable; it is apt to decay; it shrinks or swells with changes in moisture.

In wood, moderately long spans are easily provided and extremely long spans are quite practical, with the aid of trussing, or bracing. Many medieval buildings with masonry walls had wood roofs. In a few parts of the world, notably Scandinavia and Japan, wood has been used throughout, even for the most monumental buildings.

Wood is America's traditional house-building material. One authority estimates that even today almost 90 per cent of the

FHA

Small frame house, in Nashville, Tennessee. Frame construction is popular in the United States.

American people live in wooden houses. Formerly, wood construction in both Europe and America was in the form of heavy timbers, hand-hewn and joined together without nails. In the nineteenth century, a new type of construction was introduced into the United States. Slender two-inch by four-inch pieces of lumber formed the uprights of a wall; they were spaced fairly close together and tied, braced and stiffened by a continuous sheathing of one-inch boards nailed to the outside. Houses constructed thus were light, and they could be put up rapidly. This method of building, called frame construction, was regarded at first as an emergency measure for temporary building; today, with certain modifi-

cations, it is a standard practice in the United States.

The softwoods make up by far the greater part of our building lumber. "Softwood" is a technical term for lumber from any coniferous, or cone-bearing, tree, regardless of its hardness; some softwoods, such as long-leaf yellow pine, are much harder than some hardwoods like poplar or basswood.

The framework of the house is most often of Douglas fir, yellow pine, spruce, hemlock or balsam fir. The boards forming the outer surface may be white pine, redwood, cedar or cypress; sometimes yellow pine, ponderosa pine, Douglas fir or hemlock are used.

Millwork (windows, doors, their frames and all kinds of decorative moldings) is usually made of a soft, even-textured wood, with little noticeable grains: white pine, sugar pine and ponderosa pine are most popular. Doors are often made of Douglas fir or of hardwood — birch, gum, maple, oak or walnut. Flooring is almost always hardwood, usually oak, sometimes maple; occasionally it is of yellow pine or Douglas fir. Oak, walnut, mahogany, teak and many other domestic and imported hardwoods provide interior wood paneling in the more luxurious types of buildings.

Plaster and Stucco

Plaster is of ancient origin and has been used all over the world, especially for interiors. A wall may be built with considerable roughness and inequality and with no consideration for appearance if it is later to be plastered. A plastered wall or ceiling can be made to produce an unbroken, smooth white surface, which cannot be achieved by any other means. Plaster may be given various surface textures, or it may be molded or colored, as it is applied.

Elaborately molded plaster ceilings were a feature of English Renaissance mansions. Most of the extravagant ornament of the Italian and Spanish baroque periods was molded of plaster, and such molding was prominent later in the Georgian style of architecture in England. Mural paintings done on wet plaster are known as frescoes, and their beauty, as well as their durability,

The Mission San Fernando, in Los Angeles, California, is an example of fine stucco construction.

may be seen in the examples uncovered at Pompeii and in Crete.

Interior plaster has generally consisted of sand and lime, with some fiber or hair to give it additional cohesion. Gypsum often takes the place of lime, or it may be added to the mixture. Exterior plaster, usually called stucco, must be more waterproof and more resistant to freezing and thawing than interior plaster. Stucco consists of cement and sand with a small amount of lime. It was exceedingly popular in the Italian Renaissance; it has been widely applied to buildings in many lands to cover rubble masonry or poor brickwork.

There are many familiar examples of stucco work in the United States, in the Pennsylvania "Dutch" area and in the Spanish missions of the southwest. Most Latin American buildings of the colonial period are of this type. Many modern suburban dwellings are of stucco or of half-stucco. In Italy and other Mediterranean lands, beautiful effects are produced by colored stucco: reds, pinks, blues, greens and grays are popular.

Plaster may be applied directly to masonry, but in wood construction it must be applied to laths or some other special material to which it will adhere. Laths are thin strips of wood, spaced slightly apart so that the plaster can fill the space between and bulge out behind and thus be keyed securely in place. Wood laths of this type are disappearing, for it takes too long to nail each lath separately in place. Materials to hold the plaster now include expanded metal, welded wire, gypsum board and insulating board.

Hollow blocks of gypsum make good fireproof partitions and other non-load-bearing walls. The plaster is applied directly to these gypsum blocks.

Metal

Until modern times, metal was not a primary building material; but it has long served in connection with other materials. Copper and tin were the most abundant metals of ancient times. Bronze, a mixture of these two metals, was sometimes concealed in stone masonry to key the stones together; sometimes it was exposed. Doors, windows, grilles, railings, hardware, lighting fixtures, fountains and even bathtubs were made of bronze by the Greeks and Romans. Except for bathtubs, bronze is still used for these purposes, by those who can afford it, because it is a strong, handsome metal.

Copper, which can be pounded into thin sheets, has long served for roofing, gutters and downspouts. It has also provided flashing — waterproofing material used to seal a joint in the roof or between the roof and another part of the building. Copper makes fine water piping, for, like bronze, it corrodes but slightly. After years of exposure, it acquires a light, bluish green coating known as a patina.

Lead is another metal with a long and honorable history as a building material. Since ancient times it has been used for roofing, flashing, gutters, downspouts and water piping. (Our word "plumbing" comes from *plumbum,* the Latin word for lead.) It is an important ingredient in the manufacture of paint, glass and ceramic tile. Tin is another noncorroding metal of ancient lineage. Tin roofing and flashing, tin coating on steel sheets, solder (a lead-tin mixture that is part of the plumber's stock in trade) — these are some of the forms in which tin appears in building.

Zinc's most important use today is for coating steel, or galvanizing, as it is called. It is also an important ingredient in the manufacture of paint and wood preservatives. Zinc roofing, flashing and weather stripping are not uncommon.

Brass, a mixture of copper and zinc, serves principally for water piping, door hardware and lighting fixtures. It varies in color, depending upon the proportion of its ingredients, from nearly white to orange-red. Most brass is more or less like gold in color; if kept polished, it is attractive.

Iron and its derivative, steel, are so important in our modern civilization that we often call this the Iron Age. Yet they were not essential building materials until about 1850, though from ancient times iron has been important for tools, weapons and armor. In the Middle Ages, wrought iron was widely popular for door hardware, lighting fixtures, railings, grilles, dungeon bars, drawbridge chains and spikes and bolts in wood trusses.

These uses continued with little basic change down to the nineteenth century. In the Renaissance, especially in Spain and Italy, wrought-iron grillwork was developed to a fine art. Iron occasionally reinforced masonry, as in the case of the chains that were put around the base of the great dome of St. Peter's and other Renaissance domes. Wrought iron, hand-forged by skilled blacksmiths, was the only form in which iron had a place in construction until comparatively modern times.

Glass

Glass has been used for windows since Roman times, but not until the late Middle Ages did it become a major item in building. Then, in northern Europe and England, not only the great cathedrals but also royal palaces, town halls, city houses and other types of buildings were opened up with huge expanses of glazed windows. Medieval glass was in small pieces, only a few inches in each dimension. These were set in lead cames — that is, grooved bars of lead that formed a resilient and watertight frame for the precious little "lights," as the small pieces of glass were appropriately called. The glass was hand-blown and quite irregular in surface, texture and color; but it let light in. The people of

northern Europe installed magnificent stained-glass windows in their cathedrals. Glass is sometimes used nowadays for walls as well as for windows.

In the Renaissance, this brave expanse of glass gave way to the small windows of the classical styles, which had originated in the Mediterranean region, where windows were not very important. Then builders discovered the charm of glass as a material for interior decoration. They lined walls with mirrors — the wonderful Hall of Mirrors in the Palace at Versailles is literally a shining example. Small pieces of crystal cut in prisms and hung on chandeliers reflected and diffused candlelight.

Modern materials

Toward the middle of the nineteenth century, building began to feel the effects of the Industrial Revolution. Much work that had been done by hand could be done more cheaply by mass production in factories; and, as labor costs increased, it was necessary to find short cuts in building techniques. New materials were introduced; the most important were steel, reinforced concrete, insulation and sheet materials.

Iron and Steel

The first iron bridges were built in England in the 1840's. As early as 1851 the Crystal Palace, originally set up in Hyde Park, in London, was built entirely of iron and glass. Wrought iron and cast iron began to be used for columns and, occasionally, for girders and beams. In 1889, Gustave Eiffel made the world gasp by building an iron tower in Paris to the dizzy height of almost a thousand feet.

It was in Chicago in the 1880's that the complete steel frame was first used. The steel frame is a continuous cagework of steel columns and beams that forms the skeleton of a building and upon which everything else is hung. The walls do not bear the weight of the building; hence they can be quite thin and of almost any material. Windows can be as large as desired; in fact, the entire wall can be a window.

In steel-frame construction it is almost as easy to build vertically as horizontally.

As soon as the invention of the elevator made skyscrapers practical, man began to build them enthusiastically.

Steel is the form of iron most useful as a building material. It differs from cast iron in that it is not brittle, but malleable — that is, it can be shaped by hammering or pressing or rolling. Wrought iron is also malleable, but cannot be manufactured by mass production. It is, therefore, more expensive, less plentiful and less uniform in quality than steel.

Steel can withstand both compression and stretching; in this respect it may be compared to wood. Long spans are easy in steel construction. Buildings of this material look strikingly different from masonry buildings unless, as is often the case, they have been disguised to resemble masonry buildings.

Steel has two great enemies, fire and corrosion. Although steel does not burn, it loses its strength very rapidly in even a small fire. For protection from both fire and rust, the steel frame is surrounded by at least two inches of fireproof material, most often of concrete, but sometimes of brick, stone, terra cotta or gypsum.

Steel is used for many other building purposes: doors, windows, stairs, decking, roofing, gutters, siding, lath, hardware, plumbing, heating and most other mechanical-equipment items. Cast iron serves for boilers and radiators, as well as for plumbing fixtures and drainpipes.

Reinforced Concrete

Concrete was employed extensively by the Romans and can thus hardly be called a modern material. But when it was discovered in the late nineteenth century that concrete can be reinforced with steel rods, an entirely new type of material was born, unlike anything that had ever been known in building before. Concrete is as durable as any material known to man, and it is capable of supporting enormous loads. It therefore serves for heavy construction — dams, piers, docks, roads, foundations, warehouses and the like. But when reinforced with steel, it can be used, like steel, for long spans and light construction. In

Standard Oil Co. (N. J.)

A lofty iron structure: the Eiffel Tower, Paris.

Europe, where structural steel has not been so plentiful or cheap as in America, reinforced concrete may be seen in almost every conceivable type of building.

Concrete, like stone and brick, can stand very high compression but cannot withstand stretching; in other words, it has very high compressive strength and very little tensile strength. But if we embed a steel rod near the bottom of a concrete beam, the steel, which has high tensile strength, will carry the load and thus keep the concrete from cracking. The concrete, in turn, protects the steel from fire and rust. This same principle applies in reinforced brick masonry, the steel rods being placed in the mortar joints or run through holes in the brick.

Concrete is made of portland cement, sand and broken stone or gravel. Portland cement is simply a modern refinement of the various natural cements known to the Romans and later builders. It is made by burning certain limestones and shales together in the right proportions and at a very high temperature.

One of the advantages of concrete is that it is monolithic — it forms a single block.

It can be poured into forms, and it hardens in place so that the whole structure is like one massive stone; there are no joints to worry about. But this is also a disadvantage, since it requires an enormous amount of material and labor to make the forms into which the concrete is poured; besides, a good deal of valuable time is lost waiting for the poured concrete to harden. However, there are now new types of concrete that develop high strengths very quickly and permit the forms to be removed within twenty-four hours after pouring.

Another approach to the solution of this problem is to precast the concrete into blocks or slabs before they are placed in the building. These precast units resemble masonry rather than monolithic concrete. Hollow blocks of concrete are now very widely used as masonry in small buildings. Where appearance is a factor, they are usually stuccoed. Precast joists (beams) and slabs serve for floors and roofs, usually in buildings with masonry walls. Precast wall slabs are among the prefabricated home-building devices.

The difficulty with large precast concrete units is their great weight, which makes shipping and handling very expensive.

Armstrong

Insulating a room with insulating wool in roll blankets. The "wool" is made of fine glass fibers.

Therefore, a great deal of experimental work has been done in an effort to produce lightweight concrete. In one method, the stone in concrete has been replaced by lighter materials such as cinders, slag or vermiculite, a form of mica. Another method consists of omitting the stone entirely and causing the cement and sand to foam like whipped cream before the set. In both cases, the result is lighter weight and better insulation, at the cost of reduced strength and waterproofing qualities. The strength may be greatly improved by introducing wood or straw fibers.

Insulating Materials

When man began to heat his buildings, he found that heat was expensive, and so he began to develop insulating materials that would conserve the heat. These were applied first to the heater itself and to the pipes or ducts that led from it, in order to prevent the loss of heat before it was delivered at the place where it was wanted. Then the walls and roofs of buildings were insulated, so that the heat could not escape easily to the outdoors.

Boilers are insulated by means of two remarkable minerals — asbestos and diatomaceous earth. Asbestos is a rock that is fibrous, like cotton or wool, and the fibers can be woven into cloth or made into paper. Diatomaceous earth is a lightweight limestone formed by the skeletons of countless millions of microscopic plants, called diatoms, that lived in the sea ages ago. These materials are mixed with enough cement to bind them together and are then plastered to the boiler. For pipe insulation, asbestos is mixed with another lightweight mineral, magnesia, usually in preformed sections, which are held in place by metal bands.

Insulating the walls and roof of a building is very much like insulating our bodies by the clothes we wear or the blankets under which we sleep. One popular type of insulation is a sort of "quilt" made up of some woollike or cottonlike material between two pieces of very tough building paper. The stuffing may be made from rock, slag, wood or bark, which has been processed into a fibrous, woolly material, or

ERECTING A SKYSCRAPER

The construction of a large modern building is an intricate operation. It involves many steps: excavating, laying the foundations, erecting the steel framework, encasing the steel in concrete, installing elevators, air-conditioning units, plumbing and electrical systems and performing a thousand and one finishing operations. The activities of the many trades involved are so carefully planned that everything proceeds as if by clockwork. In an amazingly short time, a towering edifice arises from a yawning pit in a busy city block.

In the drawing at the right, we show certain details of the steel-skeleton construction and a few of the other building operations. The derrick used for steel erection (A) consists of a fixed mast and a movable boom, which lifts these steel pieces into place. A hoisting tower (B) is erected as the building rises. Materials are carried to various floors in a hoist (C), moving up and down in the hoisting tower. A sidewalk bridge (D) is erected over the sidewalk; it protects pedestrians and may serve other purposes. As the steel skeleton rises, work proceeds on the lower floors. Among other things, the steel framework is encased in concrete (E) and the curtain (outer) wall (F) is erected.

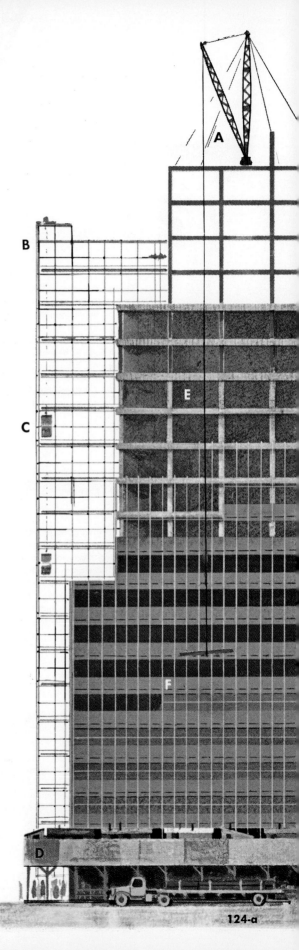

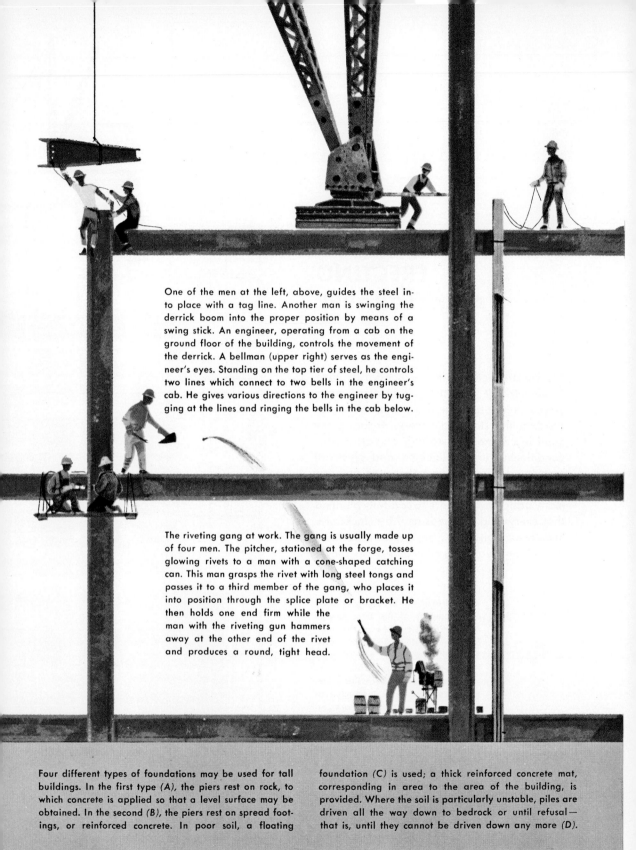

One of the men at the left, above, guides the steel into place with a tag line. Another man is swinging the derrick boom into the proper position by means of a swing stick. An engineer, operating from a cab on the ground floor of the building, controls the movement of the derrick. A bellman (upper right) serves as the engineer's eyes. Standing on the top tier of steel, he controls two lines which connect to two bells in the engineer's cab. He gives various directions to the engineer by tugging at the lines and ringing the bells in the cab below.

The riveting gang at work. The gang is usually made up of four men. The pitcher, stationed at the forge, tosses glowing rivets to a man with a cone-shaped catching can. This man grasps the rivet with long steel tongs and passes it to a third member of the gang, who places it into position through the splice plate or bracket. He then holds one end firm while the man with the riveting gun hammers away at the other end of the rivet and produces a round, tight head.

Four different types of foundations may be used for tall buildings. In the first type (A), the piers rest on rock, to which concrete is applied so that a level surface may be obtained. In the second (B), the piers rest on spread footings, or reinforced concrete. In poor soil, a floating foundation (C) is used; a thick reinforced concrete mat, corresponding in area to the area of the building, is provided. Where the soil is particularly unstable, piles are driven all the way down to bedrock or until refusal— that is, until they cannot be driven down any more (D).

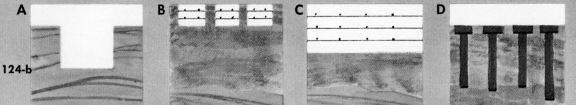

A **B** **C** **D**

JUMPING THE DERRICK

The derrick must rise with the framework—an operation called jumping the derrick. Generally it is raised two floors at a time. The platform on which the mast will rest is prepared. The boom is brought up parallel to the mast and is disconnected. It is then set on the new tier on a platform constructed specially for it, and it lifts the mast to its new location. Finally, the boom is firmly secured to the mast, and the derrick is ready for further operations.

To fireproof the columns—that is, to encase them in concrete—fitted wooden forms must be erected. Reinforcing bars or mesh are placed within the forms; so are conduits, inserts and so on. Then the concrete is poured. The reinforced concrete slabs prepared for the flooring are called arches by builders; actually, they are flat.

The outer wall, or curtain wall or skin of modern skyscrapers may be of various materials. Sometimes it consists of metal panels, applied in different ways. In the system shown here, slotted angle irons are bolted to the outer horizontal beams. Vertical metal supports are then fastened to the angle irons. Finally, the metal panels are bolted to these vertical supports.

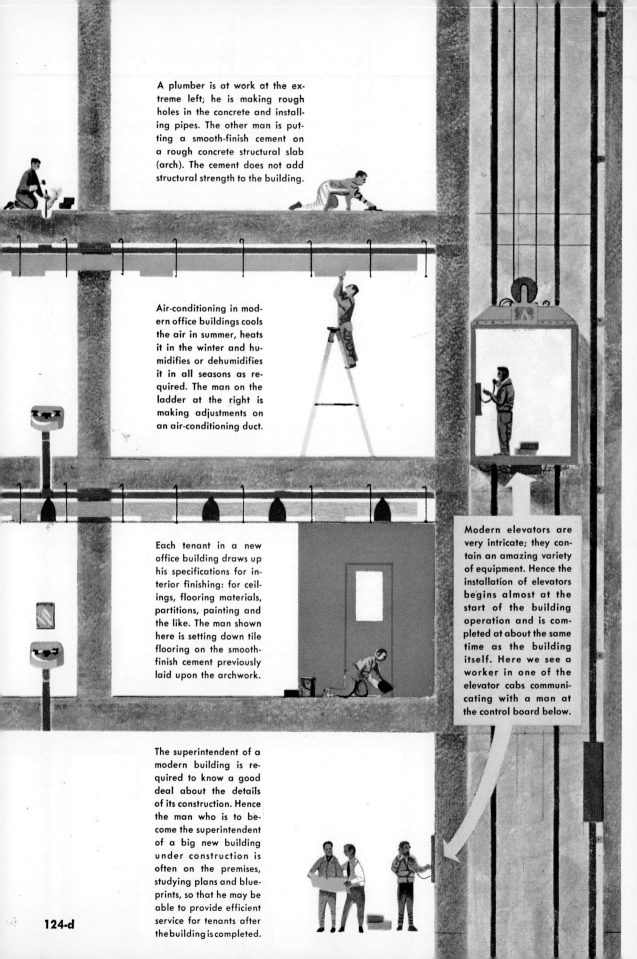

A plumber is at work at the extreme left; he is making rough holes in the concrete and installing pipes. The other man is putting a smooth-finish cement on a rough concrete structural slab (arch). The cement does not add structural strength to the building.

Air-conditioning in modern office buildings cools the air in summer, heats it in the winter and humidifies or dehumidifies it in all seasons as required. The man on the ladder at the right is making adjustments on an air-conditioning duct.

Each tenant in a new office building draws up his specifications for interior finishing: for ceilings, flooring materials, partitions, painting and the like. The man shown here is setting down tile flooring on the smooth-finish cement previously laid upon the archwork.

Modern elevators are very intricate; they contain an amazing variety of equipment. Hence the installation of elevators begins almost at the start of the building operation and is completed at about the same time as the building itself. Here we see a worker in one of the elevator cabs communicating with a man at the control board below.

The superintendent of a modern building is required to know a good deal about the details of its construction. Hence the man who is to become the superintendent of a big new building under construction is often on the premises, studying plans and blueprints, so that he may be able to provide efficient service for tenants after the building is completed.

124-d

from a natural material of this type, such as eel grass, kapok or cotton. Occasionally, the filling material in loose form is simply poured into the hollow spaces in the wall. The filler is sometimes of the mineral known as vermiculite; it comes in grains and looks much like a breakfast cereal.

Another type of insulating material does double duty. It is made in the form of a building board and can take the place of wood sheathing or lath or both lath and plaster. These insulating boards are of vegetable fiber, derived from wood, sugarcane, straw and the like. One board is made from old newspapers; it is just as good as the others, and perhaps even better!

The best insulation is a vacuum; the next best is air, provided it does not move freely. All of the insulations that we have mentioned above trap thousands of little pockets of dead air, which, since they cannot move, act as excellent insulators. An insulation that works on an entirely different principle is aluminum foil, which reflects the heat waves in exactly the same way that a mirror reflects light waves.

The weather stripping of windows and doors

A great deal of heat is lost through windows and doors, most of it through the cracks around the edges. Weather stripping is applied to stop this loss. Interlocking metal strips are the ordinary form of weather stripping, but felt or rubber gaskets are not uncommon. Heat loss through the glass itself can be reduced only by adding another piece of glass to form an insulating air space between. This is usually done in the form of a storm sash. There has recently been developed a double glass, with the air space between permanently sealed. It can be installed in any ordinary sash just like a single piece of glass. Plastic is also used for insulation purposes.

Refrigerators and cold-storage rooms must be heavily insulated, and they have a special problem of moisture condensation to contend with. In cases like these, natural cork, pressed into blocks, provides insulation. Foamed glass and foamed rubber have been developed for this purpose, and foamed plastics have also been tried.

Sheet Materials

With the thin, sheetlike materials known as wallboards, large areas of wall or ceiling can be covered very quickly, and the wall can be painted immediately afterward. About the only disadvantage of the sheet material is that the joints are sometimes unsightly. Some efforts have been made to conceal the joints in order to make the material resemble a plaster wall. It is now becoming more customary to give the joints a V-shape and expose them frankly. Most wallboards are for interiors only, but some waterproof types have been developed for the exterior of buildings. In this case, the joints are a practical as well as an aesthetic problem; they must be made weathertight, and this is not always easy.

The first wallboards were made of very heavy cardboard. Next came plaster board, an early sandwich-type board consisting of a thin core of gypsum with faces of heavy paper. Then came the fiber, insulating boards mentioned above. Another type, known as hardboard, is thin and smooth, and, as the name implies, very hard. It may be made of wood fiber compressed under very heavy pressure, or of asbestos and cement or of paper impregnated with a synthetic resin. All of these materials may be obtained already decorated, which saves another step in finishing the building. Some of the hardboards have an enamel finish of synthetic resin resembling the glaze of tile; in fact, they may be marked off into squares to imitate tile.

Plywood is a sheet material made up of three or more thin sheets of wood, called veneers, glued together with the grain of each sheet at right angles to that of the adjacent sheet. Since the strength of wood is much higher and its shrinkage is much lower along the grain than across the grain, plywood produces a stable board that is strong in any direction.

Plywood is not a new invention by any means; furniture of this material was made by the ancient Egyptians and by the fine cabinetmakers of eighteenth-century France

VERSATILE
BUILDING
MATERIALS

S. C. Johnson & Son, Inc.

California Redwood Assn.

Striking redwood ceiling, paneling and window frames in the lobby of a smart hotel in Southern California.

Aluminum Company of America

Alcoa Building, Pittsburgh, Pennsylvania. This handsome thirty-story structure is sheathed in aluminum panels, which are an eighth of an inch thick. Electrical wiring, most of the plumbing, ventilating ducts, doors, hardware and trim are also aluminum.

Alsynite Company of America

Sawing an Alsynite fiberglass building panel. It is translucent, shatterproof, lightweight and easy to use.

Glass walls in a Racine, Wisconsin, research center. Sections of glass tubing are laid horizontally, with synthetic rubber strips between them.

Slabs of pre-cast vermiculite concrete used as roof decking. Vermiculite concrete resists fire and rot.

Modern tile work in the headquarters of a building and loan company in the city of Cheviot, Ohio.

A magnificent concrete structure of advanced design: YMCA Building in Oklahoma City.

and England. But as a building material, plywood was just another wallboard until the middle of the 1930's, when waterproof plywood was developed by using synthetic resin adhesives in a hot press. This made plywood a dependable, weatherproof building material that could be applied indoors or out, structurally or decoratively. It is very important for building purposes.

Most plywood is made of Douglas fir. For decorative effect, plywood can be faced with practically any known wood, including many rare and beautiful imported ones. This has resulted in an increasing use of natural wood as an interior finish, in place of paint or wallpaper.

Two weatherproof sheet materials that have long served for exterior walls and roofs of factories and other utilitarian buildings are corrugated steel and corrugated asbestos-cement board. These thin materials have sufficient stiffness to span several feet between supports and are thus an economical method of enclosing a frame structure.

The development of sheet materials and waterproof adhesives has encouraged a more efficient use of materials in the structural engineering sense. Now we can attach a finishing material to the frame in such a way that both act together as one, and the finishing material actually carries a considerable part of the load. This type of construction is known as stressed-skin. Prefabricated stressed-skin panels are usually of plywood, though they may be of steel, aluminum or fiber board. The hollow spaces in the panels are filled with some insulating material, such as glass wool.

The sandwich-type panel represents an even more advanced step. Imagine a stressed-skin panel of the type we have just described, but with a rigid insulation, like corkboard, instead of glass wool. The faces may then be glued directly to the insulation, and the framing members may be omitted entirely. One material of this type, having a core of fiber insulating board and faces of asbestos-cement board, has been on the market for many years.

During World War II, great advances were made in the development of this type of panel for aircraft construction, and many lightweight panels of amazing strength resulted. The airplane manufacturers first used a core of balsa wood with faces of aluminum. Later, they tried foamed plastic cores, and finally honeycombs of paper impregnated with synthetic resin.

Other Modern Materials

Asphalt served the ancient Babylonians for calking and waterproofing, but modern man has given it many more jobs to do. We all know it in paving and in roofing. The commonest modern roofing material is a felt made of paper and asbestos and impregnated with asphalt. For flat roofs, this felt is applied to the roof in successive layers with hot asphalt mopped between each layer. For sloping roofs, the felt is cut into strips to resemble shingles and is applied like shingles. Asphalt is also pressed into tiles for flooring; coloring matter is added and a number of attractive designs are available.

Linoleum is a resilient, sheet flooring material made of linseed oil and ground cork. It is gaining steadily in popularity, since it gives an attractive and durable floor that has no cracks and is easy to clean. Other modern flooring materials are rubber, cork and plastic.

A number of modern materials will probably become increasingly important in building in the near future. Plastics (synthetic resins) in the form of adhesives and coatings are already very popular, as we have pointed out. In solid form, they have long been used for electrical switch plates and other small gadgets. Aluminum, more plentiful and cheaper as a result of the war, is beginning to serve for roofing, flashing and siding as well as for decorative purposes, and, in the form of foil, for insulation. Stainless steel, also popular as a decorative metal, will doubtless soon find a wider scope.

Scientific research in building materials is neither well organized nor adequately financed; nevertheless, it does go on, and it has already produced gratifying results. To an ever greater extent, the research man is helping to solve the builder's problems.

See also Vol. 10, p. 282: "Construction."

THE HOME CHEMIST III

Experiments That Entertain and Instruct

by ALEXANDER JOSEPH

IN THIS chapter of The Home Chemist we shall describe some more chemical reactions involving familiar elements. We shall show you, too, how to make certain simple food tests. Some of the apparatus and techniques you will use have been described in the first two chapters of The Home Chemist, in Volume 4. You would do well, therefore, to become familiar with these two chapters before you try any of the experiments included here.

Laboratory fun
with carbon dioxide

Carbon dioxide is one of the most important of all the chemical compounds from the viewpoint of the life that fills the earth. About 0.03 per cent of the atmosphere consists of this compound, which is a gas under ordinary conditions. Part of the carbon dioxide in the air is derived from the combustion of fuels, such as coal,

petroleum and gas. Part of it is derived also from the exhalation of animals and plants. The carbon dioxide that animals exhale is a waste product of the combustion that takes place within the cells.

Green plants use this gas in the vital process of photosynthesis, whereby they manufacture food in the leaves and certain other tissues. In this process carbon dioxide derived from the atmosphere is combined with water rising from roots in the soil and reacts in the green tissues of plants to form a simple sugar, glucose. This substance serves as a basic material from which the plant manufactures other essential foods. Animals require these basic foods just as much as plants do. Since animals cannot manufacture them in their bodies they must obtain them from plants. Some animals get the essential food elements by eating plants; others, by eating plant-eating animals. You can see,

Fig. 1.   Fig. 2.

therefore, that the carbon dioxide gas that we exhale is as vital to us as is the oxygen that we inhale.

The air that human beings expire about sixteen to eighteen times a minute contains about 4 per cent of carbon dioxide, which of course is a far greater concentration than is found in the atmosphere. You can show how little carbon dioxide there is in the air by pumping air through limewater, using a bicycle pump. Limewater turns white when it reacts with carbon dioxide. It will take about half an hour of strenuous pumping before this will happen. On the other hand, if you bubble your exhaled air through a drinking straw into limewater, this liquid will become milky white almost immediately.

How to show that fuels give off carbon dioxide

To show that fuels also give off quantities of carbon dioxide, stand a candle in a jar about one-quarter full of limewater, as shown in Figure 1. First light the candle; then cover the jar and hold it until the candle goes out. If you shake the closed jar, the limewater will immediately turn white. The burning fuel has given off carbon dioxide. You can use burning wood in this experiment instead of the candle.

The simplest source of pure carbon dioxide is dry ice, which is simply carbon dioxide in solid form. Handle this substance with gloves. Do not touch it with

your bare hands because it may cause frostbite, which will feel like a burn. To release and collect the carbon dioxide put some dry ice in a milk bottle. Close the bottle with a one-hole stopper connected to a delivery tube, and then put the bottle in a hot-water bath, as shown in Figure 2. Add some small lumps of dry ice as needed; when the water cools, replace it with more hot water. Do not put the stopper in place too tightly, and do not hold it in place by hand. This apparatus will be used in several of the experiments that we are going to describe.

In The Home Chemist II, in Volume 4, we showed you how to make a hydrogen generator, using a milk bottle, a funnel, a glass jar and some glass and rubber tubing. You can use this apparatus to generate carbon dioxide. Put pieces of broken limestone, marble or chalk in the bottle. Then put the stopper in place and slowly pour dilute hydrochloric acid through the funnel. Carbon dioxide gas will form as the acid reacts with the substances at the bottom of the bottle. Collect the gas by the displacement of water, as described in The Home Chemist II. You can use vinegar instead of hydrochloric acid, and a very small amount of bicarbonate of soda instead of limestone, marble or chalk.

Carbon dioxide is heavier than air. To show that this is so, inflate a balloon with air; inflate another balloon with carbon dioxide, given off from dry ice, as shown in Figure 3. Hang these two bal-

Fig. 3.

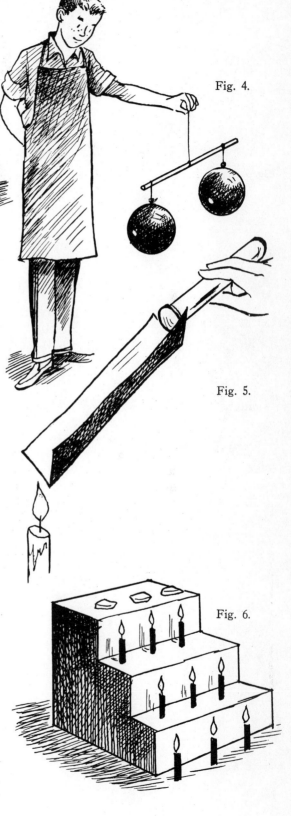

Fig. 4.

Fig. 5.

Fig. 6.

loons at either end of a balance that is made from a ruler or yardstick (see Figure 4). The balloon that is filled with carbon dioxide will pull down the end of the balance to which it is attached. You can also hang open paper bags, attached by threads, to either end of the ruler or yardstick. Let carbon dioxide flow from your milk-bottle generator into one of the bags, and watch it force the balance down. You can produce the same effect by holding a piece of dry ice with a pair of tongs over one of the open bags. Carbon dioxide gas will descend into the bag and force out the air, and the bag will sink.

Showing that carbon dioxide is heavier than air

There is still another way of showing that carbon dioxide is heavier than air. Make a trough of folded paper, as shown in Figure 5. Now put a burning candle at the lower end of the trough. Pour carbon dioxide from either of the two sources that we have mentioned into the top of the trough. The carbon dioxide will flow down the trough, displacing the air, and will smother the candle flame. There is a dramatic variation of this experiment. Pile up boxes so as to form three steps, as is shown in Figure 6. Set up a row of burning candles on the two lower steps and also at the bottom of the miniature stairway; place pieces of dry ice on the top step. As the dry ice changes into gaseous form, the gas will flow down the stairs

Fig. 7.

Fig. 8.

Fig. 9.

and extinguish each row of candles in turn. It will cause enough water vapor in the air to condense so that a mist will be readily visible.

Carbon dioxide, like any other gas under sufficient pressure, will cause a container to burst if it cannot withstand the pressure of the gas, as the following experiments will show. These experiments are safe enough, since part or all of the containers are made up of more or less flimsy substances that will burst harmlessly as the gas expands. First, place dry ice inside a balloon. Do not inflate the balloon; seal it by knotting the end or by means of a rubber band. Place the balloon and its contents in a jar of hot water. You will observe that as the dry ice changes from solid to gaseous carbon dioxide, the balloon will burst. You do not have to use hot water; simply place the balloon containing the dry ice in the sun, and it will inflate and burst (Figure 7).

Another experiment of the same type consists of placing dry ice inside a milk bottle and covering the mouth of the bottle with thin paper, held in place by rubber bands, as shown in Figure 8. Now place the bottle in hot water. As the solid dry ice changes into carbon dioxide gas, the pressure will cause the paper to burst. If you stretch rubber from a toy balloon across the top and fasten it with rubber bands, it will inflate as carbon dioxide accumulates; finally it will burst with a loud bang (Figure 9).

Making moth balls float

Fill a tall jar or bottle with cold water. Add one or two tablespoonfuls of vinegar; then a teaspoonful of dry bicarbonate of soda. Next, quickly drop in one or more moth balls. As shown by the downward-pointing arrow in Figure 10, the moth balls will sink quickly to the bottom of the container. After fifteen to thirty minutes, the balls will be covered with bubbles of carbon dioxide released by the reaction between the vinegar (an acid) and the soda. The buoyancy of these bubbles will then cause the moth balls to rise slowly (Figure 10) to the surface of the water and to float there.

How to remove
oxygen from air

Air is a mixture of various gases; it contains about 21 per cent oxygen by volume, 78 per cent of nitrogen and 1 per cent of other gases, including carbon dioxide and water vapor. You can easily remove all the oxygen in a given quantity of this mixture of gases. First make a float out of a flat section of cork or a small thin piece of wood. Put about one fourth of a teaspoonful of red phosphorus on the float. *Warning:* Do not use or handle *yellow* phosphorus, which is a highly dangerous substance. Red phosphorus is safe.

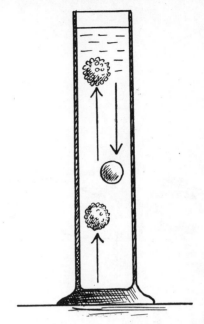

Fig. 10.

Oxygen is removed by
combining with the phosphorus

Place the float on the surface of the water in a basin or bowl (see Figure 11). Have somebody hold a quart jar in an inverted position over the float, as shown. Touch a red-hot wire (which is to be handled with tongs) to the red phosphorus. When this ignites, cover it with the inverted jar so that the mouth of the jar is just below the surface of the water. A heavy, white cloud will form in the jar. This is phosphorus pentoxide, formed by the combination of the oxygen in the air with the phosphorus burning on the float. After a while the phosphorus will stop burning; this indicates that all the oxygen in the jar has now combined with the phosphorus and that there is no pure oxygen left. In about twenty minutes the white smoke (phosphorus pentoxide) will disappear; it will have dissolved in the water. The float will rise one fifth of the way up the jar, as shown in Figure 12, because water has replaced oxygen, which composed approximately a fifth of the original mixture of gases in the jar. Almost all the gas that remains is nitrogen. To show that this gas does not support combustion, place a burning splint into the jar of nitrogen immediately after removing the jar from the water. You will observe that the flame will go out at once.

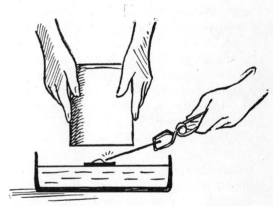

Fig. 11.

Fig. 12.

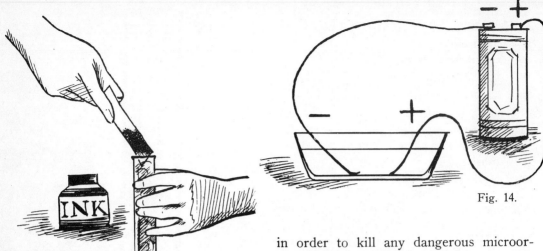

Fig. 14.

Fig. 13.

Separating table salt
into its two elements

Chemical analysis often proves that things are not what they seem. Did you know that table salt (sodium chloride), an essential food element, is made up of sodium, a soft white metal, and chlorine, a poisonous gas? We can separate salt into its components by using the electrolytic apparatus described in The Home Chemist II. Instead of acid, use a solution of table salt. At the negative pole, the sodium will separate from the salt solution and will immediately combine with water to form sodium hydroxide. If you test it with red litmus paper, the paper will turn blue, indicating the formation of a base. The test tube over the negative pole will fill with hydrogen from the water. Test it for hydrogen, as described in The Home Chemist II. The tube at the positive pole will fill with a greenish yellow gas — chlorine. This gas can be recognized not only by its color but by its distinctive odor. Its fumes are poisonous if they are present in considerable quantity; however, a tubeful of the gas is harmless. Incidentally, the chances are that you drink water containing a solution of chlorine (a very dilute solution, of course). Drinking water is often chlorinated, or treated with chlorine,

in order to kill any dangerous microorganisms that it might contain.

Turn off the current for the electrolytic apparatus as soon as you have a test tube full of chlorine; this will prevent excessive quantities of the gas from forming. To test the action of chlorine, hold the end of the tube to a piece of paper dipped in a dilute solution of ink (see Figure 13); the chlorine will bleach the paper. A compound of chlorine is used widely for bleaching purposes.

Some reactions
involving silver

Argyrol, a compound containing silver, is a solution used as a local antiseptic for the eye, ear, nose and throat. You can remove silver from Argyrol by placing two wires from a battery into a small quantity of Argyrol in a small jar or basin, as is shown in Figure 14. The silver will be deposited on the negative pole. This is an example of the process known as electroplating. You can electroplate silver from a solution by processing some well-used photographic hypo (sodium thiosulfate). Old hypo contains silver derived from silver salts left on unexposed parts of films and photographic papers. The silver will plate out on the negative pole of the electrolytic apparatus that we described above. A silver object used as the positive pole will improve the operation of the apparatus.

Silverware tarnishes because sulfur or sulfur compounds in the air combine chemically with the silver. To show this action of sulfur, heat a teaspoonful of powdered sulfur while holding a silver spoon above it. The spoon will tarnish, forming silver sulfide. A hard-boiled egg yolk will also turn a silver spoon black or brownish because of the sulfur released from the protein in the egg.

The fermentation process at work

A change that is called fermentation is brought about in materials containing starch and sugar by the action of yeast. (Yeasts are living organisms.) The starch must first be changed to sugar before fermentation occurs. As the yeast causes the sugar to decompose, carbon dioxide gas is released. If the yeast has been added to a liquid, as in the preparation of beer or wine or industrial alcohol, the escaping carbon dioxide will cause the liquor to bubble. Yeast is sometimes added to a solid, such as dough; in that case, the carbon dioxide first fills the empty spaces in the solid and then causes the solid to expand. This is why dough rises.

If you wish to see fermentation at work, prepare a 10 per cent sugar solution using two teaspoonfuls of sugar to eighteen teaspoonfuls of water; add yeast. In about one hour bubbles of carbon dioxide will be given off. After a few hours you will be able to smell the alcohol that is being formed.

Lactic-acid bacteria are involved in certain types of fermentation, as in the preparation of cheeses. For example, when skim milk sours, lactic-acid bacteria have been at work. If you slowly heat soured skim milk without letting it quite reach the boiling point, and then strain it through some cheesecloth, the resulting product is pot cheese, or cottage cheese.

Food tests made at home

You can easily carry out certain simple food tests in your home laboratory. If you wish to find out how much starch a food contains, add iodine diluted in water to a sample of the food. The sample will turn blue-black if starch is present in sizable quantities. If there is only a little starch in the food, there will only be traces of color — faint lines of blue or even green.

You can try out this test on starch that you can make at home. The best source of such starch is the common white potato. First boil a large potato or several small ones, and remove the peel. Grind up the cooked potato and place it in water; then pass it through filter paper in a funnel. Keep adding water. Allow the ma-

Some practical applications of fermentation, caused by living organisms such as yeasts and bacteria.

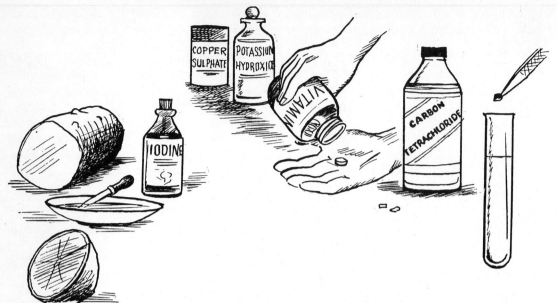

terial that collects on the filter paper to dry; this material is starch. Test it with iodine.

To test a food for fat, rub it against unglazed paper, such as mimeograph paper or newspaper. If a permanent translucent spot forms after drying, the food contains fat.

To test for protein, use the so-called biuret reaction. Add just enough copper sulfate to water to make it light blue in color. Mix strong potassium hydroxide in another bottle. As this substance is a strong alkali, be sure that it does not touch the skin or eyes. Add the copper sulfate and the potassium hydroxide to a bit of skim milk. The milk will turn blue-violet, indicating the presence of proteins. No heating is required. A very small quantity of lye (sodium hydroxide) may be used in place of the potassium hydroxide. Use the same precautions as with the potassium hydroxide.

Vitamin A is necessary for growth and health; lack of it may cause night blindness. In extreme cases blindness may result from deficiency in this vitamin. Among the best sources of vitamin A are fish-liver oil, butter and liver fat. You can determine the presence of vitamin A in food with the help of carbon tetrachlo-

ride and antimony trichloride. Handle the crystals of antimony trichloride with a pair of tweezers or tongs and not with your bare hands. Open up a vitamin-A capsule or a mixed-vitamin capsule that contains vitamin A. Add a small amount of carbon tetrachloride. After the contents of the vitamin capsule dissolve, drop in one crystal of antimony trichloride. If vitamin A is present, the colorless crystal will turn a deep blue-violet. The more vitamin A, the deeper the color. You might try this test with some of the foods mentioned above as being sources of the vitamin.

Vitamin C protects us against the disease called scurvy; it is also important in the formation of connective tissue, bone tissue and dentine. It is found in citrus fruits, berries, pears and other foods. To test for the presence of this essential vitamin, use iodine diluted in fifty parts of water. Add the iodine solution drop by drop, with a medicine dropper, to samples of citrus fruit juices. As it mixes with the juices, the brown color of the iodine will disappear. The greater the number of drops of the iodine that lose their brown color, the greater the amount of vitamin C present in the juices. Keep a record of your results with this iodine test. Try it on pure vitamin C (ascorbic acid), which you can purchase in a drugstore.

See also Vol. 10, p. 289: "Chemistry, Experiments in."

THE IMPORTANCE OF DUST

The Story of the Numberless Particles
That Fill the Atmosphere

FROM time immemorial the presence of dust particles has plagued the industrious housewife. No matter how thoroughly she cleans, a thin film of dust soon gathers upon the surface of tables and chairs and desks, and it thickens as the hours and days go by. But this is not the only grievance that man has against dust. Stores have good reason to detest it; armies of cleaners are required to combat it, and it brings about depreciation of stock. Certain kinds of dust, including sawdust, coal dust and the fine particles that accumulate in flour mills, are exceedingly dangerous. Where there is a high concentration of such dust, a single spark may set off a disastrous explosion. The reason is that the smaller a particle is, the more surface it exposes, in proportion to its bulk, to the oxygen of the air and the more readily it combines with it when ignited.

Dust is a definite menace to health. Mineral dust in mines and quarries may settle in workers' lungs and may cause the dread disease known as silicosis. Dust-laden air helps to spread the bacteria of various diseases. The sputum of a diseased person may contain millions of germs. Falling on the sidewalk or roadway of a city, sputum is soon tracked over a large area and gradually mixed with dust particles. When these particles, now bacteria-laden, are raised into the air by a gust of wind, they may be transported into the mouths or lungs of many persons.

Yet dust cannot be condemned as an unmixed evil. As we shall see, if there were no dust, there would be no blue color in the sky, no radiant sunsets, no soft afterglow. The dust particles in the atmosphere help to bring the rain that supplies moisture to our crops and fills our reservoirs. For the water vapor in the atmosphere condenses on dust as well as on particles of salt (from evaporated ocean spray), ions (electrified molecules and atoms) and other minute substances.

The dust in the atmosphere is usually made up of such small particles that most of them are invisible unless examined under the microscope; only the larger ones can be made out with the unaided vision. A ray of sunlight in a darkened room reveals the presence of these comparatively large particles of dust, which are constantly colliding with one another. They gradually yield to the pull of gravity and settle upon the nearest surface.

In spite of their exceedingly small size, it is possible to estimate with considerable accuracy the number of dust particles in a given space. The density of dust varies considerably from one place to another. The atmosphere above a mountain top and over oceans may have only a few thousand particles of dust per cubic centimeter. On the other hand, there may be as many as 5,000,000 particles, or even more, per cubic centimeter in the atmosphere over certain cities. Since a single puff of cigarette smoke sends some 4,000,000 particles into the air, the number of particles issuing from chimneys not provided with filtering devices staggers the imagination. It would be futile to try to estimate the total number of particles in the air at any one time, to say nothing of the cosmic dust particles that are found in outer space.

Dust is derived from a variety of sources; in the following pages we shall describe some of the most important ones.

Dust from the surface of the earth. Most of the dust in the atmosphere is derived from the earth's surface, particularly where exposed soil or rock has been subject to wind erosion. In places where there has been overgrazing, or where the surface has been disturbed by construction work or where there has been a long period of drought, particles of soil or rock can very readily be carried off by the wind. Sandy soils are more apt to break down under abuse than heavier soils; but even the finest-textured ones are subject to wind erosion under certain conditions.

Sometimes many tons of dust are removed by the wind from a given region. The results may often be disastrous. In the early 1930's there was a protracted period of drought in a plains area of the United States — an area including parts of Oklahoma, Texas, New Mexico, Kansas and Colorado. The winds carried off vast quantities of soil in the form of dust. In certain places wind-blown dust stopped the sun's rays so effectively that even during the day it was as dark as night. When the dust settled, it would often blanket cultivated fields and produce a barren desert. The name "Dust Bowl" was given to the region that was affected in this way. Great numbers of farms became unproductive; thousands of families were forced to move.

Dust that has been picked up from the surface of the earth may be carried long distances before it is dropped. One morning in 1918, people in Wisconsin were dumfounded to find everything covered with a thin veneer of reddish dust. The total fall came to some millions of tons. At first, the dust was thought to be the result of a distant volcanic explosion. But when some of the particles were examined under the microscope, it was found that the dust had originated in the arid regions of Mexico.

Portions of Europe frequently receive dust from the Sahara; it is estimated that

In regions where there has been a long period of drought, particles of soil can readily be carried off by the wind. Below: dust particles being blown about by a 26-mile wind in a field in South Dakota.

U. S. D. A.

The result of a typical dust storm at Childress, Texas. Topsoil blown from once fertile fields has finally come to rest, burying everything under a gray mantle of dust, several feet thick in some places.

five inches have been deposited in the last three thousand years. Dust derived from the red sands of the Sahara is responsible for the so-called "blood rains" and "blood snows" of Italy. The red color of the rain or snow has been derived from the reddish dust particles upon which water vapor has condensed. The water that falls to earth in the form of rain or snow brings the reddish particles with it.

Dust is often carried by prevailing easterly winds from the Sahara to the Canary and Cape Verde islands, in the West Atlantic. It may be so thick in the vicinity of the islands that it interferes seriously with shipping and planes. Winds from the Sahara also spread dust far and wide over western and central Africa. They produce a haze that persists, varying in density from day to day and season to season. The French call this haze "brume sèche," or "dry mist."

Sometimes wind-blown dust produces thick deposits in certain regions. Such loess deposits, as they are called, were formed extensively in the Mississippi Valley area; they are also found in South America, Europe and Asia. The most striking deposits of this kind are found in eastern China. Here the loess has been built up in many places to a thickness of several hundred feet. Since loess is easily dug and is quite dry, many families in China live in caves dug in the loess areas.

A characteristic of all true loess formations is the lack of stratification — that is, arrangement in strata, or layers. Such deposits have accumulated very slowly. As each thin layer was deposited, it was exposed long enough to the action of frost and rain so that all traces of stratification have been destroyed. The lack of strata and the enormous thickness of the loess deposits lead us to infer that wind action has been going on pretty constantly for a very long time.

Dust derived from ocean spray. Strong winds whip vast quantities of spray from the ocean. As a result, small particles of the salts contained in the sea are added to the dust of the atmosphere. Among the salts in question are sodium chloride ($NaCl$: table salt), calcium chloride ($CaCl_2$), potassium bromide (KBr) and magnesium chloride ($MgCl_2$). Traces of these salts are found in regions remote from the ocean. Naturally, toward the coast the amount of salt dust increases. Under favorable conditions it may constitute a considerable percentage of the material suspended in the air. It is estimated that about 2,000,000,000 tons of salt dust are added to the atmosphere every year.

Dust derived from volcanic activity. Volcanic eruptions throw various materials high in the air, up where the winds have great velocities and great carrying power. The coarser fragments settle down rather rapidly near the volcanic vent and aid in building a volcanic cone. The finer particles are transported great distances and they may stay in the atmosphere for long periods of time. For example, the eruption of Krakatoa, an Indonesian volcano, in 1883 added vast quantities of dust to the upper air; some of it stayed in the air for three years. It then made its way earthward, adding to the accumulation of dust on the earth. We describe the eruption, or rather the explosion, of Krakatoa elsewhere in THE BOOK OF POPULAR SCIENCE (see Index). It is said that the roar of the eruption represented the loudest sound ever heard on the earth.

Sometimes the rate of accumulation of volcanic dust is amazingly rapid. Forty hours after one eruption, some fifty inches of dust had settled at a distance of ten miles, while eight hundred miles away an accumulation of two inches quickly formed. In Oklahoma and Kansas, volcanic dust deposits fifteen to twenty feet thick have been dug to furnish the abrasive substances used in cleaning powders. They are ideal for this purpose. The deposit is made up of exceedingly fine particles; these are very

In the Chinese province of Shansi, a considerable number of people live in cave houses, dug into loess deposits, as shown above. In the lower photograph we see the living room of a typical cave dweller of Shansi.

Photos on this page, Eigner, from Black Star

angular and resistant to abrasion. Beds of this thickness undoubtedly represent repeated accumulations over a long period of time. What is even more interesting, they are several hundred miles away from any past or present known volcano.

Dust derived from forest fires. Every year millions of trees are destroyed by forest fires. These fires may be due to careless smoking, neglected campfires, the burning of debris, lightning, arson and railroad and lumber operations. In the United States alone, it is estimated that there is a total of some 13,000 fires in the national forest areas alone. Each fire adds vast volumes of ash to the atmosphere.

Dust derived from meteors. Almost everyone has seen "shooting stars" as they flash across the heavens. These are meteors which have invaded the earth's atmosphere and have been set aflame because of friction with the atmosphere. What we see is the path traced by the blazing meteor. Generally, it is entirely consumed; gas and very fine dust are all that remain.*

There is quite a good deal of dispute as to the amount of dust contributed to the atmosphere by meteors. Probably something like 2,000 tons is added yearly, though certain authorities put the figure quite a bit higher. Systematic studies have been made of the meteoric dust precipitated by rain water. Deep-sea sediments show traces of meteoric dust, deposited hundreds of thousands of years ago.

Dust derived from the burning of fuel. Incomplete combustion of fuels used in homes and in industrial establishments adds greatly to the quota of dust in the atmosphere. Smoke, soot, fly ash and other wastes combine with water particles to form a pall in many industrial areas.

* Some meteors, including a few of considerable size, fall to earth. They are then known as meteorites.

Amer. Mus. of Nat. Hist.

A considerable amount of dust in the atmosphere is contributed by meteors, which are set ablaze as they strike the earth's atmosphere and are reduced to dust and gas. A certain number of meteors, partly consumed, reach the earth; they are then called meteorites. The above photo shows a typical meteorite.

Sometimes smoke is mixed with fog to produce the type of air pollution called smog. We have discussed man-made air pollution in the article Air Pollution, in Volume 3.

Dust as the
framework of clouds

The particles of dust in the atmosphere act as centers about which moisture collects. Without these particles, water vapor would not condense in the form of water droplets; the water droplets would not form clouds; the clouds would not release their load of moisture to the earth in the form of rain, snow and other kinds of precipitation. The earth would be watered by heavy dews instead of rain; moisture would constantly condense on the ground and on foliage.

Conditions upon our planet would be quite different if there were no rainfall and if heavy dews prevailed. Certainly the geological processes connected with the transportation and deposition of sands and muds would be entirely different; so would the pattern of sediment deposition in the sea. Life would undoubtedly exist under such conditions; but it would be quite different from life as we know it today.

The fact that dust particles act as nuclei for the condensation of vapor and the formation of clouds is of the utmost importance for life upon the earth. For clouds moderate the effects of the heat rays

Maine Dept. of Economic Development

from the sun during the day and they retard the loss of heat from the earth during the hours of night. It has been said that they act like a parasol by day and a blanket at night.

Dust affects the optical
properties of the atmosphere

In a world free from dust and from any atmosphere as well, the stars would be seen clearly both day and night. The sky itself would appear jet black. We sometimes forget that the bright blue of the sky is due to the scattering of light as it strikes the particles of dust and also the molecules of the gases in its path. We have pointed out in another article (The World of Color, in Volume 7) that the white light from the sun is really a combination of different colors, representing different wave lengths of radiation. Reds and yellows pass through the dust-laden atmosphere, but the blue rays, which have a shorter wave length, are scattered, becoming visible.

The sky looks red near the horizon because, as it is viewed at that angle, the path of light through the atmosphere is longer and traverses more dust. As a result, the blue light is scattered before it can be observed by the viewer. Red is scattered too, but in time to be observed by a watcher on earth. The greater the amount of dust in the air, the greater the scattering effect and the more gorgeously the sky will be tinted. After the explosion of Krakatoa, the percentage of dust in the air was unusually large. Sunsets and sunrises were particularly brilliant at that time.

The cause
of twilight

Twilight is also due to the existence of dust particles and gas molecules in the atmosphere. If they did not exist, we would be in total darkness as soon as the sun disappeared below the horizon. As it is, even when the sun has disappeared, its rays are scattered by the particles in the part of the atmosphere that is visible to us. Hence there is a certain amount of light for some time.

THE WONDER-WORKING X RAYS

Radiations That Have Opened New Paths of Knowledge

BY HELEN MERRICK

WITHIN recent years the walls that formerly rigidly separated physics and chemistry have been crumbling away. It now seems prophetic of this trend that one of the first steps toward the discovery of X rays was taken by a scientist who was both a chemist and a physicist, Sir William Crookes, an Englishman (1832–1919).

Crookes was interested in the effects of electrical discharges through gases at low pressure. To help him in his experiments he invented a device that became known as the Crookes tube. It was a tube from which the air had been pumped; this created a vacuum, or rather a partial vacuum, since some air had to remain in the tube if it were to work. Metal wires were sealed into opposite ends of the tube. These wires served as electrodes to which a source of electric current could be attached; the current could then be sent through the air remaining in the device.

One day, after Crookes had been sending a high-voltage current through his tube, he discovered that photographic plates in a closed box nearby had become fogged. He had also observed a mysterious green glow in the tube. Crookes prevented further fogging by removing photographic plates to another room whenever the current was flowing. For some reason the strange phenomenon does not seem to have aroused his curiosity.

Our story skips a few years, to the autumn of 1895. Wilhelm Konrad Roentgen (1845–1923), the German physicist, was puttering with a Crookes tube in his laboratory at the University of Würzburg. Turning on a high-voltage current, he noticed the effects that had already attracted the attention of Crookes: the green glow; the fogging of photographic plates. Roentgen was not satisfied to let it go at that. He tried covering the tube with black paper, through which no visible light could pass, and still something came through that affected a photographic plate. He came to the conclusion that this something must be invisible rays, and he called them "X rays" because "X" stands for the unknown in science. (X rays are also called roentgen rays for their discoverer.)

The story goes that Roentgen, by a lucky accident, also discovered one of the properties of X rays. While he was experimenting with the Crookes tube, some photographic plates in a desk drawer had been fogged. Instead of throwing the spoiled plates away, he developed them. On one plate he was amazed to see the perfect image of a key appear; yet no key had been in the drawer. Then he remembered that a key had been on the top of the desk. The key had been photographed right through the desk by the X rays. Roentgen also found that the rays would pass through his hand when he inserted it between the tube and a plate. In the developed photograph, however, the bones stood out clearly as dark shadows as compared with the flesh. This indicated that the bones, more dense than the flesh, must block the rays to some extent and, therefore, that the rays pass more easily through less dense materials.

The green glow that Crookes and Roentgen saw was light waves being sent out by the glass walls of the tube when electrons bounced against them, as the result of the discharge of electricity through the gas remaining in the tube. (Electrons are atomic particles that carry a negative charge of electricity; and ordinary electric current is really a flow of electrons.) At the same time, the electrons were stimulating the atoms of the glass to send out waves of extremely short length, the X rays. Since then, physicists have learned that X rays are produced when electrons strike any atoms at high speed and, conversely, that when X rays strike atoms, electrons are hurled out.

It was finally proved that X rays are similar to those of ordinary light, though they differ in wave length. This is how it was proved:

Since the time of Sir Isaac Newton (1642–1727) it had been known that visible white light consists of a number of different colors; and that white light may be broken into bands of these colors, since each color has a different wave length. Red, at one end of the spectrum, has the longest wave length; violet, at the other end, the shortest. When white light passes through a prism, each color is bent according to its wave length, red the least and

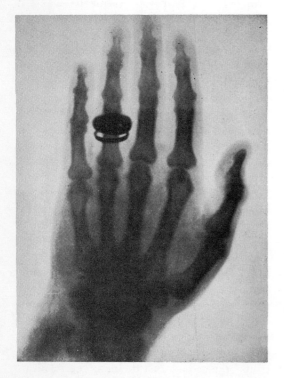

Early X-ray picture of a hand. The rings have blocked the X rays completely; that is why they appear black.

violet the most, and we get the rainbow array of the visible spectrum.

"Diffraction" is a term used for the property which waves possess of spreading out and interfering with one another when they meet a slit, a hole, or an opaque object of almost the same size as the wave length. One kind of diffraction grating consists of a system of lines drawn very close together on a glass plate. The lines are parallel, equally spaced and there are thousands of them to the inch. The spectrum obtained from such a grating is an interference phenomenon; the waves reinforce one another when their crests come together at any point, and cancel each other out when a crest meets a trough. The result is a system of bands of light and dark, visible on a screen placed behind the grating.

Von Laue proves that X rays are similar to visible light-waves

Now a German physicist, Max von Laue, was familiar with the diffraction grating. In 1912, he had the brilliant idea that a crystal, such as diamond or rock salt, might separate X rays into beams of different wave-lengths. This, of course, would prove that they do behave like light waves. A crystal came to mind because by this time physicists suspected (and later proved) that a crystal is a conglomeration of atoms arranged in a perfectly regular plan at very short distances from each other. It seemed to Von Laue that such an arrangement might act as a natural diffraction grating, with openings between the atoms small enough to diffract X rays if they really do have shorter wave lengths than those of visible light.

Von Laue's reasoning was sound. When a narrow beam of X rays was sent through a crystal, dark spots appeared on a photographic plate behind the crystal. It was found that the position of these spots depended on the different wave lengths in the X-ray beam and also on the arrangements of the atoms in the crystal.

Sir William H. Bragg, a British physicist, and his son, Sir William L. Bragg, carried Von Laue's work still further, and soon obtained clear, sharp spectrum lines with X rays. The Braggs discovered what is called the law of crystal diffraction. Using this law, if a scientist knows the wave length of the X rays, he can find out how the atoms are spaced in the crystal. The process can be reversed. If the scientist begins with knowledge of how the atoms are arranged, he can then work out the wave length of the X rays.

To describe all the experiments, theories and calculations that were involved in this work would fill volumes. Summing up, they not only furnished the key to the nature of the X rays but also to the structure of crystals. Most nonliving, solid matter is crystalline in structure.

We know now that X rays are a part of the electromagnetic spectrum. They are related to radio waves, infrared rays, visible light waves, ultraviolet and gamma rays and rays that accompany cosmic radiation. X rays belong in wave length between ultraviolet and gamma rays. (Gamma rays are given off by radium and other radioactive elements.) All electromagnetic waves travel through space at the same speed, about 186,-000 miles per second.

The angstrom unit for wave lengths

Wave lengths of the electromagnetic spectrum are expressed in terms of the angstrom unit, named for the Swedish physicist Anders Jonas Ångström (1814–74). One angstrom is 1/100,000,000 (.00000001) of a centimeter, or 1/253,999,800 of an inch. A scientist expresses this value as 1 Å $= 10^{-8}$ cm (centimeters). The symbol λ stands for wave length expressed in terms of angstroms. Thus $\lambda 5,890$ means "wave length of 5,890 Å."

Visible light has wave lengths ranging from about 4,000 Å to somewhat more than 7,000 Å. The wave lengths of X rays go from about 1,000 Å to 0.1 Å, overlapping other wave groups.

An important point to remember here is that the shorter the wave length, the greater the energy of the ray; and the greater the energy, the more penetrating the ray will be. For this reason, X rays and the tubes that produce them can be classified as "hard" or "soft." Hard rays have shorter wave lengths than soft rays and hence are more penetrating.

A descendant of the simple Crookes tube with which our story began, the modern X-

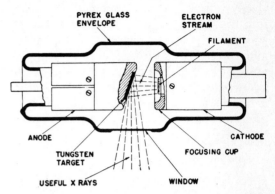

Diagram of a modern X-ray tube. Useful rays, streaming through the tungsten target, pass out through a window.

ray tube works on the same principles. It consists of a bulb containing as nearly perfect a vacuum as possible and fitted with electrodes to which high voltage can be applied. The negative electrode is called the cathode. It is usually a filament of tungsten wire wound into a spiral. The positive electrode, the anode, is called the target. It is made of some heavy metal, such as tungsten, and is welded to a copper support, which helps conduct away the tremendous heat generated. To prolong the life of the target, many present-day tubes are fitted with rotating anodes. When an electric current of high voltage is turned on, the cathode is heated white-hot and sends out a stream of electrons, or cathode rays, which are pulled through the vacuum by the high voltage until they reach speeds of thousands of miles per second. They slam against the atoms of the target and, as they strike, a spurt of X rays splashes out from the disturbed material of the target.

To explain where the X rays come from, we must go inside the atom. Whirling around the atom's nucleus, or core, are one or more electrons—in the case of a heavy metal there are many—each in a definite shell, or path. If an electron is forced to jump out of its shell, some kind of ray is emitted. Visible light is sent out when an electron is displaced from an outer shell.

If the disturbance is much more violent, an electron in an inner shell, close to the nucleus, is likely to be dislodged and removed from the atom entirely. The atom cannot remain in this unbalanced state. So an electron from the next outer shell jumps into the vacant place. The second shell is soon filled by an outer electron jumping in, and so on. Balance is restored when the atom picks up a free electron (there are always some free electrons drifting around in a metal) for its outermost shell. X rays are the result of inner-shell disturbance.

How higher voltages affect X rays

As we have indicated, it requires more energy to displace the electrons close to the nucleus than it does to displace any others in the atom. The higher the voltage in an X-ray tube, the faster the cathode rays will be pulled through it, increasing their energy. As a result the wave lengths of the X rays given off from the target will be shorter, approaching the wave lengths of gamma rays, and the X rays will be more penetrating. In giant X-ray tubes, built to stand millions of volts, the cathode rays may reach almost the speed of light (about 186,000 miles per second) and X rays are produced that will easily go through several feet of concrete.

As we have seen, one of the first properties of X rays to be noticed was their ability to penetrate solid substances such as wood, flesh and metal. However, X rays cannot be focused, or brought to a point, easily as visible light-waves can. One reason for this is that an X-ray beam of the same wave length does not penetrate all substances equally well. We mentioned earlier that the depth of penetration depends partly on the density of the substance. Ordinary X rays, produced at 100,000 volts, pass through flesh easily but through bones less so. "Filters" of aluminum or of copper are often used to refine X-ray beams. The softest rays are absorbed even by air. One of the most effective materials to block X rays altogether is lead. For this reason, glass containing lead is used on many kinds of X-ray equipment to protect those who work with it and are endangered by X rays.

Another important property of X rays is that, like light, they act on the chemicals of photographic plates, fluorescent screens and so on. This X-ray action together with the penetrating property makes it possible to take "shadow" pictures of, say, the inside of the human body or of a metal casting. Bombarded with X rays, some chemicals fluoresce: that is, they glow brightly as long as X rays strike them. One of the best-known uses to which this property has been put is in the fluoroscopic screen, which we shall explain when we take up the medical uses of X rays in more detail.

When a gas, such as the air, is exposed to X rays, it becomes ionized: that is, some of its atoms or molecules acquire an electric charge. (We explain ionization elsewhere in THE BOOK OF POPULAR SCIENCE.) The amount of ionization produced serves to measure the strength of the X rays.

Still another property of X rays is that when they fall on a substance, other rays are given off, which also consist largely of X rays. This "secondary radiation" is partly like that of the original X rays but is also dependent on the kind of substance exposed. It gives the scientist a tool with which to study the structure of matter.

Effect of X rays on living tissues

X rays also have an effect on living things. When X rays are absorbed, they change the structure of tissues, and these will be "burned" or destroyed if the exposure is long enough. Some kinds of X rays are also able to affect the genes, the elements in the germ cells of reproductive organs by which hereditary traits are passed on from one generation to the next. In this way heredity itself may be changed.

Considering the ability of X rays to probe matter and force it to yield its secrets, however grudgingly, it is hardly to be wondered at that X rays play a prima-donna role on the stage of scientific research. Practical application of new findings frequently follows so swiftly in many fields that it is difficult to separate the pure research from the everyday use.

Perhaps no field has been more rewarding to the scientist than the kind of analysis called X-ray crystallography, which began with the work of Von Laue and was developed further by the Braggs. The methods used in this field are too complicated to discuss within the brief limits of this article. To put it as simply as possible, when a narrow beam of X rays is sent through a crystal, the rays are scattered by the atoms and form a balanced pattern, which can be caught on a photographic plate. Each kind of crystalline solid has its own pattern. One result of this work is that scientists now have an accurate knowledge of the character of X rays sent out by each of the chemical elements, knowledge that has helped to determine the internal

compound by a device that shows the amount of X rays absorbed. The device is so sensitive that it can register the difference between ninety-nine and a hundred sheets of thin paper. One practical use is to measure the amount of "ethyl" in gasoline. X-ray absorptometry is possible because atoms can absorb X rays. An atom of oxygen, say, absorbs the same amount of rays whether it is in the form of the element or in a compound with other elements; and under the simplest conditions, the amount of absorption is the same whether the substance containing the oxygen be a solid, a liquid or a gas. So the amount of energy taken from a beam of X rays passing through a given mass of a substance is always the same, whether the

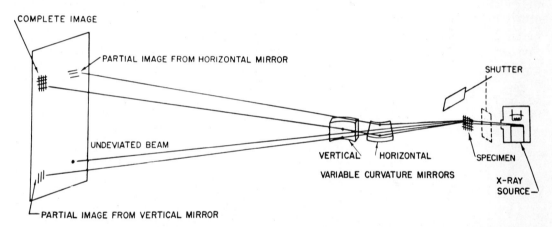

COMPLETE IMAGE

PARTIAL IMAGE FROM HORIZONTAL MIRROR

SHUTTER

UNDEVIATED BEAM

VERTICAL   HORIZONTAL

SPECIMEN

VARIABLE CURVATURE MIRRORS

X-RAY SOURCE

PARTIAL IMAGE FROM VERTICAL MIRROR

How the X-ray microscope works. The specimen is put in front of a shutter between the X-ray source and two concave mirrors with their faces at right angles to each other. Partial images are produced when rays are reflected from one mirror. The complete image results when rays bend up from the horizontal mirror and then bend out from the vertical mirror. X rays can be bent only at a low angle and only from the surface of a subject (reflection), and not in passing through (refraction). The above diagram has been greatly simplified.

structure of the atoms themselves. In fact, in what is called X-ray spectroscopy, the rays reveal information about the patterns of the electrons inside the atom.

From this kind of research, it has been a fairly easy step to investigate the arrangement of molecules in various chemical substances. They may be "fingerprinted" by the way in which X rays are scattered in passing through.

A comparatively recent development is the branch of science called X-ray absorptometry. Scientists can measure the quantity of certain materials in a chemical

substance is hot or cold, or a gas, a liquid or a solid.

Even art has benefited. X rays can be used to analyze the composition of pigments in oil paintings—to the point where sometimes the painter of what appears to be an "old master" can be identified.

The first X-ray microscope was introduced in 1950. It uses a combination of X rays and visible light-waves; and because of the penetrating power of X rays, it reveals finer detail than is possible with optical microscopes.

In order to increase the penetration of X

rays, scientists have been building ever more powerful machines. A giant among them all is the betatron—really an outsized X-ray tube. This machine was first invented in 1941 by Donald William Kerst, of the University of Illinois. He called it the "betatron," from beta ray, which is a stream of high-speed electrons.

As you may guess, a betatron is an enormously complicated apparatus. Essentially it consists of a doughnut-shaped glass tube, containing a vacuum, that is placed between the poles of a powerful electromagnet. What is called an electron gun is arranged so that it will send a stream of electrons into the tube at a certain angle. Inside the tube, two forces act on the electrons: one makes them follow the curve of the tube, and the other gives them higher and higher speeds, increasing their energy enormously.

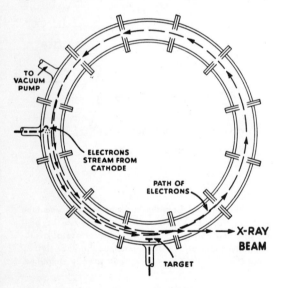

TO VACUUM PUMP

ELECTRONS STREAM FROM CATHODE

PATH OF ELECTRONS

X-RAY BEAM

TARGET

Diagram of a betatron. As the path of the electrons is bent and they strike the target, X rays spurt out.

At a chosen instant, they are bent from their curved path and directed to the X-ray target. All this happens in the twinkling of an eye; nevertheless during this time the electrons circle the tube many thousands of times. These operations are repeated over and over again.

In our daily lives, most of us are first introduced to X rays in the office of a dentist or a doctor. From what has been said, you can now understand why X-ray photography is so helpful in dentistry and medicine. Abscesses at the roots of teeth, an open safety pin in a baby's throat, a bullet in a splintered bone—all can be detected with X rays.

You may also have had your chest, say, examined with a fluoroscope, which makes use of the property of X rays to cause certain chemicals to glow. One advantage, for some purposes, of the fluoroscope over X-ray photography is that the fluoroscope allows a doctor to observe the inside of the body in action.

A fluoroscopic screen consists of a piece of cardboard (which X rays can pass through easily) coated with certain crystals, such as platinobarium cyanide or tungstate of calcium, which fluoresce under X rays. In an examination, the patient stands or lies down between an X-ray machine and a fluoroscopic screen, which the doctor can place against any part of the body. In some cases, such as an examination of the stomach, the patient usually drinks a liquid first, often a compound of barium, which blocks X rays. (Naturally, the liquid is harmless to the patient.) When the current is turned on in the X-ray machine, bones and thick organs, or organs such as the stomach that are full of the barium compound, cast shadows on the screen because they are blocking the X rays, and the rest of the screen glows. For the best results, of course, the fluoroscope should be used in a blacked-out room.

X-ray photography itself has been improved by the development of a method for taking pictures in three dimensions. Heretofore, the pictures were "flat," showing only width and length. The new method adds depth, so that the doctor may see the inside of the body in actual perspective. Another method, invented by a Swedish scientist, Dr. Arne Frantzell, permits pictures of veins and other soft tissues such as muscles, fat and the skin around bones to be taken in great detail.

X-ray photography has also made possible the mass survey of large sections of the population for tuberculosis. The mass chest X-ray units are fitted with photo-fluorescent cameras capable of taking a great

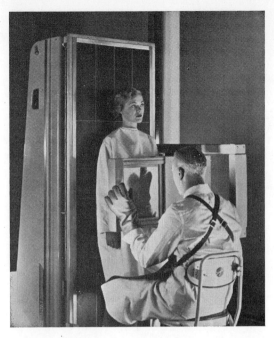

The fluoroscope at work. Special gloves and an apron protect the physician from overexposure to the X rays.

many miniature X-ray photographs. The X-ray equipment, the camera, and a dark-room for developing the films are grouped together into a truck unit. This is fitted with dressing-rooms, and it can be moved around the country. Widespread surveys of the population and the early diagnosis of cases have greatly reduced the number of tuberculosis victims.

X rays have a place in the treatment of disease, particularly cancer. A cancerous tumor is a new growth of cells which shows a tendency to spread to other areas. Now the action of X rays on living cells is mainly a destructive one. However, young cells and actively growing ones succumb to X-ray action more quickly. Also lymph tissue is more susceptible and some kinds of cancer spread by way of the lymph vessels. Cancerous cells are usually young and grow much more rapidly than healthy ones, so it is possible to adjust the strength of the X rays and limit the exposure with the result that only the cancerous cells are attacked and healthy cells are affected very little, or not at all.

Great care is necessary. The roentgenologist (a medical X-ray specialist) must be able to control the dosage precisely. X-ray machines for medical treatments are complicated affairs, with a number of precision measuring devices. Tumors on or near the surface of the body may be treated with X rays produced at 100,000 to 200,000 volts, but tumors that lie deeper may require rays from tubes of higher voltage, running into the millions. "Millions of volts" may sound alarming, but there is no danger of shock. The machines are completely insulated, and the dangerous part of the equipment is kept outside of the treatment room. One of the largest X-ray machines for treatment—2,000,000 volts—thus far built is in use at the Hospital for Joint Diseases, in New York.

A few cases of malignant tumor, considered hopeless, have even been treated with a 23,000,000-volt betatron. There were no harmful effects from the treatment on the patients, and the tumors that could be examined had shrunk.

Improvements in X-ray-treatment techniques are coming fast. An X-ray shield, developed by Dr. Hirsch Marks, of the New York City Cancer Institute, is one example. It permits extremely high-voltage

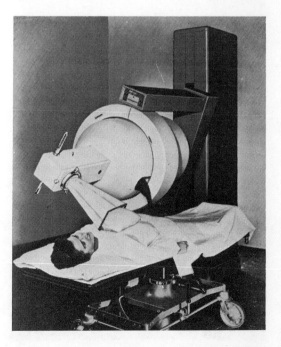

Treatment with an X-ray machine. The rays come out of the funnel and are concentrated on the area being treated.

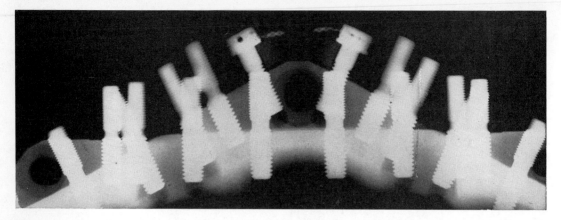

"Shadow" picture of part of an airplane crankcase, showing the position of the studs (white objects).

X rays to be trained on a tumor without damage to healthy cells. Radiation dosage is measured in terms of a unit called the roentgen, "r" for short. Without the shield, the highest safe dosage is about 200 r in air per treatment, up to a total of 3,300 r over a period of four weeks. With the shield, Dr. Marks reported, 1,200 r in air a day may be given for twenty days of treatment—in all, 24,000 r.

The industrial applications of X rays are legion, and we have space to mention only a few here. X rays are used to examine metal coatings, plastics, rubber insulation and so on, whose flaws might not appear on the surface. High-pressure boilers and pipes, for instance, are X-rayed to be sure that there are no weak places that might cause explosions. Some machines are large enough to take an X ray of an entire automobile. Most X-ray machines, however, concentrate on smaller areas.

Shadow pictures can show whether an egg is good or bad, and whether the core of a golf ball is perfectly round.

Airplane propellers are X-rayed periodically, because they show changes in their internal structure some time before they are ready to break.

Up-to-date textile mills also use X-ray equipment. Examination of fibers often shows that a better arrangement of the molecules in a fiber will make it sturdier and so will make stronger cloth.

It was only toward the turn of the century, as we have seen, that Roentgen found the answer to the mysterious green glow in the Crookes tubes and to the fogging in photographic plates. Yet already X rays have probed deeply into the closely guarded secrets of matter and have provided tools that benefit man in thousands of ways. Advances are still being made.

See also Vol. 10, p. 282: "X Rays."

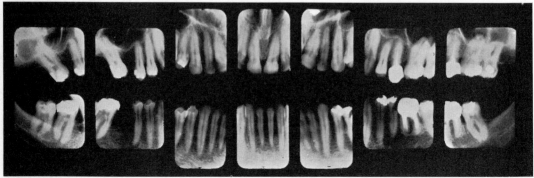

Series of X rays of a set of teeth. Such pictures give the dentist precise information about the teeth.

LIFE WITHOUT GERMS

The Answer to a Great Scientist's Question

ALMOST all living creatures, including man, are continually surrounded by a host of microorganisms (tiny living things)— bacteria, molds, yeasts, viruses and protozoa (one-celled animals). These tiny plants and animals, also known as germs, or microbes, inhabit the air we breathe, the food we eat, the water we drink and the earth upon which we live. They invade and powerfully affect the blood system, the breathing apparatus and the intestinal tract of animals.

The great French bacteriologist Louis Pasteur raised the question whether the higher forms of life could exist at all without microo.ganisms, particularly those that are found in the intestinal canal. Many biologists pondered over this question. It occurred to some of them that something more than mere theory was involved. Suppose that researchers succeeded in raising animals in a germ-free environment, so that they would be free of germs during the whole of their lives. These animals would offer the most interesting possibilities.

For one thing they would supply bacteriologists with a supremely useful tool. The bodies of animals normally contain many different kinds of germs. A bacteriologist utilizing such animals can never be certain that the conditions revealed by his microscope are due to the specific germ that he is studying. But if he could experiment with a germ-free animal, he could accurately determine the effects of a given germ, since no other microorganism would be involved.

The early efforts of biologists to raise germ-free animals were not very successful. By 1913, however, such animals could be kept alive for short periods of time. Later, the life span of the animals was in-creased and their growth rates were measured. The first successful results on a truly significant scale were produced in the bacteriological laboratories of Notre Dame University. The experiments of James A. Reyniers, director of the laboratories, and his associates opened up a new era in the investigation of germ-free animals.

Reyniers first became interested in the problem while doing research on the nature of the bacterial cell. He found that his bacteria cultures were constantly being contaminated by unwanted microbes, and he tried to find a way to exclude these uninvited organisms. He succeeded in protecting his cultures from contamination by means of containers that could be sterilized and then sealed. To test their effectiveness, Reyniers tried to raise germ-free animals in them. After much experimentation he succeeded; the glass containers he used were sterilized with germicide and supplied with air filtered through glass wool. In the years that followed, Reyniers' equipment and techniques were steadily improved. In the apparatus that is used today, microbe-free animals breathe sterile air and eat sterile food; they mature, breed and die in a germ-free environment. Sterilization is brought about by heat, chemicals and ultraviolet radiation.

The basic germ-free cage is a steel cylinder five feet long and three feet in diameter. Welded to one side of the cylinder is a smaller cylinder called a sterile lock, through which food and other materials are passed from the outside into the sterile cage. Food, instruments and other materials are sterilized in this lock before being transferred into the cage proper. Waste products, cultures and the animals themselves are removed from the main cylinder through the sterile lock.

The cage has a Pyrex observation window, through which experimenters can view the interior. Technicians, stationed outside the cage, handle and feed the animals with arm-length rubber gloves extending into the cage through ports; the gloves are sealed to the edges of the ports. The cage and its contents and the surfaces of the rubber gloves are sterilized by steam under pressure before animals are brought into the cage.

Many different kinds of cages are constructed on this basic design. There are raising cages where animals are housed. There are operating cages in which instruments and animals can be conveniently handled. A special transfer cage makes it possible to transfer animals from one kind of cage to another — for example,

containing a certain amount of glass wool.

The animals used in germ-free-life experiments include guinea pigs, chickens, monkeys, cats, dogs, rats, mice, insects and even fish. These animals must be germ-free to begin with. In the case of mammals, for example, the young are delivered from the germ-free womb of a pregnant female by Caesarean operation performed in a sterile operating cage after the mother has been anesthetized and scrubbed well with a germicidal solution. In the case of chicks the matter is simpler. Fertile eggs whose shells are sterilized by the germicide mercuric chloride are placed in a sterile cage where they hatch.

The feeding of new-born animals presents many problems. Young mammals must be fed hourly with a specially prepared formula; later they must be weaned

The new Germ-Free Life Building at Notre Dame University. This building houses specialized equipment for the raising of germ-free animals.

from the operating cage to the raising cage.

Hundreds of animals can be housed in a large tank built on the same principles of design and construction as the basic cage. The germ-free animals dwelling in such a tank are attended by a staff member, clad in a plastic diving suit; he takes a shower in a germicidal spray and a dip in a germicidal bath before entering the sterile animal quarters. All units are ventilated by air that passes through copper pipes

to a solid or semi-solid diet. They are fed by hand by means of a small capillary tube. The tip of the tube is a special latex nipple similar to the mother's nipple. Chicks are fed a sterile chick food. All foods are carefully sterilized by steam under pressure.

Once germ-free animals breed, the labor of caring for the young is considerably reduced, for the jobs of feeding and weaning no longer have to be assigned to a staff of specially trained technicians.

Each animal is frequently examined to see if it is germ-free. Samples of bodily secretions, hair or feathers, food, milk, water, urine and feces are collected. These samples are sealed in test tubes; they are then removed from the cages by way of the sterile lock and are rigorously tested for any traces of living microbes.

There can be no doubt that the experiments in the germ-free laboratories at Notre Dame have already answered Pasteur's question about the possibility of life free from germs. These experiments have proved that animals can be born, can be

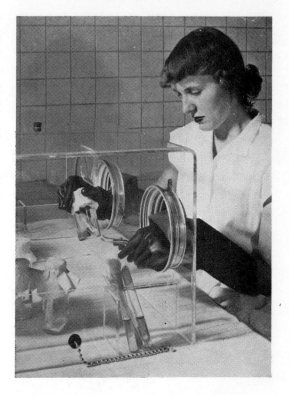

Germ-free insects (such as flies and cockroaches) are raised in simplified apparatus like that shown here. The insects are used in micrurgical studies, involving surgery at the microscopic level.

Standard Reyniers germ-free cages in the main rearing room of the Germ-Free Life Building. Each of these cages has a Pyrex observation window, through which the interior can be viewed.

This baby monkey is being raised in a germ-free cage. An attendant is feeding the animal with an arm-length rubber glove, which extends into the cage through a port.

An attendant, clad in a diver's outfit, is cleaning a cage in a large tank. He enters the tank through a germicidal bath, after taking a shower in a germicidal spray.

raised and can breed in an environment free from microbes. But they have gone far beyond this important point of departure.

For one thing, they have thrown a flood of light on the life processes of animals uncontaminated by germs. We now know that the blood serum of such animals contains few or no antibodies — substances that restrict or destroy the action of bacteria or neutralize their poisons. Germ-free animals have fewer white blood cells and less lymphatic tissue than normal animals. All of these differences are due, in part, at least, to the absence of microbes. When a dead germ-free animal is left in a sterile condition, there is no rotting, or odor or any other sign of decaying flesh.

When certain antibiotics, such as penicillin and chloromycetin, are fed to normal poultry, the birds grow considerably larger. There is no such increase in growth when these antibiotics are supplied to germ-free birds. The reason is probably that the latter, in their sterile environment, have already reached their maximum growth.

Germ-free animals have proved exceedingly helpful in the study of tooth decay. The exact cause of such decay is not known, but some think that bacteria in the mouth are responsible. Since the mouth is contaminated by different kinds of bacteria, it is difficult to single out the actual culprit. The teeth of germ-free animals are free from decay, whatever their

whether germ-free animals require the same food constituents as normal ones. Once this matter has been decided, many important dietary studies will be carried out. Researchers will be able to study, more effectively than ever before, the relation of diet to longevity, constipation, reproduction, lactation (milk production),

All photos, LOBUND Institute

A staff of technicians works around the clock feeding baby animals in germ-free rearing cages. Each cage is thoroughly sterilized by means of steam under pressure before animals are introduced into it.

diet may be. Therefore, experimenters are able to determine accurately the effects upon the teeth of specific bacteria introduced into the mouth of germ-free animals. They hope in this way to determine which germs are responsible for tooth decay.

Researchers utilizing germ-free animals are making a special study of lymphomatosis, a poultry disease resembling cancer in man. They are investigating radiation sickness. They have turned their attention to the effects of poisoning caused by antiseptics.

They are also trying to find out

growth and the development of special systems, such as blood and lymph. They will also be able to learn much about dietary diseases, such as pellagra.

New research programs are on the way in the LOBUND Institute at Notre Dame. (The name LOBUND is derived from Laboratories of Bacteriology, University of Notre Dame.) New buildings are being erected; more researchers and technicians are being added to the personnel. Director Reyniers and his associates are hopeful that tomorrow's results will overtop today's encouraging progress.

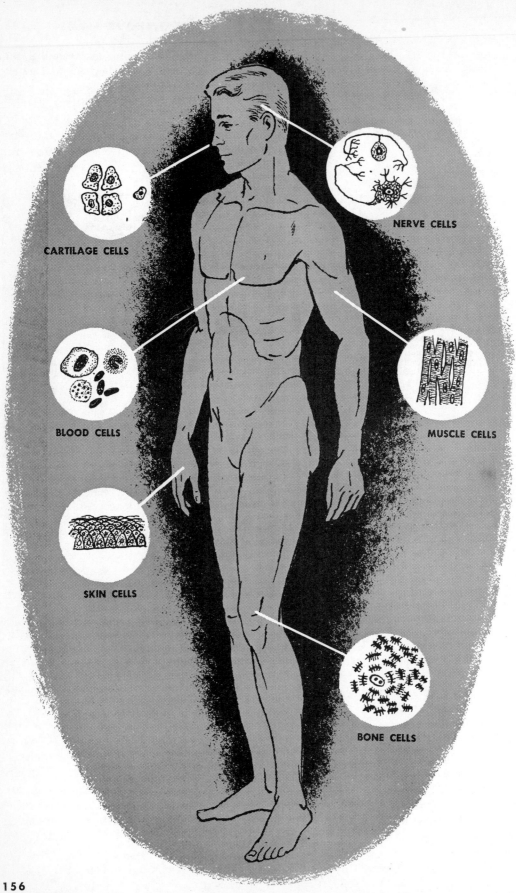

CARTILAGE CELLS

NERVE CELLS

BLOOD CELLS

MUSCLE CELLS

SKIN CELLS

BONE CELLS

SCIENCE AND PROGRESS

(1815-95) VI

BY JUSTUS SCHIFFERES

THE CELLS OF ANIMALS AND PLANTS

THE cell is the fundamental unit of living things; or, to put the matter differently, living things must be considered as communities of cells. There are certain one-celled animals, like the amoeba, and one-celled plants, like the diatom. Most animals and plants are made up of combinations of cells. Thus, in a human being we find billions upon billions of cells of different kinds — bone cells, skin cells, nerve cells, blood cells and so on. This is the cell theory of organic structure. Rather dimly foreshadowed in the seventeenth and eighteenth centuries, it was first definitely stated in the 1830's.

The seventeenth-century microscopist Robert Hooke was the first to make out the structure of the cell, or, more exactly, the cell wall. Peering through his primitive microscopes (which he had made himself) at thinly sliced sections of cork, he clearly saw that the cork was divided into numerous compartments. These seemed to fit snugly into one another, like the different sections of a honeycomb. Hooke called these compartments *cellulae,* or cells, because they suggested little rooms (which is what *cellulae* really means in Latin). The cells that he first examined belonged to dead plants; he later observed that there were similar structures in living plants.

A little later, the Italian anatomist Marcello Malpighi, studying sections of plants, observed a number of "little bodies, closely massed together and each surrounded by a wall." The Dutch naturalist Anton Leeuwenhoek and the English plant physiologist Nehemiah Grew also made out cell structures in the tissues of animals and plants. None of these investigators, however, had any idea of the nature and the purpose (if any) of these structures. We discuss these early microscopists in the chapter Science Grows Up (1600–1765) II, in Volume 3.

Probably the first man to point out that cell tissue is found in both animals and plants was the German army surgeon, naturalist and embryologist Kaspar Friedrich Wolff (1733–94). Wolff thought of this cell tissue as consisting of a mass of "cell-shaped" structures. The concept was valid enough, as far as it went, but it was expressed pretty vaguely and it does not seem to have created much stir. Wolff had other quite sound ideas. He pointed out, for example, that living creatures are not machines, as certain mathematicians had implied, since, after all, machines cannot, like living creatures, reproduce their kind. He also indicated various similarities in plant and animal development.

Unfortunately for his reputation, however, Wolff often drew conclusions that were not warranted by the facts at hand. If he could not explain something, he was wont to toss off a clever phrase in order to cover up the gap. It is said that Goethe had Wolff in mind when he had Faust remark to Wagner, his pupil, "If you do not understand a thing, simply give it a long name and then everybody will say what a brilliant fellow you are."

The early microscopists had focused their lenses not only on the minute structure of large organisms but also on the tiny creatures that they called animalcula (little animals). A generation or so later, the animalcula came to be known as infusorians, because they were to be found in different kinds of infusions. The German natural philosopher Lorenz Oken (1779–1851) pointed out a possible connection between the tiny infusorians and the higher organisms. In his GENERATION, published in 1805, he compared the infusorians to the "vesicles, or cells" found in larger organisms. "All organized beings," he wrote, "originate from and consist of vesicles, or cells. These . . . are the infusorial mass from which all larger organisms are formed or are evolved." This surprisingly acute observation was quite lost, in Oken's book, amid a flood of wild speculation.

The early investigators of cellular structure were greatly handicapped by the imperfect microscopes that they used. The chief defect of these instruments was chromatic aberration — that is, things placed under their lenses shimmered with all the colors of the rainbow, so that microscopists could not be sure of what they saw. Some of them used their imaginations pretty freely in interpreting what their microscopes revealed — or seemed to reveal. For example, some of them imagined they saw a preformed little man — homunculus — in the head of a human sperm cell.

A Swedish physicist, Samuel Klingenstierna (1698–1765), was the first to show how achromatic lenses (lenses free from color distortion) should be made. Acting under his instructions, the English optician John Dollond, in 1758, constructed the first lenses free from chromatic aberration. It was not until the nineteenth century, however, that microscopes were supplied with these improved lenses. In 1827, the Frenchman Chevalier produced the first achromatic lens system for microscopes, thus making it possible to bring very small objects sharply into view. The Italian Amici began the manufacture of achromatic lenses for microscopes at about the same time. In the 1830's, biologists began using the greatly improved microscopes and they made many striking discoveries.

These drawings of animalcules ("little animals") are from Henry Baker's treatise *Of Microscopes Made Easy*, which was published in London in the year 1785.

Thus the German botanist Hugo von Mohl (1805–72) convincingly demonstrated that new cells in plants — one-celled algae and higher plants alike — arise through the formation of partition walls in old cells. He showed that a definite cellular structure exists in bast (woody fibers), bark and other parts of plants — a point that had been denied by many previous investigators. The Scottish botanist Robert Brown (1773–1858), the discoverer of Brownian movement (see Index), showed that each cell of the orchid and of various other plants has an internal "key spot," which he called a nucleus, or areole. Johannes Evangelista Purkinje (1787–1869), a Czech naturalist, pointed out that the closely packed cell masses in certain parts of animals resemble cell masses found in plants. A French zoologist, Félix Dujardin (1801–60), observed that the life of one-celled animals is bound up with physical and chemical changes occurring in the jellylike substance of which the cell consists.

These investigations were all forerunners of the modern cell theory of organic structure. The cell theory, as we accept it today, was first comprehensively set forth by Schleiden and Schwann. Matthias Jakob Schleiden (1804–81) was born at Hamburg, the son of a distinguished doctor. He first specialized in the law. But his legal practice, in his native city, was so unsuccessful that in a fit of despondency he shot himself in the forehead. Recovering from the effects of this attempt at suicide, he determined to give up the practice of law and to devote himself to the natural sciences.

He obtained a doctor's degree in both philosophy and medicine, won fame with his scientific writings and became a professor of botany at the University of Jena in 1850. Twelve years later he resigned and, after a brief interlude as professor at

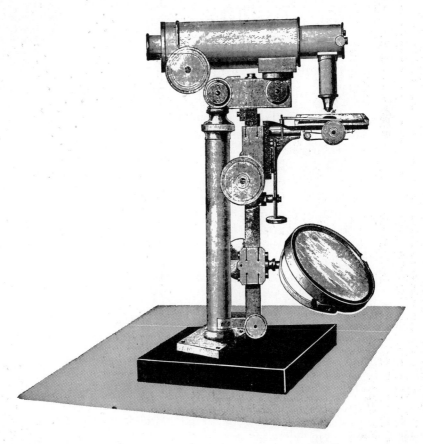

The microscope of Chevalier was the first one to use an achromatic lens system.

MATTHIAS JAKOB SCHLEIDEN

Dorpat (now Tartu), in Estonia, he gave up his scientific work. He spent the rest of his life idly wandering through the towns and the pleasant countryside of Germany.

In 1838, he wrote a paper, On Phytogenesis (origin of plants), for a scientific journal called MUELLER'S ARCHIVE FOR ANATOMY AND PHYSIOLOGY. In this paper he presented the idea of the cell as the essential unit of living organisms. He pointed out that plants that consist of but a single cell — like so many algae and fungi — are nevertheless individual, independent organisms. "Plants developed to any higher degree," he wrote, "are aggregates of . . . independent separate beings — that is, cells themselves . . . Each cell leads a double life; one which is independent and has to do with its own development alone; the other which is incidental, as an integral part of a plant. The vital processes of the individual cells . . . form the fundamental basis . . . for vegetable physiology."

All this is perfectly sound; it is accepted by all biologists. When Schleiden tried to explain the origin of cells, however, he went astray. He advanced the erroneous theory that new cells arise by budding from the surface of the nucleus.

The German naturalist Theodor Schwann (1810–82) worked out the cell theory of organic structure at almost the same time as Schleiden. The son of a Prussian bookseller, he studied at Berlin under a renowned physiologist, Johannes Peter Mueller, and after taking his doctor's degree he became Mueller's assistant. In 1838, Schwann was named professor of anatomy at the University of Louvain, in Belgium. Nine years later, he transferred to the University of Liége, also in Belgium, and here he remained until his death. He refused the professorships that were offered to him at German universities, because he objected to the constant quarreling of German professors.

Theodor Schwann's contribution to the cell theory

Schwann's chief contribution to the cell theory was contained in his treatise MICROSCOPIC STUDIES OF THE SIMILARITY OF STRUCTURE AND DEVELOPMENT IN ANIMALS AND PLANTS, published in 1839. He took as his point of departure certain striking resemblances between animals and plants. In the case of animal cartilages, for example, he pointed out that "the most important phenomena of their structure and development correspond to like processes in plants. These tissues originate from cells, which correspond in every respect to those of plants. During development, too, the cells display phenomena similar to those in plants . . . The cells — the membranes and the cell contents, as well as the nuclei [in animals] — are analogous to the parts with similar names in plants."

Schwann observed that the egg from which, when impregnated by a sperm cell, the animal body originates is really a cell. In mammals the egg (or ovum) is microscopically small in size; in other animals, like the hen, it may be quite large. Yet, large or small, all eggs are cells; in them we can distinguish the nucleus, the cell contents and the cell membrane. The egg produces a young animal by cell division. The single fertilized cell divides into two cells, the two into four and so on. The tissues that eventually arise can be distinguished from one another on the basis of the characteristic cells they contain. For example, there are tissues, like the skin, in which the cells are independent but pressed together; there are tissues, like the muscles, in which the cells are extended into fibers.

Schwann summed up his findings by pointing out that every part of every animal or plant is made up either of cells or of the substances that are thrown off by cells; that, to some extent, cells have a life of their own; that the status of this kind of individual life depends upon the life of the organism as a whole.

The cell theory, established by the

Karl Ernst von Baer (1792–1876); his treatise On the Story of the Development of Animals, published between 1828 and 1837, was long held in reverence by embryologists. Von Baer was the first to discover the eggs of mammals ripening in the ovaries and embedded in minute, fleshy nests called Graafian follicles. Examining the embryos of rabbits and dogs, Von Baer showed how the specialized tissues of the growing animal are formed in regular and orderly fashion from the fertilized egg.

While Von Baer's work marked a great advance over previous studies in the field of embryology, he did not have a clear idea of the microscopic structure of tissues. He did not realize that tissues in their

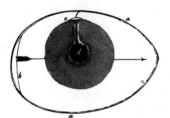 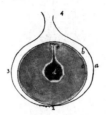 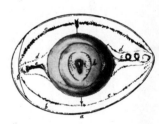

These three drawings show the development of the embryo inside a chicken's fertile egg. They are from Karl E. von Baer's book On the Story of the Development of Animals, published in two volumes from 1828 to 1837.

work of Schleiden and Schwann and sometimes called the Schleiden-Schwann theory, offers a logical and satisfying explanation of animal and plant structure and physiology. After this theory was once definitely stated, no one could successfully challenge it. It was modified in a good many details by later investigators. But the essence of the theory has remained unchanged since it was first formulated.

Out of the cell theory of Schleiden and Schwann arose the new science of cytology (the study of cells). The theory was also destined to influence profoundly many other biological sciences. For example, it revolutionized the science of embryology, which deals with the period of life between the fertilization of the egg and the moment of birth. The great name in the early development of this science is

entirety are made up of cells or of the products of cells; nor did he realize that the eggs that give rise to tissues are cells. It was not until the findings of Schleiden and Schwann were adopted by researchers in-embryology that this science was put on a firm basis. The first important work of embryology to utilize the cell theory was The Developmental History of Men and Animals, published by Albert von Koelliker in 1861.

The cell theory has become basic in many other sciences: in bacteriology, in pathology, in genetics and in histology, to name only a few. It is one of the great scientific generalizations of all time; it ranks with Newton's theory of universal gravitation and Mendeleev's periodic system of classification of the elements.

Continued on page 370.

AN ELECTROMAGNET AT WORK

General Electric Co.

A 55-inch electromagnet loading machine-shop steel turnings in the salvage department of the General Electric Company's Erie works. The magnet can lift 850 pounds of scrap like this, on the average.

MAGNETS LARGE AND SMALL

What We Know about Magnetism

by JOHN A. FLEMING

EVERYBODY is familiar with the toy magnet, that mysterious little **U**-shaped device that picks up needles or pins and that holds them indefinitely, through what seems to be sheer magic. But the magnet is far from being a mere toy. It is an essential part of a great many machines and tools and measuring devices, without which the world's work could not be done. A magnetized needle set within a compass helps the navigator to keep to his course at sea. When you hold a phone receiver to your ear, a magnet records the vibrations set up by the voice of the person talking into the mouthpiece. The electric motor, the electric generator and the ammeter could not possibly work without their built-in magnets.

Magnetism, the natural force that causes magnets to function as they do, revealed itself to men long centuries ago. A number of persons in antiquity knew that the black metallic iron ore called magnetite, or loadstone, had the property of drawing particles of iron to it. The Greek philosopher Thales (640?–546 B.C.) is said to have been the first to call attention to this property, but it may have been known long before. After Thales' time the loadstone was often mentioned in ancient writings; it was sometimes given the name of "magnet," from Magnesia, a district in Asia Minor where large magnetic deposits were to be found.

Socrates remarked, "the stone that Euripides calls a magnet . . . not only attracts iron rings but also imparts to them similar power of attracting other rings, suspended from one another, so as to form quite a long chain, and all of these derive

their powers from the original stone." The Roman Lucretius, who lived in the first century B.C., tried to explain magnetism in terms of his atomic theory.

There are many legendary accounts of the properties of magnets. The ARABIAN NIGHTS contains the story of a ship

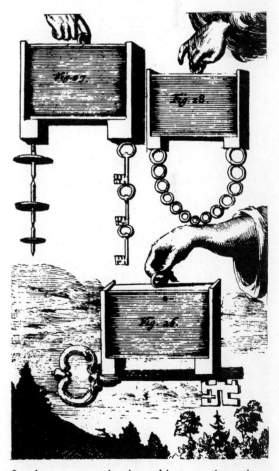

Loadstones attracting iron objects, as shown in a French treatise on the loadstone by D——, published in Amsterdam in 1687. The properties of the loadstone were known long before that time.

that approached an island made of magnetic rock; the ship fell completely to pieces because all the iron nails were pulled out of it through the attraction of the rock. Other ships, according to legend, avoided a similar fate by substituting wooden pegs for iron nails. There are stories, too, of huge statues of iron and bronze held in mid-air through the force exercised by magnetic domes.

Another tale gives a fanciful account of the origin of the word "magnet." It seems, according to this tale, that one day a shepherd called Magnes was tending his flock on the slopes of Mount Ida, in Asia Minor. Suddenly he noticed that the iron tip of his staff was being pulled toward the ground. He dug up the earth in the vicinity and found that his staff was being attracted by the loadstone in which the place abounded. Thereafter the loadstone was called a magnet in honor of the shepherd who had discovered it. Scholars have pointed out that this story originated long after the word "magnet" was in common use.

In the course of the centuries much of the mystery that once surrounded magnetism has been dispelled. In the place of out-and-out legends and pseudo-scientific speculation, we now have at our disposal a respectable amount of scientifically proved fact. Furthermore, we have been able to put the force of magnetism to work for us in a great many different ways.

Today that natural magnet, the loadstone, no longer figures prominently in the study of magnetism or in its many applications, for practically all magnets nowadays are artificial. It is easy to make such a magnet out of a steel object, such as a needle, if you have a permanent magnet. Simply draw one end of the magnet along the needle, stroking in one direction only. The needle will become an artificial magnet and will have the property of drawing various particles to it. The original magnet has lost none of its strength; it will be capable of magnetizing any number of other steel needles. Suppose, now, that we break our magnetized needle in two; each of the fragments will be a magnet.

Until about 1820 artificial magnets were made by stroking bars of steel with a loadstone or with an artificial magnet. But then a Danish scientist, Hans Christian Oersted, revealed that a magnetic field can be produced by sending an electric current through a coil of wire, called a solenoid. If this wire is wound around a core of steel, the core will be permanently magnetized when current is passed through the coil. Nowadays the manufacture of artificial magnets is based on this principle.

Every magnet, natural or artificial, produces a magnetic field in the space around it. We can map out this field in various ways. If we place a magnet under a piece of paper whose surface is covered with iron filings, the filings will arrange themselves in lines of force like those shown in Figure 1. We can also use a small compass to mark out the magnetic field about a magnet. If we put the compass in different places in the vicinity of the magnet, it will assume the direction of the lines of force at any given point (Figure 2).

In any magnet there are regions near the ends of the long axis where the attractive forces are greatest. These regions are called the poles of the magnet; the line joining the two poles is the magnetic axis. If a straight, or bar, magnet is suspended so as to move freely in a horizontal plane, the magnetic axis will assume, roughly, a north-south direction.

The pole of a magnet that is directed northward is known as the north-seeking pole, or north pole or positive (+) pole.

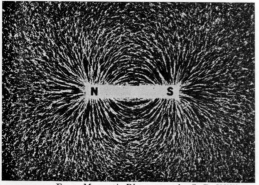

From *Magnetic Phenomena* by S. R. Williams

1. Magnetic lines of force about a bar magnet. Iron filings on a glass plate that is laid over the magnet align themselves along the lines of force.

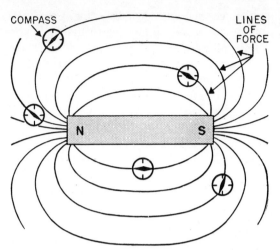

2. The magnetic field about a magnet may be marked out by a compass, which will assume the direction of the lines of force at any given point.

The pole that is directed southward is called the south-seeking pole, or the south pole or the negative (—) pole. Similar, or like, poles repel each other; dissimiliar, or unlike, poles attract each other.

When a bar magnet is bent double, it forms what is known as a horseshoe magnet. It will attract a given substance more powerfully than a bar magnet of corresponding size. The reason is that since the two poles are close together, the lines of force are crowded in a comparatively small space and exert a greater effect upon their surroundings.

Magnets lose their strength when heated; but they regain it when they become cool again. However, if they reach a temperature that is known as the Curie point, they become entirely demagnetized. The Curie point for iron is about 750° C. (1382° F.); it is different for other magnetic materials. In general, magnets tend to lose their strength as they grow older. To provide longer life, they are aged: that is, they undergo various treatments by heat, shock and repeated magnetizations and demagnetizations in weak alternating fields.

Magnetic lines of force cannot penetrate certain materials as easily as they can penetrate air or a vacuum. Such materials are called diamagnetic ("magnetic across," in Greek) because rods made of these substances tend to take up a position *across*

the field of force of a strong magnet (Figure 3). Bismuth, antimony and most of the other chemical elements are diamagnetic; so are their compounds.

Magnetic lines of force can penetrate other materials more easily than they can penetrate air or a vacuum. Such materials are called paramagnetic ("magnetic alongside," in Greek) because rods made of them tend to line up *alongside* the field of force of a strong magnet (Figure 4). Iron and liquid oxygen are paramagnetic substances.

A few paramagnetic materials are so much more easily magnetized than the rest that they are generally put into a separate class, called ferromagnetic. The ferromagnetic materials derive their name from the fact that one of them is iron (*ferrum*, in Latin). Nickel and cobalt are also ferromagnetic; so are various alloys of these

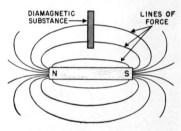

3. The diamagnetic substance takes a position *across* the field of force.

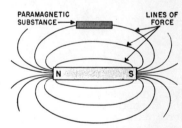

4. The paramagnetic substance takes a position *along* the field of force.

metals and a few other materials. Sometimes ferromagnetic substances are called magnetic materials, while all the rest are lumped together as non-magnetic materials.

Magnets that retain their strength for long periods of time are called permanent magnets. They have a wide range of uses in industry and in research. They serve in various measuring devices, such as ammeters, voltmeters, galvanometers, cardiograph recorders, seismographs (earthquake recorders), magnetic compasses, magnetometers and so on. They form an essential part of many kinds of scientific equipment. Temperature, pressure and traffic signals are all controlled by permanent magnets. They are used in certain kinds of lathes, chucks, conveyors, hand tools and separators, and also in various toys and novelties.

Improvements in permanent magnets are constantly being made. New alloys with superior magnetic properties have been developed. Some of these alloys are called quench-hardening steels: that is, they are

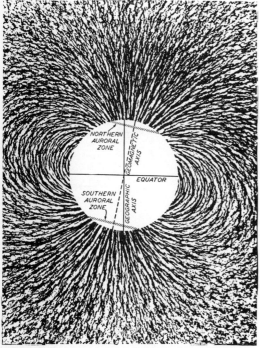

Carnegie Institution of Washington

5. Magnetic field about the earth, represented by the alignment of iron filings along the lines of force on a glass plate laid over a disc magnet.

first heated to a very high temperature and then hardened by rapid cooling, after being "quenched" in water or oil. Quench-hardening steels include alloys of steel with carbon-manganese, tungsten, chromium, cobalt and cobalt-chromium. They all contain carbon, since steel is basically an alloy of iron and carbon.

Other alloys are carbon-free. They include alloys of aluminum-nickel-iron, aluminum-nickel-cobalt-iron (trade name, Alnico), aluminum-nickel-titanium-iron (trade name, Nipermag) and cobalt-molybdenum (trade name, Comol). All these

substances require special casting, grinding, magnetizing and aging treatments.

Not all magnets are permanent. If electric current is passed through wire that is wound around a soft-iron core, the core will be magnetized but only as long as the current is turned on. If the current is shut off, the iron will lose practically all its magnetism; if the current is increased, the iron will have greater magnetic strength. This temporary sort of magnet — a core of soft iron set within a coil of wire through which an electric current is passed — is called an electromagnet.

The electromagnet is used in a great variety of modern electrical devices. It is the heart of the electric motor, and of such devices as the telegraph, the telephone and the electric bell. It serves to separate iron from its ores. Big electromagnets are frequently employed to load and unload iron and steel materials in railroad yards, steel plants and junk yards. Doctors sometimes use electromagnets to remove bits of iron or steel that have become imbedded in the eye.

We pointed out before that a magnet that is free to swing in a horizontal plane adopts, approximately, a north-south position. The reason is that it is attracted by magnetic forces arising within the earth itself. For the earth is a great natural magnet, with a North and South Magnetic Pole, a magnetic axis and a field of force that extends far out into space. Even four thousand miles above the surface of the earth, the magnetic intensity of this field is still one-eighth as great as it is at the surface. The lines of force of the earth's field are parallel to the surface near the equator; but as they approach the two magnetic poles, they bend and converge (Figure 5).

Certain substances within the earth — its magnetic deposits, for example — are highly magnetic; but that is certainly not true of most of the materials that make up our planet. As a matter of fact, a magnet of steel would be something like 10,000 times as powerful as one just as large made of typical earth-stuff. But in view of the earth's size, its total magnetism is pretty awe-inspiring. It is equivalent in effect to that produced by 800 quintillion (800,000,-000,000,000,000,000) parallel one-pound magnets if they could be placed at the earth's center or if they could be evenly distributed through the body of the earth with one magnet for about every two cubic yards.

Naturally, if the earth is a huge magnet, other magnets will be influenced by it. The north-seeking pole of a bar magnet, moving freely in a vertical axis, will evidently be attracted to the earth's North Magnetic Pole. It is true that the North Magnetic Pole does not correspond exactly to the geographical North Pole, as we shall see; but if we make certain calculations, we can determine in what direction true north lies. Of course we can then find the other points of the compass. This is the principle of that supremely useful instrument, the magnetic compass.

The south-pointing chariots of the Chinese

The Chinese have been credited with the invention of the compass. It is said that the Emperor Hwang-ti, who lived twenty-five centuries before the birth of Christ, built a chariot on which a dummy, mounted on a pivot, always indicated south. By using this chariot at a time when a thick fog had closed in upon his army during a battle, Hwang-ti was able to defeat the foe. This account is almost certainly mythical. Unfortunately it is difficult to see how much truth there is in later Chinese accounts of south-pointing chariots.

A Chinese document of the third century A.D. contains a statement that a needle enables a ship to follow a southward course. But it is not until the twelfth century that we find in any Chinese work a detailed description of the manner in which a needle is made to point to the south. The invention of the compass has also been attributed to other peoples, including the Arabs, the Greeks and the Etruscans; but there has been no authentic proof of such claims.

The earliest definite mention in European literature of the directive property of the magnet and its use in navigation appears in two Latin treatises by the English-man Alexander Neckam. These treatises, entitled OF INSTRUMENTS and OF THE NATURE OF THINGS were written toward the end of the twelfth century A.D. The first of these works describes the use of the magnetic needle to indicate the north; Neckam points out that sailors use the needle to find their direction when the sky is overcast and the stars cannot be seen. The second treatise gives a description of a magnetic needle mounted on a pivot.

An early authority on magnetism — Petrus Peregrinus

In the following century, Petrus Peregrinus de Maricourt, a soldier monk, discussed the directive property of magnets in his famous EPISTLE CONCERNING THE MAGNET. "Take a loadstone," he says, "and put it in a wooden cup or plate and set it afloat, like a sailor in a boat, upon water in a larger vessel, where it will have room to turn. Then the stone so placed in the boat will turn until the north pole of the stone will come to rest in the direction of the north pole of the heavens, and its south pole toward the south pole of the heavens. And if you move the stone away from that position a thousand times, a thousand times will it return by the will of God." Peregrinus suggested a number of improvements in the nautical compass.

In the years that followed, the magnetic compass was gradually perfected; but even the most learned had no idea why the needle always pointed in a more or less northerly direction. It was not until the year 1600 that the reason was definitely revealed.

In that year the learned English physician Sir William Gilbert published a treatise called OF THE MAGNET, MAGNETIC BODIES AND THAT GREAT MAGNET THE EARTH. This was one of the greatest scientific works of all time. In it Gilbert, who has been called the Galileo of magnetism, set forth his theory that the earth is a magnet, and that a magnetic compass needle points north because the north-seeking pole of the needle is attracted by the North Pole of the earth. Gilbert also pointed out that the Magnetic North and South Poles do not

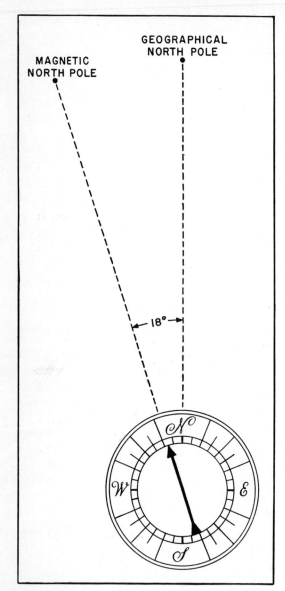

6. The compass needle that is shown above points in a direction that is 18° west of true north; hence the declination of the compass is 18° west.

the moon and radiations from outer space.

The North and South Magnetic Poles are at quite a distance from true north and south — as much as a thousand miles, or even more. They are constantly shifting their positions. The Carnegie Institution of Washington has determined their average positions for various periods, as shown in the table at the top of the following page. This table indicates that both magnetic poles have shifted in a direction that is generally north by northwest.

The declination of the compass

Since the magnetic poles of the earth do not correspond to true north and south, it is very important to know how much they diverge; otherwise a navigation officer would have only an approximate idea of his direction. We can find out how great this divergence is by examining the declination of the compass. The declination represents the angle between the magnetic needle, as it points to the North Magnetic Pole, and the geographic meridian — that is, the line passing through a given point on the earth's surface and connecting the North and South Poles of the earth. If the declination is 12 degrees west at a given point, it means that here the compass needle points in a direction that is 12 degrees west of true north (Figure 6).

Another factor that is also important in many calculations is the inclination, or dip, of the compass. To determine this, a magnetic needle, called a dipping needle, is mounted on a horizontal axis and is allowed to swing in a vertical plane. The needle will follow the direction of the earth's lines of force. The dip of the needle represents the angle between the plane of the horizon and the magnetic needle (Figure 7). The dipping needle is vertical at the North and South Magnetic Poles; it is horizontal at the magnetic equator. It occupies intermediate positions between horizontal and vertical at other places on the earth's surface.

The degree of declination or inclination at different points on the earth's surface may be shown on charts (maps) showing

correspond to true north and south. He observed, too, that a magnetic needle free to move up and down dips toward the earth at many places. Gilbert's work laid the foundations for our present-day knowledge of the earth's magnetism.

We realize now that, as a result of the magnetic field surrounding it, the earth and its atmosphere constitute a great magnetic laboratory. In this, nature continually performs her experiments, utilizing as apparatus not only the earth, but also the sun.

Five-year period * ending	Position of North Magnetic Pole	Position of South Magnetic Pole
1945.0	76.1° N., 101.8° W.	68.2° S., 145.4° E.
1950.0	73.0° N., 100.0° W.	68.0° S., 144.0° E.
1955.0	73.8° N., 101.0° W.	68.0° S., 144.0° E.
1960.0	74.9° N., 101.0° W.	67.1° S., 142.7° E.

* Note: Each date represents an official charting epoch.

the entire world or a single region. In preparing a chart showing magnetic declination, the degree of declination is determined at a number of different places. A line is then drawn through all the points where the degree of declination is 0°; another, say, through the places where the degree of declination is 5° and so on. If this method were strictly followed and if the places where the declination is determined were spaced closely enough, it would be found that the lines would form an intricate pattern of complex bends and closed loops. It is customary to smooth out the lines somewhat and to disregard irregular values.

The same method is used in preparing charts showing magnetic inclination. The charts that show lines of equal magnetic declination are called isogonic (having equal angles); those that show lines of equal magnetic inclination are called isoclinic (dipping equally). Both isogonic and isoclinic charts are called isomagnetic. On the next two pages we show a world isogonic chart.

The usefulness of the magnetic compass has been enhanced with the development of the accurate isogonic charts that we have just described. It is true that on large vessels the gyrocompass is now the chief directional instrument. In all such ships, however, magnetic compasses are held in reserve. For, after all, the gyrocompass is dependent on a source of motive power for its operation and it is also subject to mechanical failure. The magnetic compass, on the other hand, practically never goes out of order and the only "power" that is required to run it is the attraction of the North Magnetic Pole.

There are certain more or less regular changes in the earth's magnetic field, known as secular variations. There are secular variations from place to place, from season to season, from Northern to Southern Hemisphere and, for the same place, from year to year. For example, observations made at London indicate that the

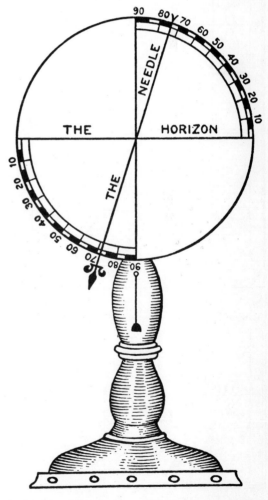

7. Sketch of the dipping needle with which Robert Norman measured dip at London in 1576. The needle, which was mounted on a horizontal axis, followed the direction of the earth's lines of force.

magnetic needle pointed 11° east of north in 1580 and 24° west of north in 1812. Since that time the needle point has shifted eastward and now points to about 10° west of north. All this indicates that this particular secular variation will complete a cycle in about five hundred years. The cycle of secular change varies from one place to another. Efforts have been made to predict such cycles on the basis of theoretical studies, just as astronomers predict eclipses of the sun and moon. It is now

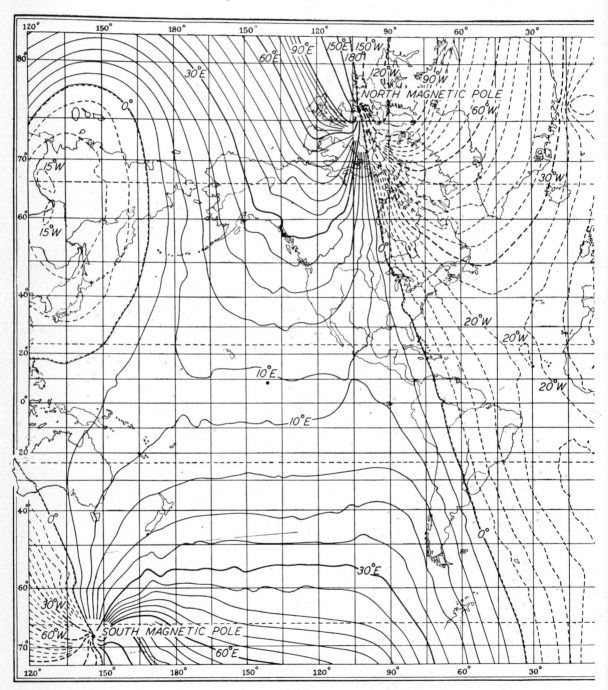

World isogonic chart for the year 1945, showing the North and South Magnetic Poles and the lines of equal magnetic declination. The magnetic declination of the compass is the same for all points on a given line in the chart. Thus the declination is 10° E. in all localities through which the line

recognized, however, that because of the many unknown factors involved, there is as yet no basis for predictions of secular change. The fact is that secular variations have not only puzzled scientists for many years but are still an unsolved mystery.

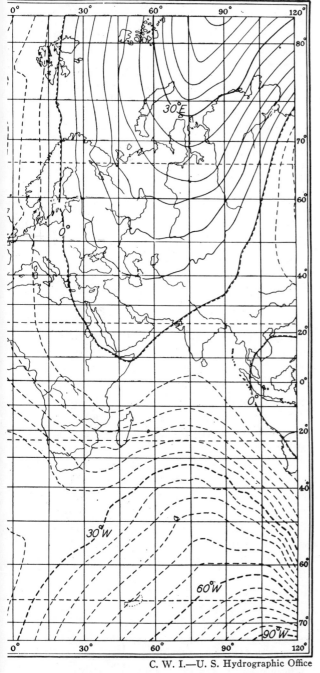

C. W. I.—U. S. Hydrographic Office

marked 10° E. passes. Navigators find charts like this invaluable; without them, the compass reading would give only a rough idea of true north.

It has been found that as the earth revolves around the sun during the year, there are corresponding fluctuations in the earth's magnetic field. The maximum variations, or crests, occur during the equinoctial months of March and September, when the sun crosses the equator in its apparent yearly journey through the heavens. (See the article The Face of the Sky in Volume 1.) The minimum variations, or troughs, come in the solstitial months of June and December, when the sun is farthest from the equator in that same yearly journey.

The presence of magnetic-ore deposits brings about anomalies, or irregularities, in the magnetic field of the earth. Such irregularities do not greatly alter the over-all picture of the whole field. They make it possible, however, to locate deposits of magnetic ores in a given area. Aerial surveys of the area are made with a magnetometer, an instrument for measuring the intensity and direction of magnetic forces. By examining the findings of such surveys, prospectors can determine the extent and depth of magnetic-ore deposits. In some cases, the soil and rocks may be only slightly magnetic, and there will be only minor anomalies in the earth's magnetic field.

Other variations in the earth's magnetism are caused by disturbances known as magnetic storms. These frequently occur simultaneously over the whole globe; they are much more violent in the polar regions because of the nearness of the earth's magnetic poles. There is not a regular cycle of quiet and disturbed days; an old disturbance may die out and a new one may occur at any time. Generally any marked disturbance in the magnetic field reappears in several successive months before it permanently disappears.

One of the principal factors in causing these magnetic storms is the existence of numberless electrified particles streaming from the sun. The earth's lines of magnetic force, extending far out into space, entrap these particles, which are made to travel in spiral paths around the lines of magnetic force. Since these lines are steepest in the polar regions, the electrified particles penetrate most deeply in the earth's atmosphere in these areas. Not only do they bring about variations in the earth's field of force but they also cause the dazzling natural displays known as polar lights.

Polar lights result from the resistance offered to the electrified particles by the earth's atmosphere — a resistance that causes the particles to glow. These brilliant lights are known as the aurora borealis, or northern lights, when they occur in the Northern Hemisphere, and the aurora australis, or southern lights, when they occur in the Southern Hemisphere. By photographing auroras simultaneously from two stations which are a measured distance apart, it has been found that polar-light beams generally do not come closer than

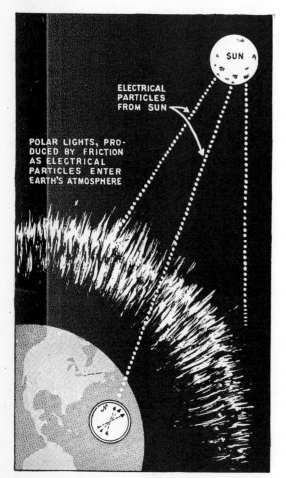

ELECTRICAL PARTICLES FROM SUN

POLAR LIGHTS, PRODUCED BY FRICTION AS ELECTRICAL PARTICLES ENTER EARTH'S ATMOSPHERE

SUN

The atmosphere's resistance to electrical particles streaming from the sun produces polar lights.

sixty miles from the earth's surface; they have been observed at heights above three hundred miles.

Observations in the United States and in various European countries have shown that there is a very close connection between sunspot activity and disturbances in the earth's magnetism. When the sunspots are the most active, there are the greatest variations in the earth's magnetic field and the most brilliant displays of polar lights.

Cosmic rays from outer space also contribute to magnetic variations. (See the article The Cosmic Rays, in Volume 8.)

Thus far we have dealt with the effects of magnetism and with its numerous applications. The question now arises: what is the ultimate cause of magnetic fields of force and of polarity? Thus far the scientists have not been able to give a definite answer. It is quite generally agreed that magnetism results from the orderly arrangement or interaction of particles of matter within magnetic materials. According to one theory the molecules in such substances are tiny magnets. When they are all jumbled together within a given substance, pointing every which way, the magnetic forces involved cancel each other and the substance is not a magnet. But if the wee molecule-magnets are lined up so that all their north poles point in one direction and all their south poles in the opposite direction, the substance becomes a magnet with a north pole at one end and a south pole at the other.

Other scientists believe that the basic units in magnetic phenomena are not molecules but much smaller particles — the electrons that revolve about the nuclei of atoms and that also spin about their own axes. According to this theory, the manner in which the electrons revolve and rotate accounts for the various magnetic effects.

None of the theories that have been advanced have provided a satisfactory explanation of the why and wherefore of magnets. When such an explanation is forthcoming, it will go far toward solving some of today's problems in magnetism.

See also Vol. 10, p. 281: "Magnetism."

THE IDIOTIC CHEMIST

A Topsy-turvy

Chemical-Magic Session

BY NELSON F. BEELER

1. The fountain that starts spouting for no good reason.

WE SHALL assume that you have gone over the three chapters of The Home Chemist in Volumes 4 and 7 and that you have carried out the projects and experiments listed there. You are now ready to put on a chemical-magic show. This will be a rather unusual type of performance, because you are going to play the part of an idiotic chemist who has no idea how his chemical magic is going to turn out. In fact, everything he tries to demonstrate will seem to work in reverse.

Of course, for any kind of magic show you must keep up a running fire of talk, or "patter." This serves to draw the attention of the audience from your preparations and to make them think what you want them to think. Part of the fun of a magic performance is working up your "patter."

In the following pages, we present a routine consisting of a number of tricks. You will find, in the case of each trick, a description of what your audience will see and some little indication of the "patter" that might be used. Then there will be an explanation of how to set up the trick (in advance of the actual performance, of course) and a brief account of the chemistry involved.

The self-starting fountain

Begin by telling your audience that there is nothing to fear from these demonstrations, as no reaction can take place until you are ready for it. While you continue to talk in this vein, move a bottle from one end of the table to the center. Keep assuring your audience that you will let them know just what is going to happen in each one of your demonstrations. As you are delivering this "patter," the bottle you have just set in the center of the table will begin to spout like a fountain all by itself (Figure 1). You will appear to be embarrassed; you will tell the onlookers that you are sure you will be able to think of some way to make the fountain stop spouting. While you are deep in thought, the fountain stops all by itself. Quickly take away the offending bottle and go on with the next demonstration on the program.

The bottle (Figure 2) is a modification of the fire extinguisher shown on page 352, Volume 4. A single-hole stopper is to be fitted to the bottle. A medicine dropper with the rubber bulb removed is forced through the stopper from the underside so

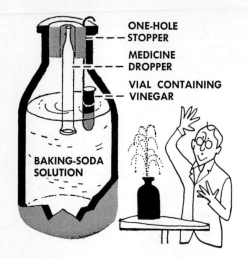

ONE-HOLE STOPPER

MEDICINE DROPPER

VIAL CONTAINING VINEGAR

BAKING-SODA SOLUTION

2. How to prepare the self-starting fountain described on the preceding page. Details are given in the text.

that the tip of the dropper is just beneath the top opening of the stopper. The bottle contains a solution of baking soda and water; the level of the solution is about halfway up the medicine dropper. A small open vial of vinegar is supported inside the bottle by means of a piece of stiff wire held in place by the stopper (Figure 2). The bottle should be covered with black paper so that the audience will not be able to see the inside arrangements.

If you tilt the bottle as you move it, the vinegar will spill out into the soda solution. The soda and vinegar will then react to produce carbon-dioxide gas, which will drive out the soda solution through the medicine dropper in a fine jet. When the solution level drops below the dropper, the fountain will stop playing. The soda solution is harmless; all you have to do is to clean up the mess with a sponge before you go on to the next act.

A powder that bursts into flame

"Liquids seem to be a little undependable," you will say. "Solid materials are more reliable." Pick up a stoppered test tube from a rack and say: "I would now like to show you a really dependable powder." Let the audience see it in the bottom of the test tube. Pour out the powder onto a piece of paper, being careful to keep the powder in a single pile. Now pick up the paper as if you were planning to show the powder to the audience again and absently tilt the paper. The powder will slide from it and will burst into flame (Figure 3).

The powder is a finely divided form of lead, called pyrophoric lead. To make it, heat strongly one-half an inch of lead ace-

A

B

3. The idiotic chemist pours out a "dependable powder" onto some paper (A). The powder bursts into flame (B).

4. To make the "dependable powder," heat some lead acetate in a Pyrex test tube, which is held with tongs.

174

tate in a Pyrex test tube, held with tongs (Figure 4), until you can observe no further change. CAUTION: The mouth of the test tube should be held *away* from you. Close the end of the tube lightly with a cork stopper. The finely divided lead will not react as long as it is all kept in a small pile; but it will ignite when sprinkled into the air. The lead bits will all burn up just as the bits of iron in Fourth-of-July "sparklers" do, so that there is very little danger of fire resulting. It would be well, however, to pour the powdered lead out over a metal cookie plate or a piece of aluminum foil placed over the table.

The magician produces
an unexpected sausage

You will now tell your audience in a self-important manner: "I should like to call to your attention the fact that many chemical reactions require the addition of heat. To demonstrate this, I shall heat a liquid in a beaker placed over a heat source [a spirit lamp or Bunsen burner]. By heating the beaker gently, I shall control the reaction so that it will proceed very gradually." Continue heating the beaker as you discuss further the control you have over this chemical reaction. Suddenly a large sausagelike growth will emerge from the mouth of the beaker (Figure 5).

The beaker contains paranitroaniline to the depth of an inch. A small amount of concentrated sulfuric acid is added to make a paste. (CAUTION: Be careful not to splash any of the acid onto your skin, as it can produce burns. If you do spill any on yourself, flush it off immediately with plenty of water.) When you heat the paste, you will produce a plastic material (the "sausage"), which is bulky as well as lightweight because it is filled with air bubbles.

When is a liquid
not a liquid?

Next pick up from the table a flask containing a liquid. Tell your audience that you would like them to pass the flask from hand to hand to observe that the liquid in it is clear and colorless. "You need not be afraid," you will add, "that the liquid will

undergo any change while you are passing it around because, as you can see, I am not doing anything to it." But, strangely enough, as the flask is passing from hand to hand, the liquid in it begins to "freeze." In a short time, the material in the bottle is completely solid, and the bottle and its contents have become quite hot, though not uncomfortably so (Figure 6).

The trick is done best with sodium-acetate crystals. Place the crystals just as they come from the reagent jar in a clean flask. Put the flask in a large container of water; heat the water and let it boil until the crystals have completely melted. Cover

5. The idiotic chemist pours liquid in a beaker (A) and lights a spirit lamp under it (B). Result: a sausage (C).

the mouth of the flask with a piece of aluminum foil to keep dust particles from falling in. Allow the flask to cool slowly, being careful not to move it. The contents will stay liquid; they make up what is called a supersaturated solution (see Index). Just before handing the flask around, drop in a tiny crystal of sodium acetate; be sure that your audience does not see you do so. The whole liquid will then begin to crystallize.

Photographer's hypo (sodium thiosulfate) can be used for this trick. It is easier to get, but its action is not so dependable. To prepare the hypo solution, add the crystals to hot water until no more will dissolve. Filter the hot solution through a small bit of absorbent cotton; be sure that no solid crystals remain. Then allow the hypo solution to cool.

Whether your flask contains sodium acetate or hypo solution, you can use the contents again for the same kind of demonstration. Simply heat the flask, and then let the solution cool slowly.

6. A flask that contains a liquid is passed around. The liquid in the flask begins to "freeze" and soon is a solid.

The not-quite-so-
colorless water

"As you all know, water is a colorless liquid. Here is some." Proceed to pour water from a pitcher into a clean drinking glass and show that it is indeed colorless. "I am going to show you," you will add, "that the shape of the container has nothing to do with the color of the water." To prove your point, pour some water from the same pitcher into a wineglass. To everyone's surprise (and yours, too!), the liquid becomes wine-colored. Next, pour the contents of both glasses and the pitcher into a milk bottle. Slowly the liquid in the bottle will turn to "milk" (Figure 7).

The "water" is actually very dilute hydrochloric acid; the solution is prepared by adding about fifteen drops of concentrated hydrochloric acid to a quart of water. Add several drops of the indicator phenolphthalein to the acid in the pitcher. The color of the indicator is not affected by the acid; hence the "water" in the pitcher remains colorless. The wineglass contains a very small amount of sodium-hydroxide solution, a base. This neutralizes the acid, and the solution will become red in color. The milk bottle has a small amount of hypo solution * in it. The excess acid from the pitcher breaks this material down and produces a suspension of finely divided sulfur, which looks like milk.

You will have to practice this trick to get just the right amount of acid and base, but once you have found the proper quantities to use, it is easy to work the demonstration successfully each time.

Some more
"colorless water"

"I don't know just what happened," you will remark, "but I still insist that water remains colorless, no matter what the shape of the container is. Let us try again." To be sure of the results this time, you discard the pitcher, drinking glass, wineglass and bottle you used in the last demonstration, and you use a new set of these containers. You add about the same amount of water from the pitcher to each of the two glasses and the milk bottle. This time you get a red liquid in the drinking glass, a white one in the wineglass and a blue one in the milk bottle. Apparently, the water will not stay colorless, no matter what you do. Admit defeat and go on to the next demonstration.

For this second "colorless-water" demonstration, the solution in the pitcher is sodium hydroxide. The drinking glass contains a few drops of phenolphthalein solution; the wineglass, a little lead-acetate solution; the milk bottle, a small amount of copper-sulfate solution. The red color in the drinking glass is due to the indicator it contains; you will recall that this indicator is red to bases, such as sodium hydroxide. The "water" in the wine glass is white because of the formation of white lead hydroxide. The liquid in the milk bottle turns blue as copper hydroxide is formed. A piece of black friction tape wrapped around

* To produce the solution, use about a half teaspoonful of hypo crystals and just enough water to dissolve them.

7. The idiotic chemist pours water into a drinking glass (A); the water undergoes no change. It becomes wine-colored when it is poured into a wineglass (B). When it is poured into a milk bottle, it turns into "milk" (C).

the bottom of the milk bottle will hide the copper-sulfate solution in the bottom. In each case, pour in just enough sodium hydroxide from the pitcher to make each color appear in turn.

Patriotic fireproofing

You are now ready to demonstrate the All-American Method for Fireproofing Paper. A piece of nearly white paper is attached to a line by means of clothespins or paper clips (Figure 8). Using an insect spray gun, you proceed to wet the paper. You explain that you are fireproofing it and you discuss the value of the process. As you continue to talk and to work the gun, an American flag appears on the paper.

The paper has been treated with two reagents made up in fairly concentrated solutions. The chemicals can be swabbed onto the paper with a piece of absorbent cotton wrapped around a wooden splint. The stripes that will later be red are put on the paper with potassium thiocyanate (sometimes called sulfocyanate). The blue background for the stars is made with potassium ferrocyanide. (Do not use potassium ferricyanide, which will not produce a blue color.) The white stripes and stars that will appear on the American flag will represent untreated paper, of course. The paper is allowed to dry thoroughly. At even a short distance, the audience will not be able to see that the paper has been treated since both of the chemicals you have used leave fairly colorless traces.

The spray gun contains a concentrated solution of ferric (Iron III) chloride. The iron compound in the spray reacts with one chemical on the paper to produce red ferric thiocyanate and with the other chemical to produce blue ferric ferrocyanide. Be sure to use as fine a spray as possible in order to keep the colors from running into the white parts of the flag.

A startling blow-up

You can produce a startling flare-up and explosion quite easily and safely, using ordinary flour. Prepare your audience by showing them a can with a press-on cover. Light a candle stub and set it upright inside the can; then tell the onlookers that you

8. Our chemist tries to fireproof a piece of paper by spraying it (A). An American flag appears on it (B).

9. What happens when the idiotic chemist tries to extinguish a lighted candle in a can by blowing into the can.

FUNNEL FILLED
WITH FLOUR
AND TILTED

10. Cutaway drawing of the can in Figure 9, showing the tilted position of the funnel used in the demonstration.

propose to blow out the candle. Replace the cover and blow forcibly into a rubber tube which protrudes from the bottom of the can. The top will fly off with a loud bang and a high flame will spring up from the can (Figure 9). Apparently your plan to blow out the candle backfired.

The can is prepared as in Figure 10. The candle can be made to stand upright in the bottom of the can if you put a little molten candle wax on a small piece of cardboard and set the candle in position on the wax. The small funnel is set in the can with its wide part uppermost and its narrow neck protruding through a hole punched in the bottom of the can (Figure 10). About a tablespoonful of dry flour is poured into the funnel before the show begins. It would be advisable to dry the flour on a cookie

sheet placed in the oven just before you are going to use it. It works best if it is very, very dry. The funnel must be tipped toward the candle so that the flour is blown into the candle flame and not into the space alongside it.

When you blow forcibly into the can, the flour grains in the funnel are distributed throughout the interior of the can as a fine dust. The candle ignites the dust near it; the flame then spreads quickly throughout the whole interior because there is a great deal of air present compared to the amount of combustible material. The heat resulting from this rapid burning expands the air in the can and forces off the cover. There is no danger from this demonstration if you will continue to blow into the can until the flame dies out, which it will do very quickly. Dust explosions of this sort occur once in a while in coal mines where there is a good deal of coal dust and also in places where flour or sugar is stored. In places such as these, great care must be taken to prevent open flames or sparks from igniting the dust.

A material called lycopodium powder will give even surer results than flour in the above demonstration. This substance is a collection of spores from the lycopodium plant, a species of ground pine. The spores, which are nearly spherical, are easily blown into a finely divided dust cloud; once this is ignited, the burning goes on very rapidly.

You can have a little fun with your audience by offering to repeat the dust explosion for them. Tell them to watch the can very carefully this time. As they are concentrating on the can, a tremendous bang is heard — a bang that does not seem to proceed from the can at all. If you have prepared the scene well, the onlookers will just about jump from their chairs.

Actually, there is nothing chemical about this particular "explosion"; it is produced by a broom hidden from view behind the table. You are to hold the broom near the end of the handle with one hand, thus leaving the other hand free for pointing at the can and in other ways diverting the attention of your audience. Place your foot on the straw part of the broom. When you let go of the handle, it will strike the floor with a resounding whack. Since your audience is tense, waiting for the explosion in the can to go off, this bang will really startle them.

The blazing handkerchief that does not burn

After everyone has recovered from the effects of the double "explosion," you are ready to talk about flames and the chemistry of oxidation. Say that you will demonstrate the chemistry of flames by burning up your pocket handkerchief. Take your handkerchief from your pocket and dip it

The final touches

It is about time to bring your routine to an end, so you proceed to wash your hands with water and a little soap, thanking your audience for having been so attentive. You rinse your hands in clear water and wipe them on a towel. Both your hands and the towel will become a dark blue.

The apparently clear rinse water is really a solution of ferric chloride. The towel has previously been dipped in potassium ferrocyanide solution and has been allowed to dry. The reaction between the two chem-

11. The chemist dips his handkerchief in a solution and squeezes it to get rid of some of the liquid (A). Then he holds the handkerchief with a pair of tongs and sets fire to it (B). The handkerchief will be undamaged (C).

in a solution which is on the table in front of you. Squeeze the cloth to get rid of some of the liquid. Then hold the handkerchief with a pair of tongs and light it with a match. It will catch fire readily enough; keep on waving it back and forth until the flame dies out (Figure 11). To the surprise of everyone in the audience, the handkerchief will be undamaged.

The solution is made from equal parts of rubbing alcohol and water. It will be the alcohol and not the cloth that will catch fire; and it will burn with such a cool flame that the cloth will not become hot enough to char or catch fire itself. It might be well to try the solution on a piece of cloth before the demonstration. If the flame does not last as long as you would like, add a little more alcohol to the solution and try again.

icals produces a blue ferric ferrocyanide. This can be removed from your hands easily with a little household ammonia and water later on.

As a grand finale, touch a match to a piece of heavy cardboard that is hanging behind the lecture table. The written words "That's all" will be seen to burn themselves slowly into the cardboard (Figure 12).

The words have been written previously on the cardboard with a "pen" consisting of a swab of cotton on a stick and with a saturated water solution of potassium nitrate as the ink. Be sure that each letter of the writing is connected to the next one. When you touch a match to the first letter, the fire will continue to burn along the entire set of words. The reason is that the potassium nitrate is an excellent oxida-

12. When a match is touched to a "blank" piece of heavy cardboard, the written words "That's all" appear on the cardboard.

tion agent; it provides a quantity of oxygen sufficient to support combustion. It is used as one of the ingredients of gunpowder for this reason.

Precautions to be taken

All the above demonstrations can be carried out safely. You should remember, however, that in many of them you are dealing with flames. You should be prepared to take care of any fire that might start. Have a large piece of heavy cardboard ready to set upon anything that continues to blaze longer than you want it to. It would be a good idea, too, to keep two large pails, one containing sand and the other water, under the table, ready for every emergency.

The routine we have given you above will be successful only if you practice each one of the demonstrations until you are letter-perfect in it. It goes without saying that you must know at all times what you are doing and why you are doing it. You must be a pretty alert chemist to give the "idiotic chemist" routine effectively.

An encore — the salt-making machine

If an encore is called for, prepare your masterpiece — a salt-making machine. Connect up, in any way you please, the glassware and rubber tubing you have used in the previous demonstrations; the only requirement is that the apparatus should look as complicated as possible. It would be a good idea to set a Bunsen burner or a spirit lamp under a beaker containing water somewhere in the assembly. At one end of the apparatus there is to be a big glass jar, with the bottom covered with sawdust or sand to a depth of three or four inches. Attach a string to a salt shaker containing salt; bury the shaker in the sawdust (or sand), letting the end of the string hang over the rim of the jar. At the end of the apparatus nearer the audience, set a flat plate with tubing dipping into it. Figure 13 gives you some idea of the way in which the salt-making machine might be set up.

Explain to your audience, as you are arranging the equipment, that you are going to make salt from the sawdust (or sand) in the glass jar. "As the salt is prepared," you will add, "it will drop into the flat plate here. I will invite you to taste it so that you will see it is real salt." Start the equipment working. It will look particularly impressive if you can heat some water and bring it to a boil. At the climax of the operation, pull up the salt shaker by the string attached to it from the big jar at the end of the salt-making machine, and sprinkle some salt on the flat plate in front of your audience. Then make your getaway!

13. The formidable-looking salt-making machine that is supposed to transform sawdust or sand into pure salt.

Actually, this device will function only if salt from the shaker hidden in the can is sprinkled on the plate!

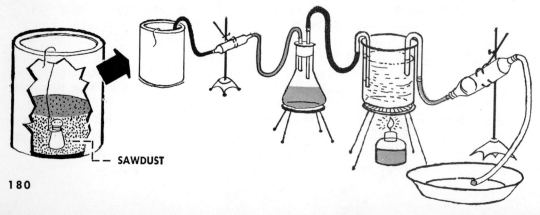

SAWDUST

MAMMALS IN THE WATER

Fishlike Creatures That Suckle Their Young

SEVERAL times in the long history of the vertebrates, or back-boned animals, species that once lived on land have invaded the seas. The marine turtles and the crocodilians (the crocodiles and their kin), descendants of land-dwelling reptiles, are now perfectly at home in the water. Many types of birds have also become specialized for this mode of life; among them are the penguins, auks, cormorants and loons. Various mammals, too, including the otters, seals and walruses, are at home in aquatic surroundings. The most highly specialized of the marine mammals, however, are the manatees, dugongs, whales, dolphins and porpoises, with which we shall deal in this chapter.

The manatees and dugongs belong to the order of the Sirenia; the whales, dolphins and porpoises, to the order of the Cetacea. Although the two groups have quite different ancestry, they are similarly specialized for the way of life that they have adopted. They have developed tapering, streamlined bodies with which they can move easily through the water in swimming and diving. Their forelegs have been greatly modified, having become well-developed flippers, used for steering. These mammals have lost all traces of external hind legs. The hind end now terminates in a horizontally flattened tail, which gives the propulsive thrust in swimming. The Sirenia and Cetacea have extremely short necks and little if any body hair. In spite of their many fishlike characteristics, however, they are true mammals, for they are warm-blooded, placental (see Index under Placenta) and suckle the young.

The Cetacea became marine mammals quite early in the history of mammalian evolution. In all probability, their ances-tors were primitive land carnivores, called creodonts. The primitive whales known as zeuglodonts swam in the sea some 45,000,-000 years ago. The nostrils of these animals had already started their migration toward the top part of the head (that is where the nostrils of modern whales are located); their hind legs had already disappeared. Their bodies were considerably slimmer than those of modern whales, so that they must have looked a great deal like sea serpents. Like the marine reptiles of previous times, the zeuglodonts had elongated jaws; their teeth were like those of early land-dwelling flesh eaters. They were quite large — up to 70 feet in length.

Today there are a number of species of Cetacea, as we shall see. Some are comparatively small; others are among the largest of all living animals. One species — the blue, or sulphur-bottomed, whale — is the largest animal that has ever existed upon the earth, as far as we know; it reaches 100 feet in length and weighs up to 150 tons. Its proportions are far more massive than those of the bulkiest dinosaurs that lived in the Age of Reptiles.

The manatees and dugongs are probably descended from land-dwelling, vegetation-eating mammals that also numbered the elephants among their descendants. The modern Sirenia, which are also known as sea cows, feed on vegetation along ocean coasts or in rivers. Their heads are not at all fishlike, like those of whales. They have heavy, bristly jowls with nostrils set at the apex of a triangular-shaped muzzle. The nostrils are fitted with valves that can be closed in order to keep out water when the animal dives below the surface. The tail of the manatee is rounded; that of the dugong is crescent-shaped.

Manatees and dugongs swim through the water by moving the tail and front flippers up and down. However, the tail and flippers undulate out of phase; that is, they do not move up and down at the same time. The animals may remain under water for some time — as much as sixteen minutes in the case of the manatee.

Both the dugong and manatee have tough, leathery skin. The brain is small. The eyes are tiny; there are no external ears and the sense of smell is limited but not absent as is true with the whales. Manatees lack front teeth entirely as adults, but have a number of molar teeth that are continually replaced from the hind end of each tooth row. In the dugongs, there are only a few peglike crushing teeth, and the lower jaw carries a horny pad; the males also possess a front pair of upper tusks formed from the incisor teeth. Aquatic plants are digested in the sirenian's complex stomach; the intestine is extremely long. Young dugongs and manatees are born in the water; they are nursed at the mother's teats located high on her chest.

It is thought that some, at least, of the fanciful legends of mermaids — creatures with a woman's body above the waist and a fish's tail — may have been based on the appearance of the sirenians. Certainly these ungainly looking animals could never have been mistaken for human beings at close range. If they were seen from a distance, however, this fancied resemblance would not be so farfetched, for the sirenians frequently stand erect in the water, supported by their tails. Their heads are often draped with loosened vegetation resembling long hair; they suckle their young in an almost human fashion.

The American manatee (genus *Trichechus*) inhabits the Atlantic coast of the Americas from Florida, around the Caribbean, to the mouths of the Amazon and Orinoco rivers. The animal is not found at sea but may occur in island lagoons as much as eighty miles from the nearest

N. Y. Zool. Soc.

Captive manatees in an aquarium. These amiable creatures browse on plants along sea coasts and up rivers.

mainland coast. This slow-moving and amiable creature has been unforgivably slaughtered almost to the point of extinction. Fortunately, efforts have been made to protect manatees in Florida so that they may not become entirely exterminated in that area. Another species of manatee lives along the coast of West Africa; it may ascend large rivers for over a hundred miles.

Manatees are able to live equally well in fresh, brackish and salt water. They travel in small bands and remain near shore, feeding on various aquatic plants. The upper lip of the manatee is split down the middle, and the two halves, which are strongly muscled, work against one another in cropping plants. The front flippers are used to stuff the vegetable material into the mouth; the food is then chewed with deliberation. If the animals are disturbed, they sink out of sight and swim away under water.

A single young manatee is born after a gestation period of about 150 days. The young is born in the water, but immediately after birth, the mother carries the little animal on her back, completely out of the water, for some 45 minutes. Then she gradually submerges the newborn during the following two-hour period.

The dugong (*Dugong*) inhabits the waters along the coasts of the Indian Ocean, from central East Africa to Indonesia and northern Australia. It may grow to a length of ten feet, exceeding the manatee by a good two feet. Unlike the manatee, the dugong is almost strictly marine and thus does not ascend rivers very far. Dugongs are sluggish animals, drifting with the tide back and forth in shallow waters where they feed. They crop seaweeds and other aquatic vegetation by wrapping the whole, extended upper lip around the material, pressing it against the lower lip and pulling it away. Apparently they do not use their flippers in feeding. Though formerly found in large herds, extensive, ruthless hunting has greatly thinned out the numbers of dugongs. Like the manatees, they are defenseless against man; dugongs, however, have been seen driving

sharks away by butting the fish with their heads.

Whales occur in all oceans and seas

There are about a hundred known species of whales, which comprise two main groups. One group — the whalebone, or baleen, whales — includes the right whales, rorquals (blue, fin and humpback whales) and gray whales. In the other group — the toothed whales — we find the sperm whales, beaked whales, white whales, dolphins and porpoises. The largest cetacean, as we have mentioned, is the blue whale; some of the porpoises and dolphins do not exceed much more than 5 or 6 feet in length. Whales occur in all open seas and oceans; some live in tropical rivers; and one species, the Chinese lake dolphin, inhabits the Tung Ting Lake in Hunan Province, China, 600 miles up the Yangtse River. Many cetaceans migrate long distances between subpolar latitudes and tropical waters.

Both small and large whales are capable of swimming at a steady speed of 15 to 20 knots. Sudden bursts of higher speeds are made in capturing food or escaping danger. A whale swims by straight up and down movements of its tail, which makes up the hinder third of the body. The tail is divided into the tail fin, or flukes, and the tail proper, a solid cone of large muscles and straplike sinews. When the cetacean begins the downstroke of the tail, the tail is at its greatest angle to the trunk, and the flukes are horizontal. As the tail moves down, the flukes are bent upward; and as the tail continues downward, the flukes gradually reach a horizontal position again. During the upstroke, as the tail moves upward, the flukes first bend downward and then again straighten to the horizontal at the end of the upstroke. The effect of moving the inclined plane of the flukes up and down is to provide a thrust, driving the animal forward. The front flippers stabilize the whale.

All whales can remain under water for considerable periods; the Greenland right whale and the bottlenose whale seem to

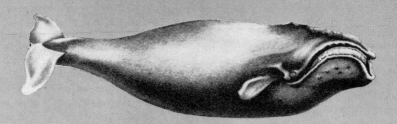

NORTH ATLANTIC RIGHT WHALE.

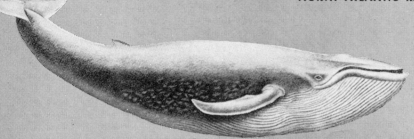

BLUE WHALE.

FINBACK WHALE.

WHALES AND THEIR KIN

Whales, dolphins and porpoises, which belong to the order Cetacea, are marine animals and look like fishes. They are really mammals, however. They are warm-blooded, breathe air through lungs, bring forth living young and suckle their offspring. Their ancestors lived on land many millions of years ago. On these two pages we show some well-known members of the order.

Amer. Mus. of Nat. Hist.

SOWERBY WHALE.

SPERM WHALE.

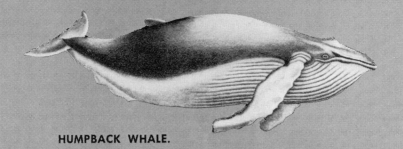

HUMPBACK WHALE.

BOTTLENOSE WHALE.

ATLANTIC KILLER.

COMMON DOLPHIN.

GRAMPUS.

HARBOR PORPOISE.

WHITE WHALE.

hold honors, submerging for 80 and 120 minutes respectively. The sperm whale, which may stay under for some 75 minutes, can sound to a depth of 3,000 to 5,000 feet. The cetaceans are capable of such feats because of special adaptations of the respiratory and circulatory systems. The nostrils, which form the blow hole, are on top of the head, and the nasal, or air, passages go directly to the lungs without joining the throat. The very elastic and extensible lungs are quickly filled with large volumes of air when the animal comes to the surface. Special networks of blood vessels throughout the body store up an extra supply of oxygen. When a dive is made, valves close the nostrils; the heartbeat slows down and oxygen is conserved by the shunting away of blood from the non-vital centers of the body.

Under the paper-thin, hairless skin of the whale lies a thick layer of fibrous tissue, which is impregnated with oil; this is known as blubber. A poor conductor of heat, blubber partially insulates the warm-blooded whale from the cold water in which it swims; but in spite of this blubber layer, most whales when at rest lose heat more rapidly than they produce it and therefore swim to keep warm. The blubber probably serves also as a food reservoir, a function supported by the fact that baleen whales have a much reduced blubber layer in winter when they eat very little. Blubber also reduces the specific gravity of the whale and probably forms an elastic covering that allows for changes in volume during deep dives.

The whale's brain is larger than that of any other animal. The sense of smell is poor or absent; the eyes are small, but under-water vision is fairly good. A whale's sense of hearing is acute, a fact that is probably correlated with the phenomenon of sound production by the animals. For example, the bottle-nosed dolphin whistles, barks and makes a snapping sound with its jaws; the white whale produces high-pitched whistles and squeals, ticking and clucking sounds and noises suggesting mewing, chirping, trilling and bell-like notes; other whales also whistle and bellow. The sig-nificance of these sounds is not yet understood. Some of them are probably means of communicating between individuals of the same species. Since many whales form herds, called gams, the production of sounds may keep a school of these animals together. Furthermore, whales may use echo-sounding to locate obstacles in much the same way as bats do (see Index under Bats); this would be especially important for deep-diving cetaceans.

During the mating season, male and female whales swim side by side, rub and nuzzle one another, give each other resounding blows with their flippers and jump clear of the water, falling back with a great splash. Normally only one young is produced after a gestation period that varies from about 1 year for the sperm and fin whales to 180 days for the porpoises. At the end of the gestation period, delivery of the young from the female's womb is rapid; the infant swims from birth and must come to the surface to breathe in the first few seconds after its emergence. Within the first 24 hours, the new-born whale begins to nurse from its mother's mammary glands, which are situated in a pocket on the rear belly and are enclosed in a compressor muscle. Nursing by the infant is very brief. The mother, swimming slowly, turns on her side; as the infant touches the nipple of the mammary gland, milk is squirted into its mouth by a contraction of the muscle surrounding this gland. The milk is rich in mineral elements (which are essential for growth of the skeleton), proteins and fats. The mother blue whale, for example, gives about 200 pounds of this rich nourishment every day to her infant, which measures 20 feet in length and weighs 2 tons. On this diet the young grows rapidly. Young cetaceans nurse from 2 to 7 months after birth. Sexual maturity is reached by most whales in their third to fifth year.

The baleen whales are clearly distinguished from the toothed whales in their manner of feeding. Baleen whales lack teeth but possess sheets of a horny material, called baleen, that hang down from the upper jaw on either side of the roof of the mouth from front to back. Each sheet of baleen is fringed along the edge that faces down into the mouth. The mouth is immense, for the jaws make up a third of the length of the whale's body. The lower jaw forms a hoop that supports a huge pouch. When the mouth is closed, the rows of baleen sheets hang down into this pouch. The lips curve upward on each side of the mouth. When the baleen whale feeds, it moves through the water with its mouth open, taking in huge quantities of the shrimplike krill and other forms of plankton (see Index). This food is swept up against the matted surfaces of the fringed baleen sheets. The whale closes its mouth and raises its tongue against the palate. Water is forced out between the baleen sheets and the loose lips, and the filtered and unchewed food organisms are pushed back by the tongue and swallowed, entering a four-

ing even the largest whales; bay dolphins feed exclusively on water plants and vegetable matter washed down from the land.

Most toothed whales are active, fast-swimming mammals; many, such as the dolphins and porpoises, possess a fin on the back. The boutu, a porpoiselike creature of the Amazon Basin, has front flippers with much flexibility in the fingers. By using these flippers, the animal can lumber over various obstacles.

Perhaps the most impressive of toothed whales is the sperm whale, which grows to a length of 60 feet. It has an enormous square snout, filled with spermaceti. This substance is highly important to the whale; together with the thick layer of blubber and the oil-containing humps along its back, it seems to cushion and protect the vital organs of the whale from the effects of the excessive pressure at great depths.

British Information Services

A catch of baleen whales being towed by a whaling vessel to the factory ship for processing. The whales are inflated with air; as a result, they float. Modern whaling methods are extremely efficient.

chambered stomach where they are slowly digested. Baleen whales will also eat small fish, in addition to plankton organisms.

The toothed whales may possess many peglike teeth in both jaws, or have teeth in only the upper or lower jaw, or have very few teeth or lack teeth entirely. Many are active predators, feeding on fish and squids. Killer whales are fierce and cunning hunters, moving in packs and attack-

Unfortunately for the whale, the same substances so vital to its deep-sea life are greatly prized by man. Because of ruthless whaling ventures, the sperm whale, and other whale species as well, face extinction. Many international agreements have been made to limit the whaling catch and protect failing species, but they are difficult to enforce, and the sperm whale and others are in great danger. Modern whaling is done on a mass scale; the mother factory-ship, in which the whales are processed, is accompanied by a fleet of small catcher-boats, the whole group remaining at sea for many months at a time.

The sperm whale inhabits mainly the southern oceans, but it ranges to subarctic and subantarctic waters. It is a fair swimmer, cruising along at four knots, and capable of three times that speed if necessary; its fabulous diving skill has already been described. After a dive that may last an hour, the whale surfaces and blows a great spray of water from its blowhole. This water is not taken in by the whale's mouth, as some people believe, but is given off as water vapor by the blood in the whale's lungs, and is condensed there as it accumulates under great pressure.

Some people wonder why a whale dies when it is cast up on the land, since it possesses lungs and breathes out of water. Actually, the whale does suffocate on land because its tremendous weight, no longer supported by the water, crushes the lungs and other organs. Occasionally, whole schools of dead whales are found stranded on the shore in what seems to be an unexplainable case of mass suicide.

Like all the toothed whales, the sperm whale feeds on sea animals much larger than the tiny crustaceans and mollusks that nourish the baleen whales. If the animal that swallowed Jonah was a whale, it must have been a sperm whale, as this is the only genus that has a gullet large enough to swallow a man whole. Though cases of whales swallowing men are reported from time to time, none has ever been scientifically verified; however, ten-foot sharks have been found in whales' stomachs.

The sperm whale's ordinary food and hereditary enemies are the giant squids. Titanic struggles between them have been witnessed by many experienced whalers, but the most spectacular ones probably take place in the natural lairs of the squids, far below the surface of the sea. The stomachs of sperm whales are often found to contain the bodies of squids six or more feet long, with tentacles up to ten or fifteen feet in length.

The hard, indigestible inner shell of the squid is responsible for the formation of ambergris in the sperm whale's intestine. Ambergris is a dark, sticky, highly unpleasant-smelling substance, usually disgorged only by sick whales. It is very rare and brings a high price (eight to twelve dollars an ounce, in 1950), for it is used in making costly perfumes.

The reproductive life of the whale shows most clearly its mammalian character; the young calf is suckled at its mother's teats for six months after birth. The newborn calf is fourteen feet long at birth; the gestation period having lasted for a year. The mothers' teats are on her underside, and suckling would be difficult if she did not turn on her side at the water's surface when nursing the calf, thus permitting it to breathe. The courting performance of a bull sperm whale is an impressive sight, granted to few; it involves tail-flapping dances, caresses, long dives and leaps that sometimes carry him completely out of the water and leave him exhausted.

Sperm whales are dark in color, sometimes black. The great white sperm whale described in Melville's MOBY DICK was an albino, a great rarity. In the winter of 1951, a white albino was caught and killed by Norwegian whalers off the coast of Peru; this was the first on record since the legendary animal of Melville's novel.

The pygmy sperm whale, together with the giant we have discussed, completes the family of sperm whales. Though called a pygmy, this thirteen-foot mammal is not quite an aquarium pet. The most unusual feature of this relatively rare whale is the proportional size of its young: the calves are generally more than half as long as their parents at birth. Sperm whales are members of the superfamily Physeteroidea (the true, or blower whales), one of the three subdivisions of the toothed whales; this superfamily also includes the curious beaked whales.

The common names of the Cetacea have become rich with confusions and misnomers, probably due to the vivid imaginations of the seafarers who named them. The cowfish is one of the beaked whales,

A bottlenose dolphin taking a between-meals snack. Dolphins rank with the more intelligent animals; they have been trained to retrieve various objects, to pull a surfboard and even to jump through a hoop.

and neither cow nor fish, but gets its name from the deep sound that it emits, like the lowing of a cow. The bottlenose whale, so called because of the shape of its head, is the most commercially important of the beaked whales; it is hunted for its spermaceti. It is also the most dangerous to hunt, as it sounds to a depth of three thousand feet in two minutes when stricken with a harpoon; a slight fouling of the line or mistake in steering the whaleboat can thus mean quick death for the crew.

The second superfamily of the toothed whales is the Delphinoidea (dolphinlike cetaceans); some of this group are called whales, some dolphins and some porpoises. The narwhal, a native of Arctic and North American waters, is twenty feet long and has only one tooth, as a rule: a nine-foot ivory tusk, sticking out in front of its upper jaw. The narwhal may have given rise to the legend of the unicorn.

The ferocious killer whale, found in every ocean, is actually a gigantic dolphin; it is the only cetacean that feeds on warm-blooded animals, including its own kind. This bloodthirsty mammal is sometimes thirty feet long, and travels in small schools. It attacks almost any animal, even the greatest whales, tearing out huge chunks of flesh with its sharp-toothed jaws.

Most dolphins, however, are mild and sociable, and run in large schools; they average six to ten feet in length, and most of them have long beaks and a dorsal fin. The most commonly seen is the bottlenose dolphin; it is very friendly and intelligent and in captivity can be trained to obey commands and perform tricks. The common dolphin is the one that is most often seen accompanying ships at sea, leaping over the waves in a graceful arc; some of them have been clocked at thirty knots. The true porpoises are members of the dolphin clan, but measure from four to six feet, and have rounded heads. The harbor porpoise is found in the shore waters of the Atlantic and Pacific Oceans.

The river dolphins constitute the third superfamily of toothed whales; they are a waning group. They include the La Plata and Amazon dolphins of South America and the Ganges River dolphin.

See also Vol. 10, p. 275: "Mammals."

ACCIDENT PREVENTION
IN THE HOME

Safety Rules for All Members of the Family

BY GLADYS J. WARD

GENERALLY we think of the home as a place of refuge, where the members of the family are safe from the thousand-and-one dangers that confront us on the highways and byways. And yet, like so many popular ideas, this one does not hold water. The average home is one of the most dangerous places in the world; within its walls lurk the hazards of broken bones, electric shock, burns, poisoning and a host of other dangers. In a recent year, 35,000 Americans lost their lives because of accidents in their own homes. Over 5,000,000 more were hurt, and of these 140,000 suffered permanent injury. A great many of the tragedies happened to small children.

The majority of accidents at home are due to carelessness; they are caused in the main by articles that are out of place or in need of repair. That means that the majority of accidents are preventable. We should seek out all possible sources of accidents and should work out a systematic plan for avoiding them.

Fires and explosions

Fire is a useful servant; it may also become a ferocious enemy. Under control, in a furnace or stove, it keeps our buildings warm and cooks our food. But when fire gets out of hand, it is a dangerous foe. It strikes quickly, spreads rapidly and may cause injury or death.

Three things — fuel, heat and oxygen — are necessary to start a fire. The fuel may be the coal or oil or gas used for heating purposes; or it may be an inflammable cleaning fluid, or waste paper stored in a basement or attic. The heat may be provided by a lighted gas burner or match or an overheated electric wire. The oxygen is always available, since roughly one-fifth of the air is made up of this gas.

We can do nothing, of course, about eliminating oxygen in the home, since we need it for breathing purposes. But we should see to it that temperatures high enough to cause fires are restricted to our cooking and heating units; we should see to it, also, that potential fuels, other than those used for heating and cooking, are kept away from extreme heat.

More dangerous fires start in the basement than anywhere else, since the furnace and the hot-water heater are usually placed there. Hence we should regularly and carefully inspect our heating equipment. Furnaces should be installed at a safe distance from wooden partitions, beams or ceilings, and furnace ducts should be at least one inch away from these. A pipe passing through a wall or floor should be encased in a metal tube an inch larger in diameter than the pipe; this tube should be provided with tight metal caps at either side of the wall or floor.

The house heating system should be large enough to heat the home comfortably. A heater that is too small may be a hazard because it may be necessary to overfire the furnace in cold weather. The heating system should be equipped with ample safety controls. Each boiler should have a device limiting the temperature of the water to a safe point. For a gas-fired boiler the principal control is the gas pilot safety device, which shuts off the gas supply if the pilot light is not burning. For oil furnaces there is a safety combustion control, which prevents the continuous pumping of oil when the flame fails to ignite. It is hardly necessary to point out how dangerous it is to let your cellar be flooded with highly combustible fuel oil.

The illuminating gas used in cooking ranges, and in certain heating units, is highly explosive. If it escapes from the range or from a leaky pipe it may set the stage for a grim disaster, for the slightest spark will explode it. Should there be the faintest odor of unburned gas anywhere in the house, put out all flames. Never strike a match in order to locate the leak; use a flashlight. Open the windows wide; then call your gas company if you can not locate the source of the smell.

on the gas and lighting the match. Keep a watchful eye on any pots or kettles containing a liquid that is about to come to a boil. If it should boil over and put out the flame, gas would be set free. If such a thing happens, turn off the gas immediately and open the windows for a few minutes before relighting the burner. Never hang window curtains too near the range, since the draft from an open window might blow them on the flame.

Fire may sometimes break out when a house is struck by lightning. Carefully installed lightning rods on the house and near-by buildings will help to prevent fires by grounding the electricity during storms.

PRINCIPAL TYPES OF HOME ACCIDENT FATALITIES IN 1961

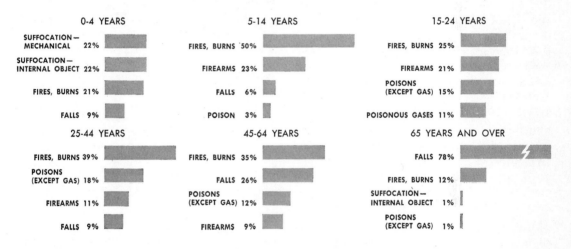

Note: These figures were compiled by the National Safety Council.

Sometimes the pilot light on a kitchen range becomes blocked up; it then goes out, allowing gas to escape. Clean the pilot light carefully, using a fine wire about the size of a hairpin. Burner cocks that become loose should be repaired or replaced at once, to avoid leaks. There is always the danger, too, that one may turn on a cock without noticing it.

Keep the surface and oven burners on gas and kerosene ranges clean and free of burned food. Light the gas oven according to the manufacturer's directions. Open the oven door to let air in before turning

If a television antenna has been installed on your roof, be sure that it has been properly grounded.

Misused or faulty electrical equipment is another common cause of fire. The trouble may come from electric wires that are exposed to a heavier current than they can safely carry; as a result, they become overheated. Fuses are inserted in each electric circuit to prevent such an occurrence. Each fuse generally consists of a strip of a bismuth alloy, with a remarkably low melting point. When the current becomes too strong for the wires, the heat causes the alloy to melt and thus breaks the circuit.

If a fuse blows (melts) on a given circuit, it may simply mean that you have been using too many pieces of electrical equipment at the same time. But if the fuse on this circuit blows again and again, it provides a warning that you should not ignore. Have a competent electrician check your wiring system; he may find it advisable to install heavier wiring or to add a circuit or two to the ones you already have.

Always have the wiring system checked before buying heavy electrical equipment. You may find it necessary to install special heavy-duty wire for the electric range, clothes dryer and ironing machine. Use electrical appliances and cords according to the written directions that accompany the equipment. Light sockets are made for lighting purposes; appliances such as irons should be plugged into appliance outlets — never into light sockets. When disconnecting any appliance, pull on the plug, never on the cord.

Worn-out cords can also lead to trouble. It is a dangerous practice to run cords under rugs, over radiators or through door jambs, since this causes undue wear. Promptly repair or replace frayed cords.

Inflammable liquids are a constant fire hazard

Inflammable liquids like gasoline, kerosene, oils, paint and turpentine are a constant fire hazard. Of course we should never under any circumstances store gasoline inside a house. If we must keep kerosene cans there, we should be sure that they are tightly closed, clearly labeled and stored far away from all sources of heat. Avoid the dangerous practice of using kerosene to start or kindle a fire in a stove or furnace. Kerosene will start a fire quickly, it is true; but you may find yourself with a much bigger fire on your hands than you bargained for.

Among the inflammable liquids that cause explosions are various dry-cleaning fluids. Keep such fluids far away from any open flame or source of heat. Open the windows wide when you clean a gar-

National Board of Fire Underwriters

This home handyman, repairing a toaster, may unwittingly remove safeguards provided by the manufacturer to prevent overheating and shock. Let an expert do the work.

ment, or, better still, do your dry cleaning out-of-doors. Never smoke or have lighted cigarettes around while the cleaning is going on. Keep the container of the fluid tightly sealed when it is not in use. Children should not be permitted to use cleaning fluids.

If you polish furniture or apply stain with an oil or wax that is inflammable, be sure that there is no flame near by. Use only a little at a time and rub it in well. Any oily cloths that are to be kept even a short time should be stored in tightly covered metal containers away from fire and heat.

A fireplace in a home is a joy for everyone if it is used properly; but it may also present a serious fire hazard. To prevent sparks from sputtering kindling or logs from igniting objects within the room, you should set a sturdy wire screen across the opening of the fireplace as soon as a fire is lighted. Do not permit children to play near the screen; they may knock it down and forget to replace it. Before you go to bed, be sure that the fire has been put out;

keep the screen in place as an added precaution. Sometimes sparks from a fireplace may issue from the chimney and set fire to the shingles of the roof. A spark arrester — a screen set across the opening of the chimney — will help to prevent such fires. So will the use of shingles made of a non-inflammable material like asbestos.

The Christmas tree is a symbol of good cheer; it is also a real fire hazard. Never set lighted wax candles on the branches; use electric bulbs that have adequate wiring. Do not decorate the tree with objects made of paper, cotton or similar materials; candy canes, oranges, nuts and apples make just as colorful ornaments and they are infinitely safer. Bear in mind, too, that the longer the tree remains in the house, the more it dries out; and the more it dries out, the more of a fire hazard it is.

Smoking has caused many a disastrous fire. The smoker should never put a lighted cigar or cigarette on a table or bureau top. He may be careful to have the lighted end protrude beyond the edge; he may intend to pick up the cigar or cigarette in just a minute or two. But

Tossing oily or greasy rags into a stack of papers and boxes invites trouble. The rags may ignite spontaneously and set the boxes and perhaps the house on fire.

suppose he forgets to do so? Smoking in bed is also dangerous. Many a tragedy has resulted when a smoker dozed off.

We have already spoken of fires starting from waste materials left about the house — stacks of old paper, oily rags, rubbish and wooden boxes in the attic or basement. Materials like greasy and oily cloths may catch fire even without an external agent, like a match or a spark; they may ignite from spontaneous combustion. Here is what happens. The inflammable materials, the oil or the grease, contained in the rags, gradually undergo oxidation — that is, they gradually combine with the oxygen of the air even at ordinary room temperature. As a result of oxidation heat is released. The oily or greasy cloths retain a good deal of this heat, and the remaining oil or grease combines more readily than before with the oxygen of the air. This raises the temperature still more, and oxidation goes on at a more rapid rate. At last the temperature is so high that the materials burst into flame. Good housekeeping will help to prevent fires of this kind. We should keep attics, basements and closets cleared of any rubbish that might form such a fire trap. We should burn all greasy or oily cloths, and all cloths used to wipe paint from our hands, unless we provide a closed receptacle for them.

Here are a few more suggestions. Use fire-resistant materials as much as possible for curtains, draperies, upholstered furniture and covers for ironing boards. Glass-fiber materials are ideal in this respect. Keep matches in a metal container with a tightly closed top and set it high enough so that the children will not be able to reach it. The metal container will also keep mice and rats from nibbling at the matches and possibly causing them to burst into flame.

Suppose that, in spite of all your precautions, fire should break out. You should always be well prepared for such an emergency. Keep fire escapes clear of obstacles such as flowerpots or boxes. Be sure that every door leading out-of-doors can easily be opened from the inside at all times. If you live in a large house, familiarize yourself with all the exits that could be used when fire breaks out, so that if flames block one exit, you may at once make for another. It might be well to have family drills, so that each person may know what to do and where to go in case of fire.

Automatic sprinkling systems, which release a spray of water when the temperature rises above a certain level, are efficient but are too costly for the average home. On the other hand small fire extinguishers are not expensive and they are effective if you catch a fire in the early stages. When you choose a fire extinguisher, be sure that you will be able to have it recharged in your own town. Have it inspected regularly to see that it is filled and ready for use. If your home has more than one story it would be well to provide extinguishers for each floor.

Falls

A bad spill may cause painful bruises or broken bones; a particularly serious tumble may prove fatal. A number of different conditions and practices may be responsible for falls; we should be on our guard against them all.

Many bad falls take place on staircases. Harmless-looking rugs at the top and bottom of stairs are responsible for many a "tail spin." Either remove such rugs or else anchor them with a non-slip device. Children should be trained never to leave objects on stairs; adults, too, might do well to keep this in mind.

Staircases are doubly perilous in the dark. Install convenient light switches at both the top and the bottom of each stairway in the house, including the basement stairs; outside stairs should also be effectively lighted. As an added safety measure, it would be well to paint the last step in the basement white.

The staircase should be carefully guarded in homes where there are very young children. In such homes, gates at the head and the foot of open stairways are a must; but remember that they give no protection at all unless they are kept closed at all times except when an adult or older child is about

to go up or down. Children, even older children, should never be permitted to slide down the banisters.

Those who are having new homes built should keep all these facts in mind. Of course one way of avoiding the problem of stairs is to build a one-story house, without a basement. If stairs are installed, there should be no "winders" or circular stairs. The risers should be not more than seven and a half inches high; six and a half inches would be better. The width of the tread or step plus the nosing or edge should be at least eleven and a half inches. One sturdy handrail or, better still, two should be provided.

A good ladder is an asset; but if it is wobbly, topheavy, worn out or placed unsafely, it is a hazard that requires your prompt attention. Be sure that your ladder can safely support your weight; be sure that you never leave it in a place where somebody can stumble on it. Above all, never use makeshift ladders made up of chairs, boxes or anything else that happens to be near at hand.

The best way to avoid a fall in walking over a slippery floor or sidewalk is to remove its slippery coating. If you spill liquid or drop fat or peelings upon the floor, clean the floor immediately. Ice on steps, porch or walk should be covered with sand, ashes or rock salt as soon as possible.

If you have to grope your way in the darkness in a room, because the switch or pull cord is set far within it, you may run the risk of tripping over a toy wagon or some blocks. Try to place switches near the doors so that you can turn on the light before actually entering the rooms or just as you enter them. In any case, keep passageways clear within the room.

There have been many jokes about the person who slips on a bar of soap in the bathtub and falls with a resounding crash or splash, but that situation is funny only in cartoons. Many a fall will be prevented if the bathtub or shower has a secure grabrail within easy reach of the bather. A rubber mat placed under the shower will also be helpful in preventing a bad fall.

How not to use a ladder! You should keep both feet on the rungs while working. Never use a shutter, window sill or other projection as an extra rung, for broken bones or fatal injury may result from such carelessness.

Inspect your yard regularly. Holes in the ground should be refilled promptly with dirt. Loose wire and boards bristling with nails should be removed. Garden tools like rakes and hoes should be kept in their proper storage place and not left in the path of the next passer-by, who may be you!

When you have checked all your equipment and surroundings, you must take still other precautions against accident. Try to avoid extreme fatigue. A tired person is particularly likely to fall; if he is not only tired but in a hurry as well, the chances of a fall are even greater. Absentmindedness is another potential cause of injury. Daydreaming is harmless when you are basking in the sun in your backyard; it is downright dangerous when you work on a high perch like a stepladder.

Burns and scalds

Among the most painful of injuries are those caused by burns or scalds. Such injuries may result, of course, when fire breaks out in the house; but they may also come about in many other ways. They may be caused when you upset pots and pans, containing hot liquids, because they have been left too near the edge of a table or because their handles protrude beyond the edge of the kitchen range. The remedy is obvious enough. Set hot coffee pots well back from the edge of the table; turn the handles of saucepans and other containers away from the edge of the stove.

Keep thick pot holders within easy reach near the range and use them in handling hot utensils on the range and in the oven. When broiling meat avoid too hot a flame; you may be spattered with burning fat when you open the oven door. If you use a cook stove, never let it become overheated. Be careful when you carry a pot or pail full of scalding water.

Pressure saucepans and cookers greatly reduce the time necessary for cooking foods, but since they contain live steam under considerable pressure, they may cause face-scarring burns if you open them prematurely. Some pressure pans can not be opened at all until all the pressure has been removed, thus insuring a large degree of safety. The most reliable rule to apply here is to follow the directions given by the manufacturer.

Electric shock

The current that runs through the electric wires in our homes does not exceed 125 volts. Since it takes from 1,800 to 2,000 volts to kill a condemned prisoner in the electric chair, the danger of electrocution in the home may seem very slight. It is true that if the average person comes in contact with the house current he will probably receive a disagreeable shock and no more, *provided that his body is dry*. The reason is that the human body normally has a high degree of resistance to the electric current. However, this degree of resistance depends upon a number of factors. If the skin is dry, the resistance is from five to twenty times as great as when the skin is wet. If a person standing in a bathtub full of water touches a defective electric fixture, he may well receive a fatal shock. Even a comparatively mild shock may prove fatal to a person with a weak heart or in a generally weakened condition. We should therefore avoid all contact with the house electric current.

We have already discussed the matter of keeping electrical equipment in good working order. Here are some further suggestions. No wall outlet should be placed within reach of the bathtub or washboard or kitchen sink — the places where water is made available to members of the family. Unfortunately, in some houses that is just where wall outlets are located. In such cases, remember that you should never touch either the outlet or any equipment connected with it with wet hands. Above all, never touch anything connected with the electric current of your house while you are in the bathtub.

Many outlets are only slightly above floor levels, so that childish fingers can probe their dangerous depths with hairpins or similar objects. It would be well to provide all such outlets with guards, which permit access to the socket holes only when a plug is to be inserted.

American Mutual Liability Insurance Co.

Household gas may be hazardous. Do not allow liquids to boil over and put out the flame, for gas will escape.

If you wish to repair a worn-out electric cord, be sure that the appliance is disconnected. Many a homeowner has absentmindedly snipped a cord above the frayed part without bothering to disconnect it, with most unhappy results. It is best to leave complicated repair jobs to a competent electrician.

Poisons and poisonous gases

One way of preventing accidents caused by poison would be to keep all poisonous materials out of the house. But poisons like iodine or ammonia or sodium arsenite may serve a useful purpose; they may guard us against infection or cleanse our garments or keep vermin out of the house. They are not the only possible sources of poison; certain medicines or drugs may be most harmful if taken in too large doses.

The important thing is to see to it that no member of the family swallows poisons, either because he mistakes them for something else or else because he is too young to know better. We should clearly label all bottles and other containers of antiseptics, medicines, insecticides and the like and keep them in a cabinet high above the reach of young children. As an added measure of precaution, it would be well to keep the cabinet locked.

It has been suggested that if pins are inserted in the corks of each bottle containing poison, anyone reaching for such a bottle in the dark will be forcibly reminded of its contents. Some people may feel, however, that this is a rather drastic measure, and they will be content with the less painful precautions noted above.

As we have seen, illuminating gas is a serious fire hazard. It is just as dangerous if it is breathed in considerable quantities. A leak in the gas pipe or an open gas cock will flood the house in time, if the windows and doors are closed, with enough gas to cause unconsciousness or death. So when we keep our gas apparatus in good working condition, we are guarding against a double danger.

Even when the gas stove is properly lighted, it can be a source of danger, for the flame uses up oxygen. Every winter

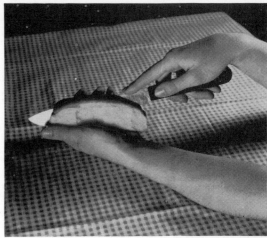

Greater N. Y. Safety Council

If you handle a knife as shown above, you will be lucky to avoid injuring your hand. Cutting strokes should always be made away from your hand, rather than toward it.

we read of people meeting death by asphyxiation because they tried to keep warm by lighting gas ovens and closing all windows. Remember that a kitchen range is *not* a heating unit. If your house is cold, bundle up in warm clothing or get into bed or visit a neighbor.

Poisonous gas fumes are sometimes given off by heating units that have become caked with soot or in which combustion is imperfect. Have your heating unit checked and cleaned by a competent service man at the start of each heating season.

Cuts and scratches

Pins or needles that are left lying on the floor or upon a chair or sofa have resulted in many scratches and infections. The writer recalls a case in which a careless person dropped a needle on a dining-room rug and left it lying there. A young girl, running barefoot over the rug, ran the needle into her foot. The foot became badly infected and required the doctor's care. It is best to do all your sewing in a special room set aside for that purpose. After you have finished sewing, put all your sewing implements — needles, pins, scissors and so on — back where they belong; and do it *at once*.

Knives, forks and other sharp tools misused or used in haste are responsible for many injuries, minor and severe. Sharp knives must be used carefully and stored safely. When using a knife, make cutting strokes away from your hand, rather than toward it, whenever it is possible to do so. Wash and dry knives separately, and store them in a knife rack out of reach of childish fingers. Children should never be permitted to use knives until they are old enough to handle them safely. Never use the sharp edge of a knife to pry open a jar. Scissors should be handled with care by adults and kept out of reach of young children.

Since so much of our food comes in cans, it is important to be able to open these containers safely. Select a can-opener that leaves no jagged edges to cut the user. An excellent type is the revolving-motion opener, which folds the sharp edges underneath. Avoid haste in opening cans equipped with a flat key for removing a metal band. Turn slowly so as to prevent breaking the band and handle the sharp edges of the band carefully.

Safeguarding young children

We can reason with adults and older children; we can win their co-operation in preventing accidents. But the very young can not co-operate in this way and therefore we should take particular pains to safeguard them.

One of the most dangerous places for little children is the kitchen. They should be kept out of this room as much as possible; if they are permitted to enter it, they should be watched every minute of the time. Needless to say, the mother should keep young children away from hot stoves or cooking ranges, and from all moving kitchen equipment, such as washing machines, clothes wringers, electric mixers and food choppers.

Play pens will protect children from many hazards while the mother is busy. For safe sleep, the baby crib with closely spaced slats will help prevent falls. So will low chairs for young children.

Since the serious business of the very young is play, we should be particularly careful about the toys that we put into their hands. Toys should be strongly built, with smooth surfaces and rounded corners. Metal toys should have no sharp point; edges should be rolled. Never give a toy so small that it can be swallowed. Rag dolls or stuffed animals are safe toys, *provided* they are washable and that they do not have eyes made of glass or pins.

Are you appalled by the long list of accidents that can happen to you or to a member of your family? Yet no accident need result in most homes if each member of the family will co-operate fully in the task of removing all hazards that may endanger young and old alike.

National Board of Fire Underwriters

Violating two safety rules. The handle of the percolator should not protrude beyond the edge of the range; the little girl should not be allowed to stay so near it.

ELECTROMAGNETIC RADIATION

Waves of Energy That Fill the Universe

by JAMES STOKLEY

IF YOU stand in front of a campfire, even though you do not actually touch the flames, something that is emitted from the blazing wood makes itself felt on your face or hands. Out in the noonday sun, on a summer's day, you get the same sensation of warmth. You know that if you stay out in the sun too long you will get sunburned, possibly painfully.

You listen to the radio or watch television. So do thousands of others, all hearing or watching the same program. Something is sent out from the transmitting station; the receiver picks up this something and converts it into the particular program that you hear or see.

These are all examples of the process called radiation, whereby energy is carried across space. Ordinarily the energy travels outward from its source, spreading equally in all directions like the radii of a circle; hence the name.

The most familiar kind of radiation is visible light, which makes it possible for us to see the world about us. Long ago men were puzzled as to just what light was, and even today we do not have the final answer. Some believed in the corpuscular theory. They held that the sun, the flame of a candle and all other luminous bodies give off tiny particles, or corpuscles, which fall upon an object in their path, bounce off and are detected when they hit the eyes. Others thought that light is something like the waves that are formed in a pond when you drop in a stone. These travel outward on the surface of the water; if they hit some solid obstruction, they are reflected — that is, a new series of waves appears.

One of the basic discoveries about light came in the year 1666 when the celebrated English scientist Sir Isaac Newton held a glass prism over a small hole in a window shutter, through which a beam of sunlight entered the room. The result, instead of a spot of white light, was a band of colors — a spectrum — red at one end, violet at the other. Thus he showed that what we ordinarily call white light is a mixture of colors. (Six are now generally accepted as the principal colors — red, orange, yellow, green, blue and violet.) Newton held up a second prism, which was turned in the opposite direction, and let the band of colors fall upon it; the light that finally emerged was white once more. He had taken white light apart and put it back together again.

Although Sir Isaac never expressed a very positive opinion as to the nature of light, he leaned toward the corpuscular theory. His high position in science did much to maintain it, even though in 1690 a Dutch physicist, Christian Huygens, published a well thought out wave theory, which later became widely accepted. According to this theory in its modern form, light is made up of waves, and their length determines the color. The longest — about 40,000 to the inch — produce the light we call red, while the shortest, which are about 1/70,000 of an inch in length, produce violet. In between are the wave lengths of the other colors of the visible spectrum.

About the beginning of the nineteenth century, scientists found that this spectrum represents only the range of radiation that the eye can see, and that beyond both ends there are other rays, which are invisible.

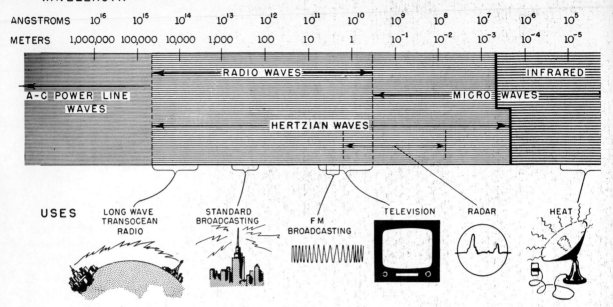

WAVELENGTH

| ANGSTROMS | 10^{16} | 10^{15} | 10^{14} | 10^{13} | 10^{12} | 10^{11} | 10^{10} | 10^{9} | 10^{8} | 10^{7} | 10^{6} | 10^{5} |

| METERS | 1,000,000 | 100,000 | 10,000 | 1,000 | 100 | 10 | 1 | 10^{-1} | 10^{-2} | 10^{-3} | 10^{-4} | 10^{-5} |

RADIO WAVES
INFRARED
A-C POWER LINE WAVES
MICRO WAVES
HERTZIAN WAVES

USES
LONG WAVE TRANSOCEAN RADIO
STANDARD BROADCASTING
FM BROADCASTING
TELEVISION
RADAR
HEAT

The first discovery of this sort was made by the English astronomer Sir William Herschel, already famous as the discoverer of the planet Uranus. In 1800, just as Newton had done, he put a prism in front of an opening through which sunlight was shining and obtained a spectrum. Then he placed several thermometers at the various colors, to see which would produce the greatest heating effect. Of all the visible parts, the red light made the temperature go up the most. Then Herschel placed one of the thermometers beyond the red end of the spectrum; the mercury went up even higher, clearly showing that invisible heat rays were present.

In this way Herschel demonstrated the existence of what we now call infrared rays. ("Infra" means "below"; the infrared rays have a lower frequency than the red rays.) He performed a whole series of experiments with these heat rays and discovered that he could reflect them with mirrors and focus them with lenses, just as visible light is reflected and focused. He also found that these rays are emitted from a bed of hot coals in a fireplace, as well as from the sun.

A year later (in 1801) a German physicist, Johann Wilhelm Ritter, discovered that there was also something at the other end of the spectrum. This did not cause heating, but was capable of producing chemical effects. Ritter knew that the white compound called silver chloride turns black when exposed to light; he put some of this substance in line with a solar spectrum, out beyond the violet part, in a place where no visible light was shining. The silver chloride blackened even more rapidly than it did in the region where the various colors could be seen.

Detecting ultraviolet rays by fluorescence

Thus ultraviolet rays were discovered. ("Ultra" means "beyond.") These rays could also be focused and reflected; but they did not penetrate glass very well. By 1852, the English physicist Sir George Stokes found that the ultraviolet rays could pass through quartz quite easily. So he made a spectrum-producing device out of quartz lenses and prisms. With the light from an electric spark, he could produce an ultraviolet spectrum ranging to a wave length about half that which produces the deepest violet light. To detect these invisible rays, Stokes made use of fluorescence. This is an effect shown by certain chemicals: when ultraviolet rays fall on them, they shine with visible light.

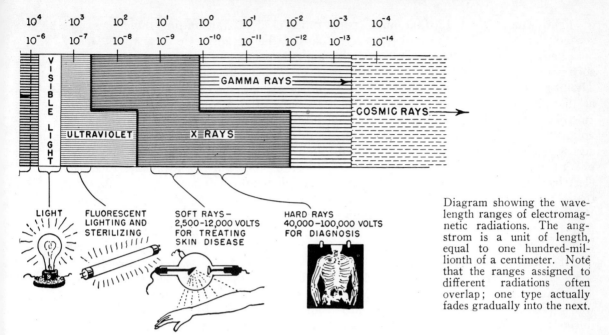

| 10⁴ | 10³ | 10² | 10¹ | 10⁰ | 10¹ | 10² | 10³ | 10⁴ |

10^4 10^3 10^2 10^1 10^0 10^1 10^2 10^3 10^4

10^{-6} 10^{-7} 10^{-8} 10^{-9} 10^{-10} 10^{-11} 10^{-12} 10^{-13} 10^{-14}

VISIBLE LIGHT

GAMMA RAYS

COSMIC RAYS

ULTRAVIOLET

X=RAYS

LIGHT

FLUORESCENT LIGHTING AND STERILIZING

SOFT RAYS— 2,500–12,000 VOLTS FOR TREATING SKIN DISEASE

HARD RAYS 40,000–100,000 VOLTS FOR DIAGNOSIS

Diagram showing the wave-length ranges of electromagnetic radiations. The angstrom is a unit of length, equal to one hundred-millionth of a centimeter. Note that the ranges assigned to different radiations often overlap; one type actually fades gradually into the next.

Since silver chloride and similar substances darken when exposed to either visible or invisible light, plates coated with these substances can be used to photograph the invisible spectrum. With refinements in this method, it was possible to extend the study of the spectrum far into the ultraviolet region. Special improved plates were used in a spectrograph from which air had been evacuated, so as to avoid the absorbing effect of the air upon ultraviolet waves.

While some photographic materials are sensitive to infrared, they do not record rays far beyond the visible spectrum; other methods must be used, therefore, in studying these rays. Some of the most fundamental work on the infrared region was done by an American, Samuel P. Langley, who for years was secretary of the Smithsonian Institution, in Washington. To study infrared rays, Langley invented an electrical device called the bolometer (ray-measurer). With the bolometer, which detects and measures small quantities of radiant heat, he studied the infrared spectrum of the sun well beyond the wave lengths of visible light.

Later, with the bolometer and other even more efficient detectors, scientists found out that still longer waves are given off from earthly sources that do not even

have to be what we commonly call "hot." Infrared rays are emitted, though rather feebly, from a glass of ice water, for example, in a long range of wave lengths, with the maximum at about 1/2500 of an inch. If the water is boiling, the emission is stronger, with the maximum wave length now slightly shorter. For still hotter objects, the total amount of radiation given off increases, and the maximum wave length becomes shorter and shorter. This is in accordance with the so-called displacement law discovered by the German physicist Wilhelm Wien in 1893. According to this law, the product of the wave length at which a radiating body gives off the greatest energy and the absolute temperature (see Index) of the body is a constant.

The first clue that there might be additional waves, still longer than those of the infrared, came from the Scottish physicist James Clerk Maxwell. In 1864 he published a series of papers dealing with the nature of light, and showed, theoretically, that it is a movement of electrical and magnetic waves; he reached the conclusion that these waves could extend far beyond the infrared. One of his suggestions was that a vibrating electrical charge could set up such waves, and that they would travel through space at the speed of light — more than 186,000 miles per second.

Perhaps the first person to produce such waves experimentally was Elihu Thomson, a brilliant young science teacher at the Philadelphia Central High School. In 1871, while experimenting with high-voltage electrical sparks which could jump across a gap of several inches, he connected one of the terminals, through which the electrical discharge passed, to a water pipe, the other terminal to a metal table top. While the sparks were jumping, he found that he could go to distant parts of the building, hold a knife blade near a metal object and draw sparks from it. After some other experiments, a few years later, he realized that he had confirmed Maxwell's prediction. The energy that produced the sparks at the knife point was transmitted from the original spark by waves that traveled through space.

As sometimes happens in scientific history, Thomson's experiments were almost entirely ignored. In 1887 Heinrich Hertz in Germany obtained similar effects with what was essentially the same equipment. The world acclaimed the German scientist, and the radiation that was produced came to be known as hertzian waves. They are now generally called radio waves.

Using radio waves
to send messages

Various people realized that these waves might make it possible to send messages between distant points, without any wires to carry them. The first to succeed was a young Italian engineer, Guglielmo Marconi. After some preliminary experiments in which he sent signals across a vegetable patch near his home, he improved his equipment and finally, on December 12, 1901, sent the letter "S" in the Morse telegraphic code across the Atlantic Ocean from England to Newfoundland. From this humble beginning came such developments as television, radio broadcasting and radar.

The waves used in broadcasting are about 1,000 feet in length. Still longer ones, as much as a mile in length, are some-times used in transmitting radio messages across the oceans. Certain waves, many miles long, are emitted from the power lines over which alternating electrical currents are transmitted. This equipment is designed, of course, to deliver electrical power; the waves are a sort of by-product.

The range of wave lengths of radiation has been extended, therefore, from those of the shortest ultraviolet to others a million million times as long. What of the other end of the spectrum? Are there any rays shorter than the ultraviolet?

Here again Germany enters our story. In November of 1895, Professor Wilhelm Konrad Roentgen, at the University of Wuerzburg, discovered that some new kind of ray was given off from a Crookes tube, an evacuated glass bulb in which an electrical discharge was produced. This ray caused some materials to glow by fluores-

A fluorescent sun lamp, suspended from the ceiling of a pigpen, subjects hogs to ultraviolet radiations.

Westinghouse

INFRA-RED

Sometimes infrared rays are used to detect alterations in paintings. When a painting by Il Tintoretto — The Doge Alvise Mocenigo Presented to the Redeemer (above) — was photographed by infrared rays, they brought out the figure of an angel hovering over the kneeling Doge. Apparently Il Tintoretto had changed his mind about the composition of his painting.

cence; it could also be recorded by its effect on a photographic plate. The most striking characteristic of the new rays was that they could pass through many solid materials — wood, paper and even human flesh; with the rays one could see and photograph the bones in one's body. Roentgen gave the name of X rays to the newly discovered and mysterious emanations, because in science X is the symbol of the unknown.

Finally it turned out that X rays — or roentgen rays, as they are also called — are radiations with wave lengths even shorter than those of ultraviolet. Their wave length depends on the voltage of the electricity used to create them; the higher the voltage, the shorter the length. For 100,000-volt X rays, most of the radiation has a wave length of about 1/1,380,000,000 of an inch, or about 1/21,400 of that of the shortest visible light.

With some of the large atom smashers that are used by physicists, X rays of hundreds of millions of volts energy are produced, and their wave lengths are still shorter. Beyond these are certain radia-

tions that occur in nature; these are associated with the cosmic rays that rain upon us from outer space.

All the radiations that we have been discussing — light, ultraviolet rays, infrared rays, X rays and the rest — are known as electromagnetic radiations. They have received this name because they are sent out into space as the result of periodic variations (variations occurring at regular intervals) in electric and magnetic fields. They are all transmitted at the same speed — 186,282 miles per second. The complete range of wave lengths of these radia-

Modern radar height-finding apparatus. Radio waves of different lengths are used in radar.

Common dog flea under the X-ray microscope. This device magnifies up to 1,500 diameters.

Using X rays to fluoroscope blueberries packed in boxes, in order to make sure that no foreign substances are present.

tions is known as the electromagnetic spectrum. (See table on pages 200-01.) The known electromagnetic spectrum extends from the cosmic rays, with wave lengths so tiny that some two trillion would have to be lined up to equal an inch, all the way up to alternating-current power waves, more than a billion billion times as long.

The quantum theory of light

So far we have been discussing the wavelike properties of light and similar radiation. The wave theory of light, which, as we have seen, had been originally proposed by Huygens, had won wide acceptance in the nineteenth century. However, about 1900 the work of Max Planck, in Germany, and later that of Niels Bohr, in Denmark, showed that there were some effects that the wave theory could not explain. Apparently, at least in certain respects, light acts as if it were a stream of tiny corpuscles of energy called photons. A photon is the smallest amount of light energy possible. Many other radiations, especially those that come from excited atoms, seem to show the same property: radiant energy is emitted and absorbed in separate, particlelike units, or quanta. These ideas of Planck and Bohr were eventually united in a revolutionary theory of atomic structure. However, the overwhelming weight of experiment still seemed to uphold the wave nature of light. Indeed, it was said that one had to believe the wave theory on Monday, Wednesday and Friday, and the quantum theory on Tuesday, Thursday and Saturday.

The wave mechanics theory

A compromise was introduced about 1925, as a result of the work of a French physicist, Louis de Broglie, and an Austrian, Erwin Schroedinger. They originated the theory of wave mechanics. According to this theory, all forms of matter and energy have both wavelike and particlelike qualities, but the two aspects never appear together under the same conditions. All "particles" have waves associated with them, whose wave lengths depend on the mass and velocity of the "particle." The bodies of our familiar world, such as a bullet fired from a gun, have relatively huge masses and low velocities, and their associated waves are therefore too small to be detected. But an electron, which is one of the "particles" of which atoms are made, has an infinitesimal mass and moves with a velocity close to the speed of light; the wave length associated with the electron is therefore great enough to measure.

Beams of electrons, and many other elementary "particles," have been diffracted to produce interference patterns just like those of a conventional wave. From these patterns, the wave length of the electron is measured and is found to agree with the theoretical prediction of wave mechanics. Therefore, the seemingly paradoxical ideas of Schroedinger and De Broglie have been generally accepted by science.

How an atom generates waves of light

For an atom to generate waves of light, changes must take place in its structure. An atom consists of a positively charged nucleus, around which are moving negative electrons; from one in the case of hydrogen to more than a hundred in the case of the heaviest element known. Normally, these electrons move in certain regions, or orbits, but when energy is absorbed they can be shifted to other orbits, farther away from the nucleus. Scientifically, they are said to be shifted to a higher energy level. Sometimes electrons may be removed completely; then the electrically unbalanced atom is said to be ionized.

After an electron has been thus shifted or removed, it tends to return to its former position; when it does, energy is given off in the form of radiation. Visible light, as well as the nearby ultraviolet and infrared, is a result of transitions of the outer electrons in atoms and molecules, while those of the inner electrons in atoms give rise to very short ultraviolet rays and to X rays. Gamma rays, still shorter in length, which come from radioactive materials, are a result of changes of energy state in the nucleus itself. In the case of the longer infrared rays, the atoms are not affected; vibrations and rotations of the molecules, consisting of a number of atoms, are responsible for the radiations.

The many
uses of X rays

Let us examine some of the applications of radiation, starting at the short-wave end. So far cosmic rays have not been used by man; indeed it is only comparatively recently, with the greatest of atom smashers operating at energies of billions of volts, that some of their effects have been duplicated for the first time in the laboratory. But X rays have found wide application. The enormous penetrating power of fifteen- to twenty-million-volt rays, which are produced by several commercial devices, have been utilized in the X-ray examination of very thick metal parts. Steel materials from eight to ten inches thick can be examined with one- and two-million-volt radiation. X rays of this same energy range are also used in treating cancer. Properly controlled, they do more damage to the cancer than to the healthy tissue.

For more general medical use, as in finding the place where a bone has been fractured, rays of about 100,000 volts may be used; these also serve in industry. X rays have been employed to analyze the structure of matter. A beam of X rays, passed through a crystal in which the atoms are arranged in regular layers, is scattered by the atoms and forms a pattern of spots, characteristic of the particular kind of crystal, on a photographic film. This is called X-ray diffraction.

The longest-wave X rays overlap into the shortest of the ultraviolet waves. One of the chief uses that man has found for the ultraviolet is in fluorescent lighting, which has now become widespread. Inside these familar tubular lamps, an electrical discharge produces ultraviolet rays, which shine on the special material used to line the tube. The invisible rays are thus turned into visible light by the process of fluores-cence. It is the ultraviolet part of sunlight that produces sunburn, in addition to forming vitamin D in the body and preventing the disease called rickets. Ultraviolet rays can also be used to kill germs; lamps producing these rays are employed in medical treatment.

The uses of rays of visible light are obvious enough; one can readily imagine what the world would be like without light. It will be remembered that in the Biblical story of the creation, one of God's earliest acts was the creation of light.

Next above the light waves in the electromagnetic spectrum are those of the infrared — the heat rays. Any incandescent lamp gives off these rays in copious amounts. There are special heat lamps in which the proportion of visible light is reduced, while that of infrared is increased. Such lamps are employed for drying purposes. For example, after an automobile body is painted, it is often run through a tunnel in which it is exposed to the rays from a battery of heat lamps; as a result it dries in minutes. The drying process would take much longer with other methods.

Microwaves and
chemical research

The shortest radio waves, which overlap the region of the longest infrared, are often known as microwaves. One use is in chemical research — in the study of the structure of molecules. Some gaseous molecules are able to absorb these tiny waves, and the energy the molecules acquire will start them spinning, much as the blades of a windmill turn in a breeze. The windmill can turn rapidly or slowly, or at any speed between; the molecule, on the other hand, can absorb only certain well-defined amounts of energy, and none that are intermediate between these. The amounts in question depend on the weights and arrangements of the atoms that make up the molecule. When the rotational energy that a molecule absorbs from microwaves is determined, it is possible to calculate the relative locations of the atoms in the molecule.

Radio waves of various lengths, down to the microwaves, are used in radar. This development makes it possible to detect ships and airplanes at night and through fog, and to determine their exact direction and distance. With radar, navigators of ships and airplanes can "view" clearly the objects and geographical features around them under adverse conditions of visibility. All this is done by sending out a short pulse of radiation; when it hits the distant target it is reflected and the echo is picked up. Even though radio waves travel nearly a thousand feet in a millionth of a second, electronic circuits determine accurately the time it takes for the echo to return, and thus give the distance. Radar pulses have even been bounced off the moon, at a distance of some 240,000 miles, and the echo has been detected and measured.

Radio waves emitted from various celestial objects have opened up a new field of research into the heavenly bodies (including many that cannot be detected by other means) and have given rise to the science known as radio astronomy. (See Index.) By studying the radio waves coming to us from outer space, radio astronomers have demonstrated, for example, the presence of clouds of hydrogen gas between the stars. The hydrogen atoms in such clouds can have two different formations; they switch back and forth between one and the other, emitting waves about eight inches in length as they do so.

Of course, the most familiar uses of radio waves are in the fields of communication: broadcasting, telecasting and radio telegraphy and telephony between ships and across continents or oceans. All these developments represent the work of thousands of scientists and engineers in many countries since that day in December 1901 when Marconi with great effort first managed to get the three dots of the letter "S" across the Atlantic.

Scientists are still at work upon the radiations that fill the universe. The result of these researches, in the centuries to come, will probably be even more fantastic than the recent developments.

See also Vol. 10, p. 280: "General Works."

Monitors in a television studio watch a baseball game as it is pictured by different cameras at a ball park. The best pictures are routed to transmitters which send them, in the form of radio waves, to TV sets in our homes.

Buttons of past generations. France, eighteenth century: upper left, left center, upper right, right center. France, nineteenth century: lower left. England, eighteenth century: lower center. Holland, nineteenth century: upper center. Place of origin uncertain, eighteenth century: lower right.

HOW BUTTONS ARE MADE

A Survey of a Flourishing Industry

TO THE average person, a button is a rather insignificant item. The word is sometimes used figuratively, indeed, to mean "something of little value," as in the phrase "I don't care a button for what he says." Yet buttons are among the most useful attachments on garments, particularly men's garments, and the button industry is a flourishing one. In a recent year, the value of its products was $58,916,000.

Buttonlike disks or knobs were used as ornaments thousands of years ago, but for a long time it did not occur to people that they could use these ornaments to fasten their clothing. It is believed that primitive man used thorns and sinews for that purpose. Later, such civilized people as the Egyptians, the Greeks and the Romans fastened their garments with pins, brooches and buckles as well as with buttons.

Some very ancient buttons were discovered by the English Egyptologist Sir Flanders Petrie (1853-1942) a number of years ago in a tomb in Coptos, in Upper Egypt. One of them went back to 2500 B.C. A hole had been pierced through a projection on the back of this button, thus forming a shank, or eye. The thread by which the button would be attached to a garment would undoubtedly be passed through this eye.

Buttons were first used in Western Europe about the thirteenth century A.D. In a manuscript that was written not later than the year 1300, there is a reference to a hero who wore buttons "from his elbow to his hand." We cannot tell whether these buttons were employed for decoration or as fasteners. In the course of time, however, buttons served both purposes. In the fourteenth century, they were used lavishly on fashionable attire. They rather lost their vogue for a time, but in the sixteenth century they became more fashionable than ever. They were worn by both sexes.

In the days of Charles I of England, in the first half of the seventeenth century, buttons attached to large handkerchiefs were much in style. Tastes had become extravagant; we find frequent mention of buttons made of diamonds and precious stones. A generation later, Louis XIV, the "Sun King" of France, is said to have had a positive mania for buttons.

By the eighteenth century, button making had become a flourishing industry in France, Germany, England and other European countries. Buttons were made of a variety of materials, including gold, silver, brass, pewter, wood, bone, glass, leather and tortoise shell. They were sometimes covered with velvet or silk cloth.

The first buttons made in America were of metal. A German immigrant, Caspar Wistar, began to manufacture brass buttons in Philadelphia in 1750; it is said that he guaranteed each one for seven years! Within a few years, several firms in New England were also producing buttons. The total production was small, however. In the American Revolution (1775-83), many of the buttons used on the uniforms of American soldiers had to be imported from France.

Button imports in the United States were cut off during the War of 1812 as a result of the British blockade. Aaron Benedict, a button maker of Waterbury, Connecticut, profited by the emergency. He bought all the brass pots and pans that he could and rolled the metal in his own rolling mill; then he made buttons of it. After the war, the American metal-button industry made rapid progress.

About the middle of the nineteenth century, the Frenchman Emile Bassot devised a method for making buttons from the softened horns and hoofs of cattle. The vegetable-ivory button was developed in Austria at about the same time. It was manufactured from the tagua nut, the fruit of the tagua palm, found near the west coast of South America, from southern Panama to northern Peru. Tagua nuts could be readily sawed, carved and turned in lathes in a great variety of sizes and shapes.

Pearl buttons won great popularity in the last half of the nineteenth century. The first buttons of this type were made from the shells of pearl oysters; they became known as ocean-pearl buttons. In 1891, the pearl-button industry in the United States received a great impetus because of the discovery, in the Mississippi River, of mussels whose shells could be made into buttons. Because of the abundance of suitable shells in its vicinity, Muscatine, Iowa, was selected as the site of the first factory in the United States devoted to the manufacture of pearl buttons from mussel shells. Many other factories were established in the years that followed.

In the present century, buttons made of various plastics have come to the fore. Plastics, as we point out elsewhere (see Index), are synthetic materials that can be molded or pressed into a desired shape. They lend themselves excellently to mass production; and there is a minimum of waste. Today more buttons are produced from plastics than from any other material. They can be made to resemble vegetable-ivory, pearl and horn buttons so closely that only an expert can tell the difference between the real and synthetic articles. As a result, they are taking over the markets formerly served by natural products.

The manufacture
of plastic buttons

Certain plastic materials, such as Bakelite and Plaskon, are made into buttons by the process known as compression molding. The manufacturer receives the plastic material in the form of a dry powder. This powder is converted into preforms, or pellets, each of which is about the size of a finished button. The preforms are molded by compression in cavity dies. The buttons are then tumbled in a revolving machine to remove excess material; finally, they are filed, drilled and polished.

A process called injection molding is used to manufacture buttons from cellulose-acetate plastics, such as Tenite. The plastic material, in powder form, is dumped into a hopper in the injection-molding machine. It then goes into a heating chamber, where it acquires the consistency of paste. A piston pushes it out of the heating chamber through a nozzle; from the nozzle it is squirted into a mold. This has a number of cavities, each large enough to contain material for one button. The cavities are connected with each other by little channels. The plastic paste runs along these channels and into the cavities.

When it has hardened, the contents of the mold are ejected. We now have a number of buttons connected by little stems (the plastic material that has remained in the channels). The stems are broken off from the buttons by means of buffing machines; they are then reground and used as materials for other buttons.

Celluloid buttons are made in various ways. In one process, blanks (pieces about the size of the finished buttons) are cut from large rods. These blanks are sometimes placed in holders, which are called chucks, and then turned; sometimes the buttons are held stationary while a revolving blade shapes them. In making die-pressed celluloid buttons, round disks are cut from celluloid sheets and are pressed into shape by means of dies. Twisted celluloid buttons are made from thin rods.

A favorite plastic material for buttons is Galalith. This trade name, which comes from two Greek words meaning "milk stone," is quite appropriate, for Galalith is produced by the action of formaldehyde on milk casein, producing an insoluble, tough, hornlike substance. It is interesting to note that Galalith was created by two German scientists, Wilhelm Krische

COMPRESSION MOLDING

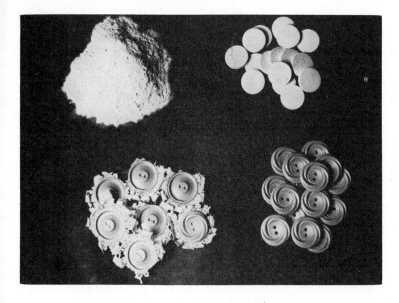

Powdered plastic, upper left, the material from which many buttons are made. Button-size pellets or preforms, upper right, ready for compression molding. Molded buttons, lower left, before excess material is removed. Finished buttons, filed, drilled and polished, lower left.

A pellet-making machine of the same type as that used in making various kinds of pills, such as aspirin. Here, dry powdered Bakelite or Plaskon, plastic material, is formed into button-size preforms on the machine. At left, preforms are expelled automatically from machine.

INJECTION MOLDING

Merit Plastics

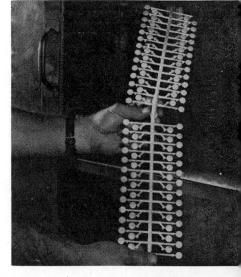

Injection molding is accomplished in a very different way from compression molding. At the right of the picture you can see a pattern of holes that are part of the stationary button mold. A movable mold is brought up against the stationary one to fashion the buttons.

"Trees" of button shapes, in branch formation, after they have been removed from the ejector unit of the injection mold. The stems will be broken off the buttons in a buffing machine and will be reground later for material to be used in the making of still other buttons.

and Adolph Spitteler, who were trying to develop a composition material to replace slate for blackboards.

Galalith is mixed with coloring matter and various other ingredients. The mixture is then fed into a machine that operates on the general principle of a meat grinder. When the mixture comes out of this machine, it is in the form of plastic rods. Sheets of plastic are sometimes made from these rods by pressing in heated presses. Disks are sliced from the rods or punched from the sheets; they are then hardened in formaldehyde. The disks are turned, carved, drilled and polished. Galalith makes a sturdy, hornlike kind of button.

Buttons from other materials

Pearl buttons are used in considerable quantities for shirts and undergarments. Shells for ocean-pearl buttons come principally from the waters around Australia and the East Indies. The shells are soaked in water for several days to make them less brittle. Then they are cut into blanks by tubular saws that revolve at high speed. The blanks are split lengthwise two or three times, depending on how thick the finished button is to be. The back of each blank is ground flat or else is rounded under an emery wheel. After the completion of the grinding process, the blanks are placed in chucks on an automatic pattern machine. As each blank is carried around the machine, it is processed by a succession of tools. These shape it into the desired pattern, round its edges and drill it with two or four holes. The button is put in a barrel of acid to give it luster and is then tumbled in a barrel of sawdust. Fresh-water pearls are manufactured like the ocean-pearl variety, except for the fact that the blanks are not split.

Most of the nuts that are used for vegetable-ivory buttons come from Panama, Colombia and Ecuador. The nuts are first dried under intense heat in order to crack their shells; then the shells are broken off by means of rolling in a metal drum. The nuts are now cut by circular saws into slabs, which are dried in kilns until all moisture has been removed. Blanks are cut from the slabs, turned on lathes and drilled. The buttons that are to have a solid color are dyed in vats. To produce a mottled effect, various color shades are sprayed upon the buttons through a series of stencils; then the buttons are dipped in vats. Sometimes vegetable-ivory buttons are embossed — that is, ornamented with raised work. Vegetable-ivory buttons are used principally on men's suits, army shirts and women's fine tailored clothing.

The evolution of pearl buttons from fresh-water mussel shells to the exquisite finished product.

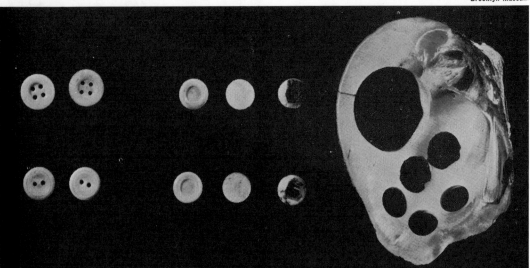

A native boy, standing beside a tagua palm, proudly displays a sliced fruit from the tree. The cross section of the large husk shows the egg-shaped nuts from which vegetable-ivory buttons are made.

Metal buttons are used mostly for military uniforms and work clothes and as ornaments for women's coats and dresses. They are produced mostly from rolled sheet brass. Circular blanks are first punched out of the brass. An automatic machine then depresses the center of each button and forms the edge. The holes are also punched by machinery. When shanks are used, they are formed of wire by a separate machine, which cuts off the pieces and bends them into loops of the required form. After these are soldered on, the buttons are trimmed on a lathe. They are then finished by gilding or burnishing.

Fancy metal buttons are sometimes manufactured from so-called white metal, which has a tin base and is alloyed with copper and antimony. The metal is heated until it is molten; it is then poured into rubber molds. After being molded, the buttons are plated with gold, silver or copper.

Horn buttons are now made chiefly from cows' hoofs, ground to a powder and mixed with color ingredients and water. Since the hoofs contain natural glue, no binding material is added. The mixture is poured into dies containing about a hundred cavities. The dies are heated and then they are chilled. After the buttons are removed from the dies, they are filed, drilled and polished. Bone buttons are generally made from the shinbones of cattle. They are used chiefly on undergarments or the inside flaps of garments, where the buttons will not be visible. First, the bones are softened by soaking; then they are cut into slabs by circular saws. Automatic machines cut buttons from the slabs, turn them and then drill them as desired.

Covered buttons are generally manufactured by covering metal parts with the same textile materials as those from which the garment itself has been made.

The unit of measurement for buttons

The unit of measurement for buttons is the line; there are 40 lines to the inch. For example, a 30-line button is ¾ of an inch in diameter; a 20-line button has a diameter of ½ inch.

213

EXPERIMENTS IN MECHANICS

Moving, Lifting and Holding Things

BY NELSON F. BEELER

THE science called mechanics is concerned with the motions of bodies and with the forces that bring about such motion. It also takes up the forces that act upon bodies in such a way that they remain at rest.* In this chapter, we shall illustrate some of the principles of mechanics with a series of simple experiments that you can easily do yourself.

For a long time, men have known how to move things more effectively by using the device called the lever, or pry. It consists of a rigid piece that can be made to turn about a point called the fulcrum, or pivot. To show how the lever works, first try to lift up from the floor a friend who is about as heavy as you. You will find it hard, of course. Now put a long, strong board on the floor. Set a small piece of wood underneath the board, about a foot from the end; this will be the fulcrum. Have your friend stand on the end of the board nearest the fulcrum (Figure 1). It will be easy to lift him when you push down on the other end of the board. You have successfully applied the principle of the lever. You exerted your effort at a greater distance from the fulcrum than the distance from the fulcrum to the place where your friend was standing. In so doing, you multiplied the effect of your effort, so that you had to apply only a fraction of the force that had been required to lift your friend from the floor without the use of a lever. To make up for this, in order to lift your friend only a few inches, you had to move the board down at your end perhaps a foot or so.**

* The study of such bodies is the province of statics, which is a branch of mechanics.

** The ratios involved are given in the article Simple Machines, in Volume 3.

1. How a lever works. When the person on the right gets on the board, his end of the board will go down. It will be lifted up when somebody gets on the other end.

2. You can illustrate the three classes of levers with the above apparatus: a spring balance, a can filled with sand and a light piece of wood into which screw eyes are set.

Every lever has three parts — the fulcrum, the effort and the resistance (also called the load). In the lever you just prepared, the fulcrum, as we pointed out, was the small piece of wood under the board. The effort was applied at the place where you pushed down on the board. The resistance was at the place where your friend was standing. His weight, pressing down on the board, *resisted* your attempt to lift him and this resistance had to be overcome by exerting an effort.

There are three possible combinations of fulcrum and forces, or three classes of levers, as a scientist would say. In the first class, the fulcrum lies between the effort and the resistance. In the second, the resistance is between the fulcrum and the effort; in the third, the effort is between the resistance and the fulcrum. You can illustrate these three kinds of levers with the simple apparatus shown in Figure 2. The lever can be any light piece of wood, into which you set screw eyes as shown. You will also need a small can; to this you will attach a wire hook that can fit into a screw eye. The can is to be filled with sand. You will apply the effort by pressing up or down, as the case may be, on a spring balance, attached to one of the screw eyes. The marker on the spring balance will enable you to measure the amount of force required for the different classes of levers. Finally, a nail is to be driven into a vertical post. The screw eye which will serve as the fulcrum will be passed through this nail. By changing the position of the fulcrum, effort and resistance, as shown in Figure 3, you can produce any one of the three classes of levers.

In Figure 4, we show certain familiar devices, all based on the lever principle. To which class of lever does each of them belong?

Not all levers are straight. The tool called the claw hammer is a bent lever. If you hammer a nail about halfway into a piece of wood, you will find it difficult, if not impossible, to remove the nail with your fingers. However, you can easily get it out by applying a claw hammer to it, as

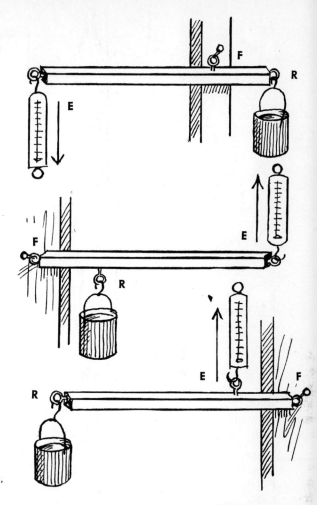

3. The three classes of lever can be produced (above) by changing the positions of the fulcrum, effort and resistance. *F* represents the fulcrum; *E,* the effort; *R,* the resistance. 4. Below: devices based on the lever principle. Can you identify the class of lever in each case?

215

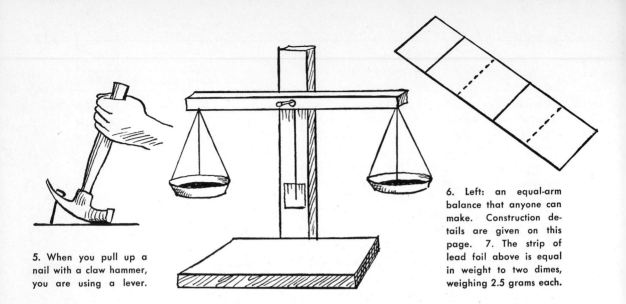

5. When you pull up a nail with a claw hammer, you are using a lever.

6. Left: an equal-arm balance that anyone can make. Construction details are given on this page. 7. The strip of lead foil above is equal in weight to two dimes, weighing 2.5 grams each.

illustrated in Figure 5. Can you tell whether this hammer is a lever of the first, second or third class?

The lever principle is applied to the equal-arm balance, which is used to find the weight of things (Figure 6). The balance arm, which is the lever, is made of a uniform piece of wood about a foot long. Drill a hole as near the exact center of the piece as you can. Attach a vertical piece to a horizontal stand, and hammer a thin finishing nail part way into this vertical piece, near the top. The balance arm is to rest on the nail as shown; it should be able to pivot freely. Suspend two pans, made of jar lids, from the two ends of the balance arm. Attach a knitting needle to the balance arm at its center; this will serve as a pointer. Glue a piece of paper to the upright, at the place where the end of the knitting needle swings to and fro. Place marks on the paper so that you can tell when the arm is swinging to one side or the other of a middle line. If the needle does not swing evenly about this middle line, add a small piece of modeling clay to the light side of the balance arm. Take off or put on a bit at a time until the pointer moves as far to one side as to the other when there is no load on either pan.

If an object is placed on one pan, the force of gravity will cause it to move downward. If you place suitable weights in the other pan, you can bring the first pan back to its original position. In this case, the part of the balance arm through which the nail passes is the fulcrum; the effort is applied to the pan on which you set the weights, and the resistance that must be overcome is the object in the other pan. You can find out the weight of this object by comparing it with the known weights that caused it to return to its original position.

Most weighing in scientific laboratories is done with the units of the metric system, in which the gram is the unit of weight. It so happens that a dime weighs 2.5 grams. Get a piece of lead foil or other heavy metal foil. Find out how much of it is needed to just balance two dimes. This will represent five grams. Divide the piece of foil in five equal parts, and cut it as shown in Figure 7 so that you will have a one-gram weight and two two-gram weights. With these, you will be able to weigh different objects in grams.

The lever is called a simple machine; it is one of six basic types, discussed in the article Simple Machines, in Volume 3. Another simple machine is the pulley, which makes it possible to lift things up in the air by pulling downward. A single fixed pulley does not increase one's force; it merely changes the direction of the pull. You can easily make a single pulley from a wire coat hanger and a thread spool, as shown in Figure 8. Attach a small pail or box to a string and pass the end of the string over the pulley. When you pull down, the box or pail will go up.

Now prepare a second spool pulley and put the two of them in place as shown in Figure 9. Notice that you will still pull down on the string and that as you do so the weight will move up. However, you will find that it requires less effort than before. What do you notice about the amount of string you pull down compared to the distance you lift the weight?

By passing a rope over several pulleys in turn, it is possible to produce an enormous force with very little effort. In this case, you will have to apply the force over a great distance in order to move the weight up a very short distance. Construction crews use a block and tackle, which is a pulley system using several pulleys. Very heavy weights can be lifted with this device. An even more complicated system, called a "chain fall," is used in garages to lift engines from automobiles. You may have seen one of these devices in operation.

With two broom handles and a length of clothesline, you can rig a pulley system that will show you how much this device can multiply the force that you exert. Attach one end of the line to one of the broom handles. Wrap the line around the handles as shown in Figure 10. Have two grownups grasp the handles firmly; let a child hold the end of the rope. As the latter pulls, the two adults will be surprised to find that they are forced to move toward each other. A child who can pull with a force of twenty-five pounds can produce a force of a couple hundred pounds with this device. Try adding more turns of line around the broom handles. What happens when the child pulls?

Sometimes we want things to move round and round. Suppose we have a motor which is turning the way the hands of a clock move — clockwise — and we want to produce motion which is just the reverse — counterclockwise. We can bring this about by means of gears. A gear is a variety of the simple machine called the wheel and axle; it consists of a wheel with teeth cut into its outer edge.

Bottle caps can be set up to work like gears. Get three caps that have been re-

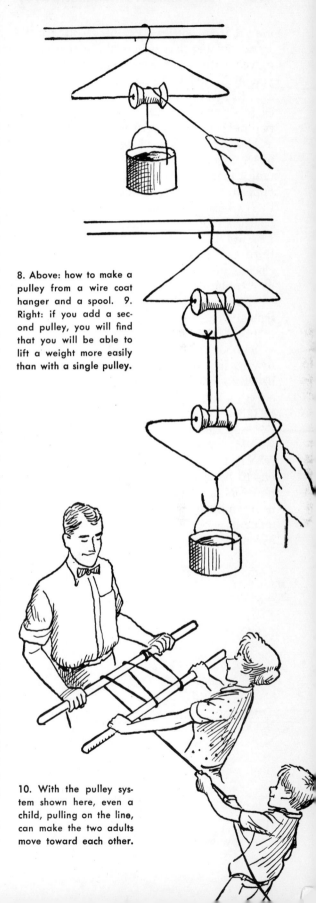

8. Above: how to make a pulley from a wire coat hanger and a spool. 9. Right: if you add a second pulley, you will find that you will be able to lift a weight more easily than with a single pulley.

10. With the pulley system shown here, even a child, pulling on the line, can make the two adults move toward each other.

moved from bottles containing soft drinks. Press on the place where the cap was removed with a bottle opener; in this way, you will make the cap perfectly round again. Make a hole in the exact center of each of the caps with a nail. Now mount two of them on a board by means of tacks, as shown in Figure 11, so that the "teeth" mesh together. Turn one cap; you will notice that the second one will move in the opposite direction. Add a third cap to the two you have already installed on the board, in order to make a "gear train" (Figure 12). In what direction does the third cap move? Gear trains are used in clocks, mechanical toys and in some kinds of computing machines.

An important kind of motion for scientists and engineers is that of pointers across dials. The pointer readings in the cockpit of a modern airplane supply a great deal of vital information to the pilot. Workers in radio and television stations, atomic-energy plants, weather stations and a host of other scientific operations use information supplied by pointers and dials.

Many times the motion which is being recorded is an up-and-down motion, which has to be changed to an across-a-dial motion. Many times, too, the motion that is being measured is small and it must be made larger. The operation of an aneroid barometer is a good example.

The heart of an aneroid barometer is a small metal can with the air pumped out of it. As the pressure of the atmosphere increases, the can is squeezed smaller; if the pressure decreases, the can expands. The amount of change in the can in either case is small. The small up-and-down motion is made larger and turned into a pointer motion by means of a lever-and-gear system, known as a linkage.

In Figure 13, we show a linkage consisting of a fan-shaped piece, called a sector, which fits into the teeth of a round gear. A pointer is attached to the gear and moves with it. If you make an enlarged copy of the diagram on heavy cardboard and then cut out the parts, you can mount this linkage on a board and show how it works. Note that the up-and-down motion at A is made larger by the sector so that the part at B is moving much faster. Also note that the up-and-down motion finally becomes round-and-round motion as the teeth of the sector turn the gear. More complicated linkages have extra levers and gears.

A belt may also be used to transfer motion from one wheel to another. In the early factories, a number of machines were operated by a single steam engine. This engine turned a long shaft on which pulleys were fixed; a belt of rope or leather ran from each pulley to a machine. Belts are still used to transfer motion. Your refrigerator may have a compressor which is connected to a motor by means of a belt. The fan in front of an automobile engine is turned by a belt.

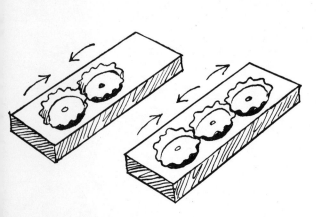

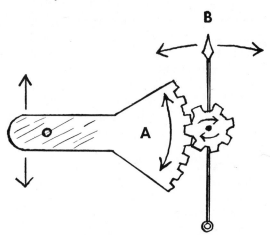

11. Left, above: gears made out of two bottle caps.
12. Right, above: a "gear train" made out of three caps.

13. A linkage consisting of a fan-shaped sector, which fits into the teeth of a round gear. Note the pointer.

To see how the belt principle works, set up two spools from a wooden construction toy set as shown. The large one is to be the driver. Use a rubber band as a belt between the spools (Figure 14). When you turn the larger spool, the smaller spool will turn in the same direction. It will rotate a number of times while the larger spool is rotating only once. If you wish to have the smaller spool move in the opposite direction, cross the belt, as shown in Figure 15.

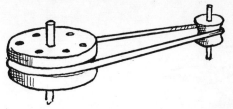

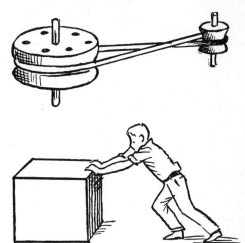

14. Above: a rubber band fitted between two spools shows how the belt principle works. 15. Below: to make the two spools move in opposite directions, cross the band.

Whenever things are moved, some work disappears. This is because heat is produced whenever one object rubs against another; a part of the mechanical energy is dissipated, or lost, in the form of heat energy. Try to push a heavy wooden box along the floor (Figure 16). The bottom of the box touches the floor and rubs against it. Some of the work you put into pushing the box heats the floor and the box because of moving friction, or rubbing, and it is not available for the motion of the box. Now put some dowels or other rollers under the box and try again (Figure 17). You will find that it is now much easier to move the box. When it is on rollers, only a small part of the surface of each roller comes in contact with the floor. Not so much rubbing results; less work is lost as heat.

16. Because of moving friction, it is hard to move the heavy box shown above along the floor. 17. If you put a pair of dowels under the box (below), it can be moved quite easily, because there is less friction than before.

If balls are substituted for rollers, there will be even less surface to be rubbed against and less heat will be lost because of friction. To show how well things move on balls, set up the following simple apparatus. Get two paint cans or similar cans with press-on lids, which fit in a "track" around the top of the can. Set one can right side up on the table and fill the track on the top with glass marbles, all of the same size. Put the second can bottom side up on the marbles. Place a board on the top can and put several books or other heavy objects on top of the board (Figure 18). The board will turn very easily on the ball bearing you have just made. If you put a little machine oil on the marbles, the board will move even more easily. Ball bearings are used in a great many machines in order to reduce friction and wear.

18. How to make a set of ball bearings with two books, a board, two paint cans and some marbles.

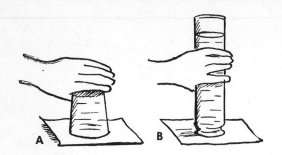

19. If a piece of cardboard is held on top of a glass tumbler partly filled with water (A) and the tumbler is then turned upside down, the water will not be spilled. The cardboard will also be held in place even if one uses a tall jar (B), in which the column of water is higher.

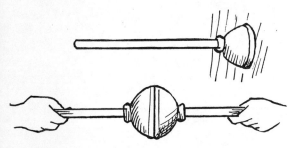

20. In the upper drawing, above, a rubber plunger, whose lip has been wet, adheres to a wall against which it has been pressed. 21. If the lips of two plungers are wet and forced together, it will be difficult to get them apart.

22. When you blow into a paper bag that has been placed under a book, the compressing of the air contained in the bag will cause the book to be lifted, as shown above.

The pressure of the air can be made to move things. We are constantly being subjected to such pressure; the reason is that air around us presses upon all things with a force of about 14.7 pounds per square inch at sea level. (The pressure decreases with distance above sea level.) To show how strong this atmospheric pressure is, partly fill a glass tumbler with water, and cover the top with a piece of light but stiff cardboard. (A 3 × 5 filing card will do.) Hold the card firmly on top of the tumbler and then quickly turn the

tumbler upside down (Figure 19A). Remove your hand from the card. The air pressing upon the card will hold it in place despite the downward pressure of the water due to the force of gravity. Try the same thing with a tall bottle such as an olive bottle (Figure 19B). The air will hold up this taller column of water too. As a matter of fact, the air could support the weight of a column of water about 34 feet in height, if you could find a jar as tall as that and if you could arrange to have the jar turned upside down.

Get a plumber's force cup, which is a rubber plunger used for opening drains. Wet the lip of the cup and press it firmly against a hard surface such as a wall (Figure 20). The air pressure will keep the cup attached to the surface. A small object can be raised from the ground by attaching a force cup to it and then lifting up the cup.

Provide yourself with two plungers of the same size and wet the lips of each; then press them firmly together. Have a friend hold the handle of one of these plungers and hold the other handle yourself. Now try to pull the two cups apart (Figure 21). You will find it very difficult.

Ordinary air pressure is great, but even greater force can be obtained by compressing the air — that is, by squeezing it into a smaller space. You can show how compressed air can be put to work by means of a simple experiment. Place a paper bag under a heavy book. When you blow into the bag, thus compressing the air in it, the book will be lifted up (Figure 22).

Compressed air can be made to move water out of a bottle, producing a miniature fountain. Set up a flask and a one-hole stopper as shown in Figure 23A. A glass tube from a medicine dropper is forced into the stopper so that the small end of the tube comes out of the top of the stopper. A second piece of glass tubing is attached to the first tube, as shown, with a short length of rubber tubing. The end of the second glass tube should extend almost to the bottom of the flask. Put about an inch of water in the flask. Hold the stopper in place and blow into the tube (Figure 23B).

Close off the open end of the tube with your finger before taking your mouth away. When you release your finger, the air compressed in the flask will force the water out in a fast-moving stream (Figure 23C).

Compressed air has many uses. You have probably seen men operating a pavement breaker, which is a drill used to break up concrete or other solid materials in a street or pavement. A rubber hose brings air under pressure into the device, and it is this air that operates the drill. The force exerted is greater than could be produced by men using sledge hammers. Compressed air in a garage is used to fill tires. Torpedoes are operated by compressed air; so are certain types of locomotives.

Liquids do not squeeze together much when a force is applied. Because of this fact, pressure applied to any part of a liquid in a closed vessel is transmitted throughout the vessel with equal force. This is called Pascal's law, because it was discovered by the French scientist and mathematician Blaise Pascal (1623-1662). This principle is applied in the toy called the Cartesian diver, which will seem very mysterious unless one knows what makes it work.

Get a gallon glass jug, a medicine dropper and a tall drinking glass. Fill the jug and the glass with water of the same temperature. Draw up water into the medicine dropper from the glass until it holds just enough to allow the dropper to float in the glass with its tip barely touching the top. Carefully transfer the dropper to the gallon jug without losing any water from the dropper. The dropper will now just float in the water at the top of the jug (Figure 24).

Press the fleshy part of your hand on the open mouth of the jug. The dropper will go down. When you ease up on the pressure, the dropper will rise. It will go down again if you press down with your palm as before. Thus you will be able to make the dropper rise or fall at will.

How does the Cartesian diver work? Remember that pressure on the water at the top of the jug will be felt equally throughout the liquid. Water will be pushed up into the open end of the medicine dropper even though the pressure you apply is in a downward direction. The water entering the dropper will compress the air within it. Since the dropper just barely floats at the start, the added water will make it heavier and it will sink. If the pressure is lessened, the compressed air in the dropper

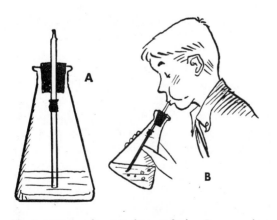

23. A miniature fountain from a flask. Force a glass tube from a medicine dropper into a one-hole stopper, which is inserted in the flask (A). Hold the stopper in place and blow into the tube (B). The air compressed in the flask will force out the water in a fast-moving stream (C). 24. The Cartesian diver, described above.

25. A Cartesian diver made with a flat-sided bottle. The medicine dropper within it will sink when you press the sides of the bottle and will rise when you relax the pressure upon it.

26. How to make a hydraulic lift with a hot-water bottle, a funnel, a glass tube and some boards.

will push some of the water out. The dropper will therefore become lighter and it will rise.

If the dropper is carefully adjusted, it will sink when only a very small quantity of water is added to it. In fact, you can prepare a Cartesian diver that will work when you merely press on the outside of a bottle of water in which you have put a dropper. The glass will "give" and will force enough water into the dropper to make it sink. Get a flat-sided pint bottle. (One kind of laundry bluing is sold in such a bottle.) Boil some water to get rid of the air it contains. When the water is cool, fill the bottle to the top. Get the smallest medicine dropper you can find. One kind of vitamin concentrate used for babies comes in a tiny bottle with a dropper less than an inch long. A large dropper will work but the result will not be so spectacular. Float the dropper as you did before in a glass full of water and then transfer the dropper carefully to the pint bottle.

Force the cork into the bottle (Figure 25). This cork should fit tightly enough so that no air can enter. The dropper should float at the top of the water; if it does not, pull out the cork a little to cut down the pressure on the liquid. Now grasp the bottle firmly between your thumb and fingers and squeeze strongly. If the

diver is properly adjusted, you can move it first down and then up by alternately squeezing it and then relaxing the pressure on it. Most people will be baffled to see the diver apparently acting on the command of the operator.

The force transmitted by a thin column of water will be strong enough to lift you up. First lay a hot-water bottle on the floor. Place two thin boards to the left and right of it, as shown in Figure 26. Set a third board over these so that it rests on the hot-water bottle. Put a right-angled glass tube through a one-hole stopper set in the bottle. Attach a length of glass tubing to the right-angled tube, using a small piece of rubber tubing to make the connection. Attach a small funnel, by means of another piece of rubber tubing, to the top of the second piece of glass tubing, as shown in the illustration. Have someone hold the stopper firmly in the hot-water bottle. Stand on the board and pour water into the hot-water bottle through the funnel. Soon you will find that the water pouring into the hot-water bottle will lift you up. The hydraulic lift that raises a car in a garage works in much the same way. Air pressure supplied by a compressor is used to push oil under a large cylinder. The cylinder then rises and lifts the car.

See also Vol. 10, p. 289: "Physics, Experiments in."

THE MOTIONS OF THE STARS

Apparent and Real Movement in Space

FOR many centuries, the stars were known as the fixed stars because they did not seem to change their relative positions in the heavens. They were contrasted by early astronomers with the planets. The latter could be seen to follow definite paths in the skies with reference to other heavenly bodies and therefore they were called "wanderers." (The literal meaning of the Greek word *planetes,* from which "planet" is derived, is "wandering.")

If the individual stars did not change their relative positions, they did apparently move around the earth, which was thought of as a fixed body set in the center of the universe. The different constellations and other star groups rose and set just as the sun did. They had to be looked for in different places in the heavens according to the time of day and also the season. These apparent motions of the stars have been known and recorded ever since the heavenly bodies were first studied.

The rising and setting of the stars and their positions in the skies have, from time immemorial, caused certain star groups to be associated with various events of human life. Thus the ancient Greeks began their navigation season when the Pleiades were first seen to climb above the horizon before sunrise in the month of May. To them, consequently, the Pleiades were the "sailing stars." Other peoples associated these stars with the planting season.

Chaucer and other medieval writers often used astronomical terms, based on apparent star motions, in referring to the time and season. This practice shows that in the Middle Ages the motions of the stars were not only the concern of special stu-

dents but were familiar to educated people in general. The men of the medieval period marveled at the wonderfully consistent patterns the stars formed as they wheeled majestically through the skies. No wonder that even learned men thought that these glittering celestial bodies were closely linked with human destinies.

The Copernican theory, which was advanced by the Polish astronomer Nicolaus Copernicus (see Index) in the sixteenth century, dispelled the old belief that the earth stood fixed in the center of the heavens. Copernicus pointed out that the earth revolved around the sun and that it held no more exalted place in the scheme of things than did Venus, or Mercury, or Mars or the other planets. Astronomers came to realize that the revolution of the stars around the earth was only apparent. It was an optical illusion, based on the rotation of the earth about its axis.

Yet modern science has shown that the stars in our galaxy (the only ones that can be made out individually even with the most powerful telescopes) actually move, so that they gradually change their positions with respect to one another. The British astronomer Edmund Halley, after whom perhaps the most famous of all comets was named, was the first to call attention to these individual motions of stars. He reported that the stars Arcturus and Sirius were both definitely south of the positions noted by the Alexandrian astronomer Ptolemy in the second century A.D.* With the introduction, in the nineteenth century,

---

* These positions were given in Ptolemy's *Great Composition,* which is better known as the *Almagest* (the name given to it by the Arabs).

of the photographic method for determining star positions, astronomers have been able to measure the individual motions of many thousands of stars.

The study of these motions is one of the most difficult branches of astronomy. There are several reasons for this. In the first place, the distances between stars are so immense that one can make out changes in their relative positions only with the most precise instruments and the most painstaking methods. Another problem is the lack of a reliable frame of reference in measuring stellar motion. Since we live on the planet earth, we have to use it as a reference point. Obviously, it is not very satisfactory for this purpose. Not only does it make its annual journey around the sun, but the entire solar system, as we shall see, is also moving constantly through the skies. Despite all obstacles, however, astronomers have made a great deal of progress in analyzing the different types of stellar motion.

The arrow in the photograph below points to Barnard's star. The arrow in the right-hand photograph points to the same star. Note the evident shift in position with respect to the other stars in the vicinity. Barnard's star has a greater proper motion than any other star.

The proper motions of stars

We determine the position of a star in the heavens at a given time by observing its right ascension and its declination. The right ascension of a star is its east-west position along the celestial equator.* The star's declination is its position north or south of the celestial equator. We explain these two terms in greater detail elsewhere in THE BOOK OF POPULAR SCIENCE. (See Index, under the entries Declination; Right Ascension.)

When we take observations of a star's right ascension and declination over a period of years, we find that its position gradually shifts. To a certain extent, this change is only apparent. It is partly due to precession (the wobbling of the earth on its axis), which causes apparent changes in the locations of all the stars in the skies. It is also due, to a certain extent, to aberration. This is caused by the changing direction of the earth as it revolves around the sun and the resulting apparent change in speed with respect to a given

---

* This is the extension of the earth's equator into space until it touches the imaginary outer limit of the heavens.

star. All the stars in a given region are equally affected by precession and aberration. The changes in position that result from these phenomena are known as common motions. After astronomers take precession and aberration into account, there still remain certain individual motions that can be traced in the heavens. These are called proper motions, because they are proper to (characteristic of) given stars.

We may define proper motion as the apparent motion of a star across the sky, with relation to neighboring stars. It is given in seconds of arc. To understand what this means, we must think of the heavens as a hollow sphere, with the celestial bodies appearing on its inner surface. Any two objects in the sky are to be thought of as lying on a circle passing through the center of the hollow sphere and bound by the inner surface of the sphere. Each circle is divided into 360 units, called degrees of arc. The distance between the two heavenly objects mentioned above is expressed in degrees (°) of arc, if the distance is considerable. Smaller distances are measured in minutes (′) or seconds (″) of arc. There are sixty minutes in a degree and sixty seconds in a minute. The proper mo-

tions of stars are so small in all cases that they are given, as we pointed out, in seconds of arc.

We determine proper motion by comparing star catalogues compiled at widely separated periods of time, or by examining photographs made with the same telescope at intervals of several years. The change in the alignment of a given star with respect to other stars in its vicinity will represent its proper motion.

The star with the greatest proper motion known to us is Barnard's star, which was discovered photographically by the American astronomer Edward E. Barnard at the Yerkes Observatory in 1916. This faint tenth-magnitude star,* which is in the constellation Ophiuchus, has a proper motion of 10.3″ a year. The annual proper motion of Kapteyn's star (ninth magni-

* "Magnitude," in this case, is a measure of brightness. In the catalogues of the ancient astronomers, the stars were grouped in six classes, called magnitudes, on the basis of their brightness. The first magnitude was assigned to the brightest stars, the second to the next brightest and the sixth and last to those just visible with the naked eye. (Of course, there were no telescopes in those days.) Our system is based on this classification, though with certain modifications. For one thing, with our powerful telescopes we have extended the scale of magnitudes; we can make out stars of magnitude 23. The scale is such that if one star is five magnitudes brighter than another, it is a hundred times as bright. Certain stars have been found to be brighter than first magnitude; these stars are assigned negative magnitudes. Sirius, for example, is of magnitude −1.52.

Both photos, Yerkes Observatory

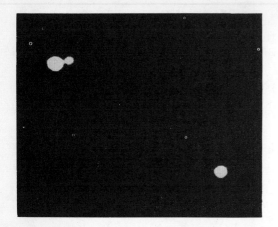

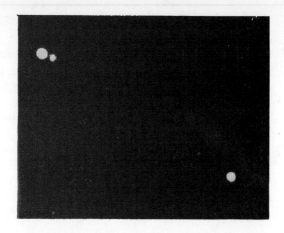

tude) is 8.7″; of Groombridge 1830 (sixth magnitude), 7.04″. It is interesting to note that the stars with the greatest proper motion are faint stars which, like Barnard's star, can be made out only with the telescope. The only bright stars with a proper motion of more than 2″ a year are Alpha Centauri (3.7″) and Arcturus (2.3″). These motions are so slow that it takes hundreds of years before the change of position of the star with relation to its neighbors can be made out with the naked eye.

The radial
motions of stars

Stars not only move across the heavens, as seen from the earth, but they also move toward us or away from us. They may move at an angle (other than a right angle) to our line of sight, or they may travel directly along the line of sight. In either case, motion toward or away from the earth is involved. It is called radial motion and it can be detected by means of the spectroscope. This device, described elsewhere in THE BOOK OF POPULAR SCIENCE (see Index), analyzes the colors of the spectrum. Those in the visible spectrum range from violet to red. If a star is approaching the earth, all the lines of its spectrum shift toward the violet end. If it is speeding away from the earth, the lines of the spectrum shift toward the red. Thus we can determine whether a star is traveling toward the earth or away from it; by analyzing the amount of the shift in the spectrum, we can determine the velocity of its motion.

Radial velocity is indicated in kilometers per second (km/sec). When the number of kilometers is preceded by a plus sign, it indicates that the star is moving away from us; the radial velocity is then said to be positive. A minus sign before the number of kilometers shows that a star is coming toward us; the radial velocity is said to be negative. For example, when we say that the radial velocity of Van Maanen's star is + 240 km/sec, we mean that it is receding from the earth at the rate of 240 kilometers per second. The radial velocity of Helsingfors 956 is — 325 km/sec; this means that the star is coming toward us at the rate of 325 kilometers per second.

The space
motions of stars

Most stars have both proper and radial motions: that is, they are moving across our line of vision and at the same time they are receding from the earth or coming nearer to it. Hence we must know a star's proper motion and its radial motion before we can determine its actual motion in space — its space motion, as it is sometimes called. But to find out the space motion of a star, it does not suffice to know its proper and radial motions; we must also know how far distant it is from us.

To show the importance of distance in calculating the actual motion of a body in space, let us consider the flight of a jet plane across our range of vision. If the plane is only a few hundred feet above the

The double star called Krueger 60 is made up of a pair of stars that describe distinct orbits around each other — orbits that can be made out with the telescope. The three photographs at the left show different stages in the orbital motion of Krueger 60.

Yerkes Observatory

surface of the earth, it will zoom past us in a matter of seconds. Suppose that the plane flies at the same speed at an altitude of five thousand feet. In that case, it will move across our range of vision much more slowly. The "proper motion" of the plane for a given period of time when it is near the surface of the earth is much greater than its "proper motion" at a considerable height.

The factor of distance from the earth is just as important in analyzing the velocity of stars and the distance they cover in space in a given time. If the proper motion of two stars, A and B, is the same and if A is much nearer to us than B, the distance actually traversed by B in a year, say, will be greater; so will its velocity.*

Fortunately, it is possible to determine the distances of stars with considerable accuracy. In the case of certain stars, we can determine distance from the earth by working out the parallax (see Index). In other cases, indirect methods are used.

Once we know the proper motion of a star, its radial motion and its distance, we can determine its space motion. It is expressed in kilometers per second. Among the stars with the most rapid space motion are BeB1366 (425 km/sec); Groombridge 1830 (348 km/sec); Kapteyn's star (287 km/sec); and Van Maanen's star (247 km/sec).

* To have large proper motion, a star must not only be comparatively close to the earth but it must also be moving rapidly. Some stars that are relatively near us have a small proper motion. Sirius, with an annual proper motion of only 1.38", is a good example.

The space motion of the sun

The sun, traveling through space with its retinue of planets, asteroids, comets, meteors and other members of its family, has a space motion of its own. The determination of this motion is difficult because we upon the earth share in it, and also because the stars that might serve as frames of reference are moving in varying directions and at varying speeds. However, in spite of the apparent confusion of these motions, astronomers have been able to determine the sun's space motion with considerable accuracy.

In 1783, the great German-English astronomer Sir William Herschel, taking into account the proper motions of thirteen reference stars, came to the conclusion that the sun was moving toward a point in the constellation Hercules. He called this point the "apex of the sun's way." Since he did not know the distances of these stars and their radial velocities, he was unable to give even an approximate idea of how rapidly the sun was moving through space.

In the course of time, the proper motions of more and more stars became known to astronomers and they acquired more information about stellar radial velocities and distances. Hence it became possible not only to map out the sun's course in the heavens but also to determine its velocity. Today it is generally agreed that the sun and its satellites are moving at the rate of about 12 miles a second in the general direction of a star in Hercules — Mu Herculis. In comparison with other stars, the space motion of the sun is quite slow. As we shall see, in addition to its space motion, the sun is carried, together with all the other stars in our galaxy, around the center of the galaxy.

Star groups with
common space motion

There are certain groups of stars whose members move through space at equal speeds and follow parallel courses. In 1915, Lewis Boss, an American astronomer, discovered what came to be known as the Taurus moving cluster. The stars making up the group (they are now known to be about a hundred in number) were seen to move at the same velocity and along paths that seemed to converge at a point somewhat east of Alpha Orionis. This apparent convergence is really an effect of perspective — an effect that we see in parallel railroad tracks that apparently merge in the distant horizon. Actually, the stars of the Taurus moving cluster follow a parallel course. Some of the stars now within the cluster are probably interlopers, which do not share in the motion of the cluster as a whole. In time, these stars will no longer form part of the group.

Another group of stars with common space motion is the so-called Ursa Major (Great Bear) group, which includes five of the seven stars that make up the Big Dipper. The group also includes Sirius, Alpha Coronae, Beta Aurigae, Delta Leonis and Beta Eridani, which are found scattered in the heavens. The Ursa Major group is moving toward a point southwest of Altair, the brightest star in the constellation Aquila. There are other moving groups of stars in the constellations Perseus, Orion, Scorpius and Centaurus. The stars that make up the Pleiades, in the constellation Taurus, constitute still another group of stars with common space motion.

The orbital
motions of stars

Certain stars, occurring in pairs, show a different type of motion, called orbital

motion: they describe distinct orbits around one another.* If they can be made out with the telescope, they are known as visual binaries. On page 229, we show three such visual binaries — Mu (μ) Draconis, 70 Ophiuchi and 61 Cygni. Other well-known ones are Krueger 60, Alpha Centauri and Castor.

Certain star pairs are too far away to be seen as separate stars even when viewed through powerful telescopes. In such cases, the orbital motions can sometimes be made out by the use of the spectroscope. Unless the orbit of such a pair is at right angles to the line of sight, the revolving stars alternately approach the earth and recede from it. Their orbital motion can be detected by analyzing the shift toward the violet or toward the red of the spectrum, as described previously. Star pairs whose orbital motion is revealed in this way are called spectroscopic binaries. Among them are Mizar (Zeta Ursa Majoris), Spica (Alpha Virginis) and Beta Aurigae.

In certain cases, we can detect orbital motion because two stars periodically eclipse one another. Such stars are called eclipsing binaries. They include Algol (the "Demon Star"), UX Ursae Majoris, Epsilon Aurigae, Zeta Aurigae, Y Cygni and GL Carinae.

The number of star pairs displaying orbital motions is very great. A great many stars which were once thought to be single stars are really binaries with orbital motion. Among them are Sirius and Barnard's star.

Orbital motion is not limited to star pairs; it is also found in the more or less spherical groups of stars called globular clusters. There are over a hundred of them in our own galaxy and they are undoubtedly to be found in other galaxies. Each globular cluster contains many thousands of stars, which are in particularly dense swarms toward the center. We can make out only the brightest stars in each cluster even with our most powerful telescopes.

---

* Not all star pairs in the skies display orbital motion. Two stars that appear close together when viewed through the telescope may simply happen to lie in the same general direction; they may have nothing else in common. Such stars are called optical pairs.

Astronomers agree that the stars that make up a globular cluster move in orbits about the cluster's center of gravity. Very little is known about the orbits of the individual stars of such a group. It is evident, however, that these orbits must lie in different planes, to account for the spherical shape of the cluster as a whole.

The rotation of stars about the galactic center

In addition to the different kinds of stellar motion that we have already mentioned, the stars in our galaxy are all moving about its center. This is called the galactic rotation. Of course, our own star, the sun, takes part in this type of movement; it travels at the rate of about 120 miles per second as it circles the center of the galaxy. Its circuit is completed in something like 200,000,000 years, a period of time sometimes known as the cosmic year. (The ordinary year, of course, is the time taken by the earth to make a single circuit of the sun.)

The speed of rotation of the stars around the center of the galaxy depends upon their distance from it; the closer they are to it, the more rapidly they revolve. Since the sun is located about two-thirds of the way from the center of the galaxy to its outer edge, its rotational movement is slow compared to that of the innermost stars. *See also Vol. 10, p. 267: "General Works."*

Visual binaries — μ (Mu) Draconis, 70 Ophiuchi and 61 Cygni — in two different stages of their orbital motion. Visual binaries are orbiting star pairs that can be made out with the telescope.

Yerkes Observatory

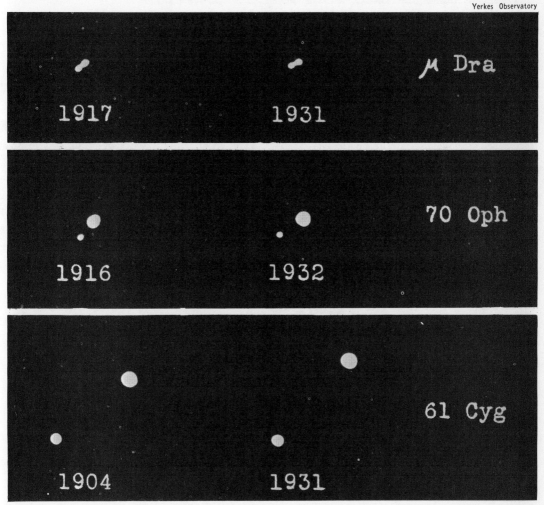

Canadian fur trapper
hanging a muskrat pelt
on a stretcher to dry.
The animals are usu-
ally skinned the same
day they are caught.

"Fleshing" a sealskin
(removing the flesh)
is a familiar task to
this Eskimo woman
from one of the small
islands in James Bay.

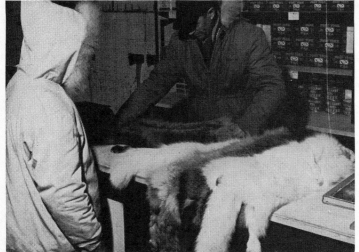

An Eskimo of the Bel-
cher Islands, in Hud-
son Bay, trades furs for
provisions at a store
operated by the Hud-
son's Bay Company.

Photos, Nat. Film Board

The Fashioning of Leather and Furs

From Animal Pelts to Valuable Industrial Materials

by KENNETH E. BELL

A FINE pair of shoes, a sturdy brief case, a gorgeous mink wrap, a richly ornamented saddle, an exquisite morocco bookbinding, heavy transmission belting — all these articles and numberless others are products of two of the oldest industries known to man. For the conversion of animal pelts into leather and into furs goes back to the early days of recorded history.

Animal skin, which yields such varied products, is a living tissue, with its own system of blood vessels, nerves and glands. In the outer layer, called the epidermis, we find great numbers of hairs growing in little depressions in the skin. The part of the skin beneath the hair roots is made up of a vast number of matted fiber bundles. This skin layer is known as the corium; it is the part that is converted into leather. Corium is made up largely of collagen, a gelatinlike protein; it also contains fats, carbohydrates, mineral matter and water.

In the pages that follow we shall describe in turn the processing of leather and of furs from animal pelts.

THE PROCESSING OF LEATHER

In the Boston Art Museum there is a beautiful coat of white antelope that was made in Egypt some five thousand years ago. Evidently the Egyptians were skilled workers in leather at that early period. The ancient Hebrews also made numerous articles of leather. The Greeks, Romans and Anglo-Saxons used leather for body armor, shields and bowstrings, as well as for gloves, belts, shoes (or sandals) and clothing. Parchment, prepared from sheepskin, was used for manuscripts long before the birth of Christ.

During the Middle Ages workers in leather formed powerful trade guilds. Apprenticeship was long, and stern measures were taken to ensure the quality of work turned out by guild members. Medieval craftsmen produced many useful and ornamental leather articles: shoes, buff jerkins, belts, gloves, scabbards, inkhorns, caskets and the like.

Methods of preparing leather changed little for centuries thereafter. Since the equipment required was relatively simple, tanning was essentially a home industry. Toward the end of the eighteenth century, however, the Industrial Revolution brought about a great change. Ingenious machines were invented and new techniques were developed; large factories gradually replaced small establishments.

Today leather ranks as a highly important industrial material. Yet, curiously enough, most leather is a by-product, derived from the skins of animals, such as cattle and sheep, that are killed chiefly for their meat value.

Like every other industry, the leather business has its own special vocabulary. The word "hide" refers to large, heavy skins. "Skin" generally applies to smaller and more delicate skins, like those of calves, sheep, goats and kids; however, the word sometimes refers to all animal pelts. When we speak of the "grain" of a skin we have in mind the outer or hair surface; the term "flesh" refers to the inner surface. Most leathers are finished on the grain surface, because this can stand the greatest amount of abrasion and wear. In cases where a softer texture is desired, the skins are finished on the flesh side; skins

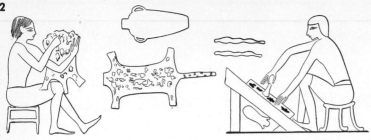

Ancient Egyptian tanners preparing hides.

prepared in this way are called suèdes. The name "suède" comes from the French word for Sweden; it is thought that the process of finishing skins on the flesh side originated in that country.

The leathers used today are divided into two groups — upper leather and sole leather. The term "upper leather" applies primarily to leathers used for shoe uppers; but it also includes lightweight leathers used for gloves, bookbinding and so on. Sole leather serves as sole and heel material for shoes and for other heavy-duty materials, such as industrial belting, harness, luggage and upholstery.

Upper leathers come chiefly from calfskins (the skins of calves), kips (the skins of somewhat older cattle) and the lighter weight hides of fully grown cattle. Sheepskins are also used as upper leather; so are the skins of goats, pigs, kids, kangaroos, antelopes, ostriches, alligators, lizards and snakes. Sole leather is made almost exclusively from heavyweight cattle hides; however, the hides of buffaloes, horses and other large animals are sometimes used.

Skins are sprinkled with salt or immersed in brine (water saturated with salt) within a short time after the animals are killed. This process preserves the skins by partially dehydrating them (freeing them from water) and by killing certain bacteria. The salting process, however, does not preserve the skins indefinitely; they spoil if they are not used within a few years. After salting, the skins are made up into bundles for the tanneries; or if they are destined for foreign markets they are packed in large casks.

The tanner begins by cleaning the skins thoroughly. They are put in soaking vats — wooden boxes containing cold, clean water — or in water-filled pits. The soaking process takes from one to seven days, during which the salt is dissolved, the dirt is loosened and the skin is restored to a soft condition. Next, the skins are usually rinsed to remove the loosened dirt.

Most skins or hides have flesh or fat adhering to the flesh side as a result of imperfect flaying. Such materials prevent the chemicals used in later operations from penetrating uniformly. Hence they are removed at this point by a fleshing machine, consisting of spiral knives set in a revolving cylinder. This machine leaves a comparatively smooth surface; any flesh still remaining is removed by a knife.

The next step is unhairing, or removing the hair. The skins are placed in vats containing a solution of slaked lime to which other chemicals, such as sodium sulfide, are added. The purpose of liming, or treating with lime, is to loosen the outer layer of skin in which the hair is embedded so that the hair may be easily removed. After from two to seven days, the skin is removed from the vat and is fed into an unhairing machine. This is quite similar in construction to the fleshing machine; only in this case the knife edges are dull, and it is the hair side of the skin that comes in contact with the knives. The squeezing action of the dull knives removes practically all the hair and its roots. The skins are then carefully inspected; any hair that still remains is pulled out by hand.

In order to remove the lime used in the unhairing process, upper leathers are placed in a bating bath — a solution containing an extract of pancreatic glands and an ammonia compound. The bating bath removes most of the lime; it also dissolves certain proteins, making the skin soft and pliable. After bating, the skins are placed in a solution containing common salt and sulfuric acid — a process called pickling.

Sheepskin is unhaired, limed, bated and pickled before it is sent to the tannery.

Sole leathers are generally placed in a lactic-acid bath after liming instead of in a bating bath. The lactic acid does not remove as much of the hide substance as the bating bath — sole leather, of course, must be as heavy and thick as possible.

For thousands of years all hides were converted into leather by the process called vegetable tanning. This method is still used in the preparation of sole leather. It is based on the fact that tannin, a bitter ingredient found in vegetation, will combine with the proteins of the skin, forming a chemical compound that resists decay. The tannin used in vegetable tanning is derived chiefly from the bark of the oak, hemlock and mimosa, from the wood of the quebracho tree, from acorns and from the abnormal growths, called galls, found on oak trees. Hemlock and oak bark

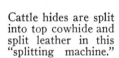

In this big lime vat the hides are carefully processed for the removal of the hair.

Cattle hides are split into top cowhide and split leather in this "splitting machine."

Photos, Les Cooper

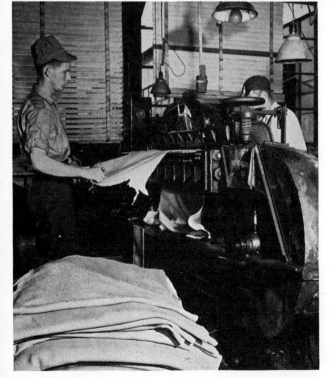

liquors and extracts are used more than any other tanning preparations.

Before tanning, the hides that are to be used for sole leather are cut into two parts, called sides. They are placed in pits containing weak solutions of tannins and are kept in slow, constant agitation by a rocker mechanism. This flexes the hides so that the tannins can penetrate more rapidly; it also keeps stirring the tannin solution. Next the hides are placed in pits containing high concentrations of tannins. Often, ground tannin extract is dusted directly on the surface of the hides. It takes up to two months for leather to be completely tanned by the vegetable tanning process.

Most upper leathers are prepared by a different method, called chrome tanning. The essential ingredient in this process is chromium — the metal used for bathroom faucets and automobile grill work. Before chromium can be used for tanning, it

Conner-Geddes

These hides are drying under infrared lamps after dyes have been applied.

must first be dissolved in acid to form the compound known as basic chromium sulfate. The skins are placed in revolving drums, or cylinders, which contain the tanning solution. The drums are usually raised several feet above the floor so that, when the tanning process is completed, the skins can be dumped into wooden boxes set under the drums. Leather can be tanned by the chromium compound in less than twenty-four hours.

Vegetable tanned leathers are brownish in color, while those that are chrome tanned are greenish blue. (Of course this is before the dyeing process.) You can sometimes see these colors when you examine the cut edges of finished leather articles.

For white leathers alum is used in place of the chromium compound. In other cases formaldehyde or rare metals, such as zirconium, are employed. Chamois is tanned by the process called chamoising. In this, oil is rubbed into hides or skins that are stretched out on a frame.

Many cattle hides are split by a machine into two parts after tanning: a grain split and a flesh split. The grain split is commonly used for shoe uppers; the flesh split serves for gloves, shoe tongues and medium-grade luggage.

The skins are now ready to be dyed. Some important tanners' dyes are made from the wood of certain trees; others are aniline colors, created by chemists from coal tar. Leather may be colored, with either natural or synthetic dyes, to any shade of the rainbow that happens to be fashionable. Lubricants like cod oil, neat's-foot oil or castor oil are generally added to the coloring solution. The coloring operation is carried out in drums similar to those used in chrome tanning.

When skins are removed from the drums, they are dripping wet. They are first put in machines that wring out most of the moisture and are then placed in driers. Most modern tanneries employ conveyer driers; the skins are carried on conveyers through drying chambers in which the temperature and humidity are carefully regulated. By the time they have dried out, the skins have become stiff; they are now softened in machines that flex the fibers.

The finishing processes come next. In the case of upper leather, solutions containing casein, synthetic resins or lacquers are applied to the grain surface. This is done either by means of spray guns or by seasoning machines, in which a rotating brush applies the finishing solution to the grain surface. Sole leather is finished by applying wax emulsion to the grain together with some filling material. A varnish of linseed oil is used to finish patent leather.

Leather is polished by lustering or burnishing machines in which cylinders of glass or metal rub or compress the surface. Ironing machines, like those found in laundries, are sometimes used for this purpose; in other cases machines operated by hydraulic pressure give the desired luster.

Any desired pattern can be imprinted on the surface of leather by means of engraved plates applied by hydraulic pressure. For instance, skins can be given an ostrich print so that they resemble genuine ostrich in appearance. Such skins are never sold as genuine ostrich skins; they are called ostrich printed skins.

Special effects can be obtained by applying gold leaf or by stenciling. Vegetable tanned leather can be tooled so as to produce the designs one sees in saddles, or wallets or book covers. Such work is usually done by craftsmen who make leather articles and not by tanners.

Finished leather is sorted by experts who grade it according to appearance, feel and freedom from defects. The individual skins are then passed through an ingenious machine that measures their area and records the measurement. Finally the skins are made up in bundles for shipment.

Today about 85 per cent of the leather processed by tanneries is used for shoes. Genuine leather shoes are attractive in appearance and durable; they are comfortable, too, for air permeates the fiber substance of the leather so that the foot can breathe. Leather has literally hundreds of other uses. It is a favorite material for

jackets, belts, gloves, luggage, bookbindings, harnesses, saddles and upholstery. Power is often transmitted from motors to machines by means of leather belting. Leather washers are often used in pumps where, unseen and unattended, they withstand hundreds of thousands of flexings and give years of service.

Leather is not the only product of the processing of skins and hides, for the material that is trimmed off is processed into valuable by-products. The flesh is converted into gelatin, used as a food and also in photography. The hair goes into the manufacture of felts, carpets and rug anchors; it serves as a binding material in plaster or wallboard and for insulating purposes. Large quantities of oils and greases are recovered from hides and skins and are converted into soaps, beauty creams and lubricants. Other types of leather scrap, because of their high nitrogen content, make valuable fertilizers.

Plastic materials are sometimes used today in such articles as shoe soles and heels, ladies' bags, belts, brief cases and upholstery. Yet the demand for genuine leather articles is as great as ever.

HOW ANIMAL PELTS ARE CONVERTED INTO FURS

It is probable that prehistoric men made garments out of the furs of animals long before the first written records appeared. The Chinese held furs in great esteem as early as 1,500 years before the birth of Christ; the Egyptians seized them as tribute; the Phoenicians traded in them.

Greek and Roman writers often referred to the use of furs. In ancient times the fur-trading center of the world was the great plain of Armenia, called the Plain of Taurus.

In Medieval Europe, furs were worn by all classes of society. The nobility and

Canadian mink ranch. Many striking new varieties of mink have been developed in ranches like this one.

Fur Trade Journal

the higher clergy wore expensive Russian sable, vair and gray squirrel; they often bordered their cloth garments with beaver, fox and other furs. Lesser folk satisfied themselves with such furs as rabbit, cat, sheep and lamb. In England the wearing of ermine was restricted to royalty by King Edward III (reigned 1327–77). This beautiful white fur also came to be part of the insignia of judges; as a result the name ermine became a symbol of the judge's office. In those days furs were obtained chiefly from northern and central Europe.

The modern age of the fur trade began with the discovery of America. The quest for fur-bearing animals was a powerful spur to the exploration of Canada by the French. The Hudson's Bay Company entered upon the scene in the last part of the seventeenth century; in time its trading posts extended from the east coast to the west and northward as far as the Arctic Ocean. The fur trade became fabulously lucrative; John Jacob Astor, for example, laid the foundations of the Astor family fortune by trading in furs at the turn of the nineteenth century.

Today furs are as popular as ever, at least among the feminine part of the population. As in the Middle Ages, costly furs, like precious gems, are often held to be the badge of wealth. But furs are not restricted to any one class of society; people of modest means can buy attractive and serviceable furs at prices they can afford.

The heaviest, thickest fur is generally found on animals living in regions that are very cold for at least part of the year; naturally such animals need more protection against cold than those living in warmer climates. An animal's coat is at its best in midwinter; the fur is not only heavier but it is also silkier, finer, more lustrous and of better color.

The most valuable furs today come from North America and Siberia. North America supplies beaver, mink, skunk, raccoon, various kinds of foxes and other furs; northern Europe and Siberia yield ermine, wolf, bear, badger, kolinsky, mink, marten and Russian sable. Furs also come from other regions of the world. Thus

Nat. Film Board

Inspecting and tagging choice beaver pelts flown in from Amos, in Quebec.

China provides kid, goat, lamb and marmot; Australia, rabbit, opossum, red fox and wallaby; South America, skunk, chinchilla and nutria.

In general, fur-bearing animals are taken by trappers in the wilds by means of snares or traps; this method is preferred to shooting, which would damage the pelts. The skin is removed from the carcass as soon as possible after the animal is killed. It is lightly tacked out on boards, with the fur side outward, and permitted to dry. All raw skins are subjected to this drying process with the exception of sealskin, which is salted. The pelts are collected by dealers or large companies and are then sold, generally at auction, in New York, St. Louis, London, Leipzig or other world fur markets. The prices paid depend on the rarity and beauty of the pelts as well as on the fashion demands of the season.

An increasing number of pelts now come from animals raised on farms or ranches; these animals include foxes, sables, minks, beavers, skunks, raccoons and chinchillas. The furs of animals raised on farms are free from the scars that are acquired in the wilds through fighting. Moreover, the scientific breeding of animals for their furs has resulted in the development of striking new varieties. Mink, for example, is now available in a great number of beautiful shades.

The pelts of true fur-bearing animals have a double coat of hair. There is a soft, silky undercoat of fur fibers, called the pelage; from this there protrude straight and rather coarse hairs known as overhairs or guard hairs. In nature, guard hairs serve to shed water and to protect the fur fibers, which provide most of the warmth. The guard hairs of certain fur-bearing animals, like the seal and the beaver, are plucked in the course of manufacturing; in other cases the guard hair is retained.

Nat. Film Board

Workers carefully cutting and matching furs in a large Winnipeg factory.

We can give only a very general account of the methods of dressing and dyeing furs, since furriers jealously guard their trade secrets. This much is certain: costly furs, like sables, chinchilla, ermine and mink, are still largely processed by hand. First, fat is scraped off the pelts by dull-edged knives; then the pelts are soaked. Next comes the fleshing operation; a skilled worker removes the flesh from the underside of the skin with a razor-sharp knife. The pelt is then put in a pickle solution, containing alum, or a bite, consisting of either hydrochloric acid, or sulfuric acid plus salt. The furs are now carefully lubricated with natural oils. They are often dyed black or brown or any other desired color.

After being dyed, the pelts are dried and stretched; certain types are plucked or sheared. The sheen, or luster, of furs is improved by tumbling them in drums together with hardwood chips or sawdust. They may also be run through combing machines, where wire bristles or combs straighten out the fibers.

Certain furs are pointed in order to improve their appearance — that is, overhairs are inserted in the fur by hand. For example, to produce the fur called pointed fox, the fox pelt is dyed black and the silver hairs of a badger are inserted. Each hair is generally attached directly to the surface of the skin by means of a tiny blob of cement.

This is only the barest outline of the dressing and dyeing processes; there may be as many as a hundred different operations in the case of certain furs. In dressing inexpensive furs, ingenious machines cut down the time required.

There are over a hundred different kinds of furs in the fur trade; we name and describe some of the more important ones in the table on the next page. These furs are made into coats, wraps, jackets, capes, scarves, stoles and muffs. They are also used as linings for cloth garments and gloves as well as for trimmings of women's millinery, cloth garments and slippers.

Synthetic furs have appeared on the market in recent years. In some cases plastic filaments of different colors are applied to a base of shorn sheepskin to produce cheap imitations of costly furs. In other cases "furs" are made of various woven materials, both synthetic and natural. They may imitate broadtail, chinchilla, ermine and other expensive furs; or else they may have exaggerated markings, and they may be dyed in all sorts of fantastic colors. These synthetic products are not quite so warm as real furs.

See also Vol. 10, p. 283: "Furs."

LIST OF IMPORTANT FURS

BADGER. Pale, fluffy fur, with long silky guard hairs. Best quality comes from western Canada. Guard hairs are often plucked out and used to point other furs.

BEAVER. Untreated peltry has ugly guard hairs, which are plucked; thick underfur is sheared. Fine beaver skin is bluish brown, with silvery belly. Canada is largest producer of beaver.

BROADTAIL. Pelt of a baby Persian or karakul lamb. No curl; watered pattern.

CHINCHILLA. Animal lives in Andes (Peru, Chile, Bolivia). Fur is delicate; lustrous slate blue in color. Chinchilla farming is carried on in North America.

CIVET. True civet lives in southeastern Asia. In fur trade, "civet" usually refers to coat of little spotted skunk of North America; animal is black with white markings.

CONY, see Rabbit.

ERMINE. A weasel, which changes from brown in summer to white in winter; in far north may stay white all year. Southwestern Siberia supplies most beautiful ermine.

FISHER. Dark, blue-brown peltry, with long guard hairs and full underfur. Eastern Canada produces best skins.

FITCH. Animal is a wild ferret; member of weasel family. Best yellow fitch comes from central Europe; best white fitch, from Siberia.

FOX. Red fox is one of the most important furs. Black, silver and cross fox are color variations of red. North America and Kamchatka Peninsula, in eastern Asia, produce finest red fox. White and blue fox are color variations of Arctic fox. Platinum silver fox was first discovered in Norway in 1935.

HARE. Used to imitate white fox, Alaska seal and other furs.

HUDSON SEAL. Muskrat, in which guard hairs are plucked, underfur is sheared and skin is dyed black to resemble Alaska seal. Most muskrat skins treated in this way come from eastern Ontario and around Great Lakes.

KARAKUL LAMB. Flatter and more open curl than Persian lamb. Best comes from Kirghiz region, western Asia.

KOLINSKY. Name given to China mink; see Mink.

KRIMMER. Trade name for lamb pelt; mixture of gray with some white and black; open curl. Best source is Circassian lambs of Caucasus.

LEOPARD. Somali leopard, from Somaliland, is choicest. Himalayas, mountains of China and jungles of Africa and India yield fine furs.

LYNX. Long and silky guard hairs and fur fibers; most valuable natural color is blue-gray. Hudson Bay region and Alaska produce best skins.

MARMOT. Finest skins come from Mongolia and Central China. Usually dyed to imitate mink; is then called mink-dyed marmot.

MARTEN. Beautiful full-haired fur with fine guard hairs and silky texture. Canadian marten is often called Hudson Bay sable; Baum marten, which comes from Europe, resembles Siberian sable.

Stone marten, also from Europe, is the most durable.

MINK. Best peltries from North America. Mink farming is important in the United States and Canada. Blue-brown is most desired color; silver-blue and snow white are being bred. China mink is yellow; Japanese mink is yellow-brown. Both are always dyed.

MOUTON. Trade name for skins from fine wool sheep. Usually dyed to imitate beaver or Alaska seal.

MUSKRAT. Huge quantities of this skin, supplied mostly by North America, are used in the trade. Skins may be used in natural brown color, but are often dyed or blended to resemble more expensive furs.

NUTRIA. Native to South America; introduced into North America and Europe. Long, ugly guard hairs are plucked. Best quality underfur is dark bluish brown.

OCELOT. Central and South America produce most of skins. Tawny, with black, brown and white oblong spots.

OPOSSUM. Skins obtained chiefly from North America and Australia. Generally is dyed brown, blue or black, and may be used to imitate skunk, a more expensive fur.

OTTER. Peltries obtained from all continents except Australia. Fur is short, lustrous and dark brown.

PERSIAN LAMB. Covering of animal is not woolly, but silky, with tight, lustrous curls. Usually dyed black; some natural grays and browns are also used. Comes from Persia and east of Caspian Sea; best quality produced in region around Bukhara.

PONY. Finest skins come from Poland and Russia; flat short-haired pelt is usually dyed.

RABBIT (Cony). Caught everywhere; peltries used in North American trade come mostly from Australia and New Zealand. Used chiefly to imitate other furs.

RACCOON. Caught from southern Canada to Gulf of Mexico. Best quality is silvery, with thick gray-brown underfur. Sometimes sheared to imitate beaver.

SABLE. A choice fur. Peltries come chiefly from Siberia; some from northern Canada. Finest peltries—dark bluish brown—are caught northeast of Lake Baikal.

SEAL. Obtained mostly from Pribilof Islands, off Alaskan coast. Fur is soft, short, fine and thick; is dyed black or brown.

SKUNK. Most desirable peltries are dark smoky black; poor qualities are dyed black. Minnesota is chief source of supply.

SQUIRREL. Europe and Asia are largest producers. Best skins are steely blue and full-furred. Inferior skins are dyed.

WEASEL. In fur trade means American animal closely allied to ermine. Fur is soft, lustrous, with thick brown underfur. Inferior skins may be bleached or dyed.

PLANTS AND
THEIR PARTNERS

The Pollination of Flowers by Insects

PLANTS and animals alike have a good many special adaptations for damaging or destroying other plants and animals in the fierce struggle for existence. Carnivorous mammals use claws and fangs to seize and kill their prey; carnivorous plants, such as sundews and pitcher plants, possess leaves modified for trapping insects. The vegetation-eating mammals have teeth that crop the leaves from a plant or strip trees and shrubs of their bark. Leaves arrange themselves so as to receive the maximum amount of sunlight; in so doing they may cut off the light from other plants and stunt or kill them.

A great many plants and animals, on the other hand, are specially adapted to serve one another. The reproduction of many plants, for example, is made possible through the help of insects. The fungus known as the stinkhorn thrusts forth a stalk whose head is covered by a slimy substance smelling like decaying flesh. Contained in this slime are the spores, or reproductive cells, of the fungus. Carrion beetles, attracted by the foul odor, crawl over the spore-bearing slime and eventually help to spread the spores from which new fungi will grow.

There is an even more intimate relationship between plants and insects: that in which certain insects, such as beetles, flies, bees, moths and butterflies, cross-pollinate flowers. In cross-pollination, the pollen from the male organ of a flower is transported to the female parts of flowers of entirely different plants. The insects that perform this essential task benefit by it; they obtain food in the form of nectar and pollen.

Botanists have found that in the case of many plant species, cross-pollination is more effective than the process of self-pollination, in which pollen from the male organ is transferred to the female organ of the same flower. Plants that are grown from cross-pollinated seed tend to be stronger, grow to greater size and produce more and better seeds than plants grown from seed of self-pollinated flowers.

Since plants cannot transport their own pollen, they have developed special adaptations to ensure such transportation, including the very effective one of attracting insects. For their part, insects must be specially adapted for penetrating the internal parts of certain flowers. Sometimes insects and plants have become highly specialized in maintaining their partnership. Yucca flowers, for example, are pollinated only by a certain species of moth; pollination is effected in the flowers of fig trees by small wasps. If the particular insect that the plant requires for pollination becomes scarce, the plant will be hard put to maintain its species. If its plant partner dies out, the insect that pollinated it may also soon disappear.

Early seed plants ensured cross-pollination by producing great quantities of pollen. The winds, however capricious they might be, would be sure to take up at least a certain amount of pollen, carry it and spread it over the flowers of other plants of the same species. Many plants, such as pine trees and corn, still depend upon this method of pollen distribution. The flowers of such plants, unlike those pollinated by insects, are inconspicuous in size and color and have no odor.

Wind pollination is a very wasteful process. Formerly, when the pines in the forests around the Great Lakes were discharging their pollen, the water would be yellow with it for miles from the shore — all wasted because the wind happened to be blowing in the wrong direction.

The "essential"
parts of a flower

The "essential" parts of a flower are the pistil, in which the seeds are matured, and the stamen, in which is developed the pollen that makes the seeds grow.

The pistil contains a seed box, or ovary, in which the seeds grow, and the stigma — a spongy and sticky surface which catches the pollen. The pollen grains pass down pollen tubes to reach the ovules and enable them to grow into seeds.

The stamen always has an anther, or pocket, in which the pollen is developed. Sometimes the anther has a stem, called the filament, which holds the anther up. Anthers open in different ways to let the pollen out — like books, boxes, bags or even like pockets.

The petals, which are usually colored to attract the living pollen carriers, are varied in form. The sepals, which protect the flower while it is in bud, also have various forms.

Insects are generously rewarded for carrying the pollen. They have all that they need for food; the bees have plenty to carry home and feed to the young. Some flowers give sweet nectar for the insects to drink; the bees make this into honey. Nectar is the chief food of butterflies and of many species of moths.

Flower colors and
odors attract insects

Flowers attract the attention of insects both by color and by odor. Because insects have a very keen sense of smell, they are probably more attracted by odor than by color. Their organs of smell are in their antennae and they can detect odors at great distances. Even the extremely disagreeable odors of some plants attract certain kinds of insects.

Insects and flowers are wonderfully adapted to help each other. In some cases the scent of the flower becomes strong at the exact time when the flight of certain insects begins. Some of the honeysuckles and petunias, for instance, which have a very faint smell, or none at all, during the day, are powerfully scented in the hours of the evening when the particular insects that visit them are on the wing. Other flowers, acutely scented during the daytime, when butterflies and bees are on the wing, have no odor after sunset.

Convenient flower
"doorsteps" for insects

Some flowers, such as the sweet pea, pansy and mint, offer a convenient alighting place to the short-legged insects such as the bees. Usually one or more petals have a specialized form to make a "doorstep" on which the insect may rest while probing for nectar. In some flowers there are guide lines of different colors which converge and point like so many fingers to the opening that leads to the nectar well. Often the margins around the opening to this well are of contrasting color and show where the sweetness lies. The nectary is often adapted in length and size to the tongue of the insect which carries the pollen. This is true of the long, deep nectar tube of the Nicotiana, perfectly adapted for the long-tongued sphinx-moth which carries its pollen. Some flowers save the time of bees by changing color as soon as they are pollinated, like asters and the lupines. The flowers of the white clover not only change color but bend downward as soon as they are pollinated.

Devices that
prevent self-pollination

The most common of the devices to prevent self-pollination are as follows: (1) the stigmas ripen before the pollen of the flower is ready to be shed; or the pollen is shed and gone before the stigmas ripen, as in the case of the common garden pink and the hollyhock. (2) The flowers that produce the pollen grow on one plant and those producing the seeds grow upon another, as is the case with the pumpkins, the cucumbers and the willows; or they grow on different parts of the same plant, as is true of the maples.

(3) Two forms of flowers are developed; the stigmas of one form are especially placed so that one receives the pollen of the other. Primroses and bluets use this method. If we look carefully at bluets, for example, we find two forms of flowers: (a) those where the throat of the flower tube seems to be closed by four anthers, which join like four fingertips pressed together and (b) those with a two-lobed stigma protruding from the opening of the flower tube. In the first type (a), the stigma may be found about halfway down the tube. Just the reverse is found in the second type (b). In this, the anthers are attached halfway down the side of the tube and the stigma is at the top.

An insect visiting the second type (b) would get its tongue dusted with pollen

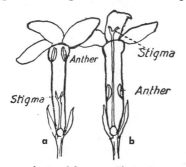

Bluets have two forms of flowers. Their stigmas and anthers are so arranged as to ensure effective pollination.

from the anthers at the middle of the flower tube. This pollen would then be applied to the stigma of the first type (a) as the insect entered the next flower. On the other hand, a bee visiting a bluet of the (a)-form would receive pollen at the base of its tongue, where it is conveniently placed to be brushed off by the protruding stigmas of the (b)-form.

In looking for the two types of flowers among garden primroses, we may find that most of the flowers are of only one type. It should be understood that this is caused by the fact that these flowers were artificially multiplied. By perseverance, it is usually possible to find at least a few of the other type even in an artificially cultivated garden.

Flowers employ a number of devices to make the pollen carriers work effectively

for them. The most important device is the positioning of the nectary (the nectar-producing gland) so that a visiting insect or hummingbird must, in order to reach the nectar, brush against the anthers or stigmas or both during the process. The nectar-producing glands of the flower are usually located near the bases of the petals. The nectary may be in a modified petal, as in the violet or columbine; or it may be in a sepal as in the nasturtium. In every case, the path to the nectar-producing gland is barred by the anthers, the stigmas or both. In this way, the flower makes certain that a visiting insect that takes nectar will aid in the pollination process.

The nectary may not be accessible to all insects

The positioning of the nectary, its depth and its shape may play an important part in determining who the pollen carrier will be; only certain insects or birds will be able to enter the nectary. Red clover, for example, has adapted its nectaries so that the tongue of the bumblebee can just reach them; the nasturtium allows access particularly to the probing beak of the hummingbird. In the case of orchids, many species are so specialized that they are suited to only a single pollen carrier; in this case, a partnership exists between just one plant species and one insect. Similarly, the yucca plant and its partner the yucca moth are completely dependent upon each other.

Devices that secure cross-pollination

Plants have a wide variety of mechanisms to secure cross-pollination. In the nasturtium, five sepals are united at their base. One of them is extended into a long spur, which forms a tube leading to the nectary. The five petals are arranged around the mouth of the tube. Two of the petals stand up like a pair of fans with converging lines pointing toward the mouth of the tube. Similar road signs are found on the sepals. The lower petals stand out, making a landing platform, or doorstep, upon which the pollinator alights.

But it requires a large insect or a hummingbird to do the work of pollinating this flower. If a bee, fly or other comparatively small insect should alight on the petal doorstep by chance and try to steal into the cave, it is prevented entrance by the structure of the flower's lower petals. Each of these lower petals narrows to a mere insect foot bridge at its inner end; and this foot bridge is rendered quite impassable by being studded with irregular little spikes and projecting fringes, sufficient to discourage any small insect from crawling that way.

But why all these guiding lines and guarded bridges? If one watches the same blossom for several successive days, it will reveal this secret. When a

Flower of the spring beauty, clearly showing the five stamens and the style. You will note that the latter is split in three parts at the tip.

flower first opens, the stamens are all bent downward, but when an anther is ready to open its pollen doors, the filament lifts it up and places it like a sentinel blocking the doorway to the nectar treasure. Then when the pollinating partner comes, whether it be a butterfly or a hummingbird, it gets a round

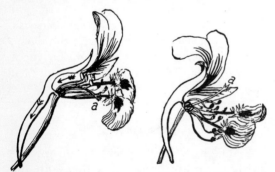

(1) A nasturtium flower in the early stage of blossoming. The closed stigma is shown deflected at a in the diagram. (2) A later stage; in this one, the stigma (a) is elevated.

of pollen ammunition for its daring. Perhaps there may be two or three anthers standing guard at the same time, but, as soon as their pollen is exhausted, they shrivel and give room for fresh anthers. Meanwhile, the stigma has its three lobes closed and lying idly behind and below the anthers; after all the pollen is shed, the style rises and takes its position at the cave entrance and opens up its stigmas, like a three-tined fork. These rake the pollen — from another nasturtium flower — from any visiting insect or hummingbird, thus robbing it of the precious gold dust that will fertilize the seeds contained in the flower's three-lobed ovary and that in this particular manner will assure the continuation of the species.

The bee larkspur has a similar story except that the beautiful color of its blossoms is entirely in the sepals, and the petals form nectaries and give an alighting place. By their contrasting color they also show the bees where the nectar wells lie hidden.

In barberry and laurel (*Kalmia*), the anthers are held under tension in pockets formed in the petals. The filaments of the stamens are elastic. When an insect alights or bounces on the petal, the petal bends and releases the anther from its position in the pocket. The springing anther dusts the insect with pollen. Stamens and pistils mature, or ripen, at different times to prevent self-pollination.

The iris blossom has a strange appearance, and this is because in it nothing is as it seems. The style of the pistil is divided into three broad branches, which look like petals; together with the sepals, they form a tunnel for bees. Apparently the petals serve only to deceive butterflies and other insects that seek nectar in the center of the flower. If we look directly down into the flower of the iris, we see ridges on the broad styles and purple veins on the petals. They all point plainly to the center of the flower.

"Road signs" that lead insects astray

An insect would naturally seek the nectaries to which all the lines seem to lead. Unfortunately for these insects, the nectar-producing glands are not in the center of the flower but in the sepals. In his admirable study of the blue flag iris and its insect visitors, J. G. Needham observed that the little butterflies called "skippers," the flag weevils and the flower beetles all fell victim to this deceptive appearance. Thus they provided evidence that the nectar guide lines on flowers are noted and followed by various insects. The nectary may be bordered with contrasting colors.

What insects does the iris deceive?

The iris deceives insects such as butterflies and beetles because they do not help to pollinate it; they are interlopers at best. It is the bees that are the pollinating agents. They are never deceived into seeking the nectar in the wrong place for they know exactly where it is. A bee will alight on the lip of the sepal and go directly into the tunnel leading to the nectary. The tunnel's floor is formed by the sepal; its ceiling, by the stigma and style. At the very entrance, we find the stigma's receiving surface; it curves up so that it faces outward. Within the tunnel, attached to the roof, are the anthers with their load of pollen. When the bee enters the tunnel, its back brushes against the receiving surface of the stigma. If the bee had visited another iris before coming to this one, it probably picked up some pollen on its back. This pollen is deposited on the stigma as the bee enters the tunnel.

The bee proceeds down to the nectary, where it feeds on the nectar, and then makes its way upward again. In doing so, its back brushes against the pollen-laden anthers on the tunnel roof. In this way, the bee receives a new supply of pollen to carry to the next flower. As it leaves the tunnel, the bee again makes contact with the stigma, but because the stigma's receiving surface faces outward, no new pollen is deposited on it. This ingenious arrangement of the iris ensures that it will be fertilized by the pollen of a different flower and that only the pollen carrier, the bee, will get the nectar.

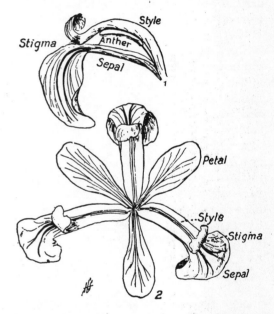

Blossoms of the common iris flower. (1) Side view of passage to anther. (2) Looking directly into the flower.

The violet has five sepals and five petals. Of these petals, two are located above, one on either side and one below; the lower petal offers a doorstep to visiting insects. This petal forms a long spur which holds the nectar. In order for the insect to reach it, it must thrust its tongue through a little door guarded by both anthers and stigma; in this way the pollen is spread and stigmas pollinated. Many species have fringes on their side petals, forming an arch over the door of the nectary. While this structure is generally thought to be a guard to keep out undesirable insects, such as ants, it is possible that it is also useful in brushing the pollen from the tongues of the insect visitors.

Two unusual and closely related plants

The Jack-in-the-pulpit and the calla lily are members of the same family. They are characterized by a long, fleshy spike that points straight up and that is called a spadix. Surrounding the spadix is a circular cuplike leaf known as a spathe. The spathe of the Jack-in-the-pulpit is striped green and purple; in the calla lily, it is all white. Many people mistakenly take the spathe of the calla lily for a petal, although it is really a modified leaf.

The flowers of these plants are located on the lower portion of the spadix, hidden within the spathe. There are two kinds of flowers — staminate (stamen-bearing) and pistillate (pistil-bearing). Generally the same plant does not have both pistillate and staminate flowers. Where both types are present in a plant, the staminate flowers are below the pistillate ones. As a result of this arrangement, insects migrating up the spike will carry pollen from the stamens to the pistils. In most instances, the insects must travel from one Jack-in-the-pulpit or calla lily to another in order to bring about pollination.

The petunia is a very popular garden flower in the United States and Canada, particularly because it stays in bloom from early June to September. When the flower first opens, there is a green stigma and five anthers in the throat, or bottom, of the flower cup. The filaments are attached to the bottom of the cup and also to the side of the corolla about halfway up their length. From this second point of attachment, they bend inward toward the center of the flower and they cluster about the stigma. The filaments of the stamens are of different lengths. If we look into a newly opened flower, we see a pair of anthers directly above the stigma, a second pair slightly below it and a single anther below them. In this way, the petunia is able to have a pollen-shedding anther at several different times during the season.

How cross-pollination is brought about in the petunia

The clustering of the anthers in the center of the flower is the petunia's device for bringing about cross-pollination. The nectary is located directly below the stigma in the cup of the flower. An insect seeking the nectar must pass its tongue down the center of the flower and through the clustered anthers. In the process, its tongue is dusted with pollen for another flower. The stigma, surrounded by its anther "guards," looks somewhat like a closed fist;

(1) Jack-in-the-pulpit. (2) Spadix with pistillate flowers. (P) Pistillate flower enlarged. (3) Spadix with staminate flowers. (An) Staminate flower enlarged, showing four anthers. Right: cross section of petunia blossom.

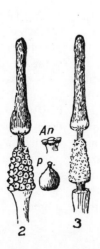

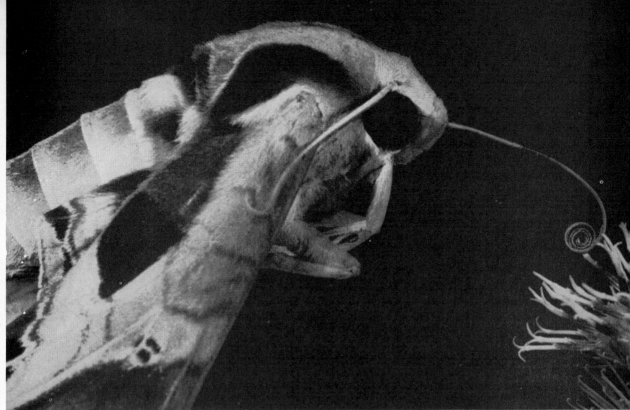

Lynwood M. Chace, from Nat. Audubon Soc.

The coillike tongue of the tomato sphinx moth is used to extract nectar from flowers. This moth pollinizes long, tubular flowers, such as the jimson weed. It ranges over much of North and South America.

its structure is such that it cannot be pollinated by anthers that hang above it. As the flower grows older, the anthers lose their pollen and are soon empty. The stigma then opens into two lobes that are ready to accept pollen that is brought in from other flowers by visiting insects. In this way, the petunia insures its cross-pollination.

The moccasin flower, one of the most

Andrena bee pollinating the oddly shaped flower of the mocassin plant. For details, see the text on this page.

attractive members of the orchid family, has a particularly interesting method for insuring that visiting bees will participate in the cross-pollination of the flowers. The two sepals and two side petals are formed into streamers, which attract attention to the flower and direct bees to the lower petal. This petal is in the shape of a puffed-out sac that opens above, with its edges incurved. The coloring of the edges of the sac is particularly attractive and entices bee visitors. Once a bee enters the sac, it is kept a prisoner by the incurving edges. At the bottom are delectable vegetable hairs for the bee to eat. Soon the bee will search for a means of escape. It will observe two small openings near the stem of the plant, quite far from the original opening through which it entered. Extending into each of the openings is an anther containing a plasterlike pollen preparation. As the bee crawls through the opening, the anther smears this preparation on its back so that the pollen will be carried to the next flower. There, the bee will most probably scrape it off upon the stigma

245

against which she has to push in order to get out. She may then move from this flower to another where she will no doubt brush off some of this pollen. At the same time, she will acquire, by the same method, new pollen which she will leave on another flower and so on ad infinitum.

The number and variety of devices employed by flowers to accomplish cross-pollination is limited only by the number of observations that one can make. Each flower has its own method. For example, the bleeding heart, the sweet pea, the columbine, the sage, the hollyhock and the laurel all have very interesting devices for securing cross-pollination. The composites have an equally amazing array.

How flowers conserve their pollen for desirable carriers

Flowers show an extremely high efficiency in distributing their pollen to carriers that best serve their purposes and conserving it when these carriers are not available. Most flowers will close during storms, dark days and nights because the insect pollen-carriers are not flying around at these times. The hepatica, for example, closes its petals and hangs its head during a snow storm. The dandelion does not open until late in the morning and closes early in the afternoon. The tobacco flower will open only in the evening when the sphinx moth, its special partner, is in its period of maximum activity. The petunia, which also depends on the sphinx moth, opens at about the same time as the tobacco flower. The evening primrose flares its petals wide in the later afternoon when its special partner is most active. Examples such as these could be multiplied almost endlessly.

Flowers also discourage undesirable guests from partaking of their pollen and nectar. The devices they use usually involve some method of barring the entrance to the nectary to intruders that do not regularly visit the flower. The obstructions used are quite effective against small insects, but are rarely strong enough to resist the larger ones, which can force their way through obstacles set up by the flower to stop them.

The devices we have just described are adaptive ones developed by the flower in its fight for survival. This development has also brought to the flower a much higher degree of specialization. The other half of the partnership, the insect carrier, has been profoundly affected by the development of the flower and has developed a greater degree of specialization itself because of it. The best case is that of the bee, the most important pollen-carrier of all. It carries more pollen than all other agencies combined. The bees

Below, we see a worker bee gathering nectar. Note the pollen basket that is attached to the hind leg.

N. Y. State Coll. of Agriculture

Lynwood Chace, from Nat. Audubon Soc.

The bumblebee helps ensure an abundant fruit crop by pollinating apple blossoms as it forages in the orchards for nectar and pollen. This bee is an efficient pollinator.

ing nectaries especially fitted for their sucking proboscis but also visit those flowers not meant for them — such as nasturtium, scarlet sage and columbine, which are adapted for hummingbirds or butterflies. Bees carry pollen for the flowers that do not invite them and cross-pollinate many of the flowers that produce no nectar, such as the hepaticas, the poppies and many others.

To gather this pollen, the honeybee has a hairy body like a brush to which the pollen clings when the bee brushes against it. Long hairs on the front pair of legs remove this pollen from the mouth parts and the head. With the second pair of legs, the bee scrapes its thorax and takes the pollen from the front pair. Then the stiff hairs of the third, or hind legs, comb the abdomen, collect the mass of pollen from the second pair and put the accumulation into pollen baskets attached to the hind legs. When the bee arrives at the hive it uses the long spur at the tip of each front leg to push the pollen out of these baskets and into the cells of the honeycomb.

It appears that the evolution of flowers and of insects must have progressed hand in hand, for the flowers needed the insects for pollination and the insects, bees in particular, needed the pollen and nectar for their survival. As a result, flowers and insects developed special features. The flowers have bright colors, sweet scents and delicious nectar to entice the bee, which has a long tongue for sipping the nectar and a special honey stomach separate from its own food-digesting stomach. The bee carries the nectar in this special honey stomach where the sweet fluid is ultimately transformed into honey.

The moths and butterflies also have been much modified by their partnership with flowers. In their case, one pair of jaws has been greatly elongated, each jaw being grooved on the inner side, and the two joined together lengthwise to form a long tube through which nectar is sucked. This long tongue is carried coiled in a spiral when not in use.

use pollen for rearing their young through larval and early adult stages. In maturity they feed on honey or the nectar from which honey is made. In seeking pollen and nectar from flowers, the body of the bee becomes literally coated with pollen grains which are transferred later to other blossoms. Thus the insect performs the important service of distributing pollen. As a pollinator, the bee is very efficient. This is partly due to its habit of visiting only one plant species and none other at certain seasons or times of day. The pollen of one species usually cannot fertilize the immature seeds of another species; for example, larkspur pollen can never fertilize a poppy. The bees not only pollinate those flowers possess-

The butterflies frequent the deep-throated flowers that open by day, and the moths visit those that open in the evening. The sphinx moths are the most efficient pollen-carriers of the moth family. In both butterflies and moths, hairs and the fringes and scales around the base of the tongue brush off pollen from the ripe anthers, or male parts of the flowers, and carry it to the stigmas, or female parts.

Another common pollen-carrier is the diminutive hummingbird. This tiny creature gains its food more in the manner of an insect than a bird, drawing it from the interiors of flowers. It is the only bird known that can fly backwards and sideways as well as forward. It can also move straight up and down or just hover like a helicopter. It accomplishes all this by fluttering its wings so rapidly that the human eye cannot follow them. They appear as a dim haze or halo encircling the tiny body. The rapid motion of the wings produces the characteristic hum from which this bird gets its name. The hummingbird darts rapidly from flower to flower. It hovers for only a brief moment at each one and dips its long slender beak into it, extracting the nectar, as well as the insects that were trapped in the flower while seeking the sweet liquid.

As a matter of fact, the hummingbird is particularly interested in the insects in the flower rather than the nectar. They form the main part of its diet; the nectar makes up a sort of "dessert." The hummingbird's long slender beak encloses a tongue that is well adapted for the bird's method of obtaining its food. The end of the tongue is split into two parts which can be turned upward and inward to form parallel tubes or "straws" through which the bird is able to draw up insects and nectar. The end of each tube is equipped with a minute brush that sweeps insects into the opening of the tube.

Like pollen-carrying insects, the hummingbird helps in cross-pollinization. As it darts about, dipping into one flower and then another, it is unwittingly carrying pollen on its beak.

Thus nature has set up two opposite systems of survival. The law of the jungle says: "Kill or be killed"; plants and their partners co-operate to survive.

See also Vol. 10, p. 273: "Insects and Plants."

The black-chinned hummingbird, shown below, is found in most of western North America. It is here about to dip its beak into the flower in search of insects trapped while seeking nectar.

A. M. Bailey & R. Niedrach, from the Nat. Audubon Soc.

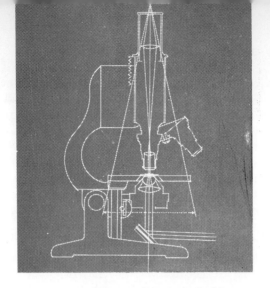

AN INTRODUCTION TO OPTICS

Some Striking Effects of Light and How They Serve Man

BY J. H. RUSH

MEN of science have realized for years that light is a form of radiant energy. Light, X rays, radio waves and gamma rays are all electromagnetic radiations. They are so called because they are sent out into space as a result of variations in electric and magnetic fields.

Not so many years ago, scientists were quite sure that light and other radiant energy consisted of electromagnetic waves rippling through space. They distinguished sharply between the wave character of radiation and the particle character of matter. Today the picture is not so clear. In some respects, at least, light and other radiations behave like streams of particles. What is even stranger, perhaps, electrons and other very small particles sometimes behave like waves.

The peculiar dual nature of radiation and of material particles is the subject matter of wave mechanics. It would take a great deal of space to explain this theory, even nonmathematically. It is perhaps enough to say here that neither radiation nor material particles ever exhibit both wave and particle characteristics at the same time. In transit — that is, as it moves from place to place — light behaves like a system of waves. It has a fixed speed in empty space; its wave lengths can be measured by a variety of methods. In many ways its behavior is entirely consistent with the theory that it is an electromagnetic wave disturbance. But in its emission from electrons in an atom and in its absorption or other reaction with atoms in its path, light behaves more like a stream of very small particles of energy — photons.

The old argument over the nature of light seems to have ended in a draw. Instead of saying that light (or any other kind of radiation) consists of waves or particles, physicists recognize that it behaves like waves in certain respects and like particles in others.

The speed of light plays an important part in astronomical and physical research. It is extremely important, therefore, to measure it as accurately as possible. For nearly three centuries scientists have gradually refined and improved their techniques, until now the speed of light is known to within about one part in a hundred thousand. It is very nearly 299,792 kilometers, or 186,281 miles, per second.

Strictly speaking, light means any radiation whose wave lengths excite a sensation of brightness, or illumination, in the retina of the eye. These wave lengths range from about $^{16}/_{1,000,000}$ of an inch to $^{30}/_{1,000,000}$ of an inch. Certain types of radiation — ultraviolet light and infrared light — are similar to visible light, though their wave lengths are somewhat shorter or longer than those of the visible range. In the following pages we will use the word "light" in the sense of all wave lengths that behave similarly to those that excite vision.

What happens when
light encounters matter

When light encounters some form or other of matter, it behaves in different ways. For example, when it strikes an opaque object (one that does not transmit light), the object casts a shadow. Much of the early argument as to whether light is made up of waves or particles hinged on observations of shadows.

If light consists of a shower of flying particles, those striking an opaque barrier in their path would be stopped. Other particles would miss the edge of the barrier, as shown in Figure 1a. If the latter struck a screen set beyond the obstacle, a geometrically sharp shadow would appear on the screen. That is apparently just what happens. In this respect light seems to

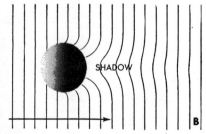

1. Shadow formation. In A we assume that light consists of particles; in B, that it has a wavelike nature.

act unlike waves of water or sound, which bend around obstacles. That is why Sir Isaac Newton and other thinkers came to the conclusion that light does not consist of waves, but of tiny flying particles.

Careful observation, however, shows that light *does* bend around obstacles (Figure 1b). Short waves of any kind bend less than long ones, and light waves are very short indeed. Yet under special conditions this bending of light at edges — a phenomenon called diffraction — produces striking effects. A certain amount of light makes its way into the shadow area. Ordinarily it is not noticeable. However, in the case of a small ball or disk, all the light diffracted around the edge crosses at the center of the shadow, and a bright spot appears. Immediately outside the shadow cast by the sharp edge, the light is broken up into a series of bands, or fringes, that alternate in intensity.

When light falls on a material barrier, some of it is reflected; it bounces back from the surface, like a ball from a floor. If the material is not quite opaque, some of the light is transmitted through it at reduced speed. In its course it is refracted, or bent. Usually a large portion of the light is absorbed in the material. It disappears as light, and its energy is changed into other forms.

The most common effect of absorption is the change of light energy into heat. The rapidly oscillating electric and magnetic fields involved in the light wave set the electrons in the material to vibrating. As a result, all the molecules of the material vibrate more energetically and the material becomes hotter. To understand why this is so, one must bear in mind that heat is simply the kinetic energy, or energy of motion, of the molecules of a substance. The more vigorously its molecules vibrate, the hotter the substance is.

Light energy is not always transformed directly into heat. If it strikes a clean metal surface in a vacuum, the light energizes some electrons so violently that they jump out of the metal and fly through the adjacent space. This so-called photoelectric effect is used in television cameras and many other electric-eye devices. (See Index, under Photoelectric cells.) Light also causes chemical changes in many substances. It can bleach various dyes, or cause intense irritation of the skin or pro-

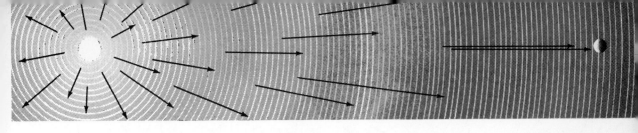

duce subtle changes in the silver chloride of a photographic film. From the human standpoint, the most important photochemical effect of light takes place in a substance in the retina of the eye, called visual purple. This pigment is partially bleached by light, and the chemical change that takes place is an essential part of the visual process.

Light affects matter chiefly by acting upon the outer electrons in atoms. When light energy sets the electrons dancing, each electron acts like a tiny radio antenna, radiating back part of the energy it has received. Some of this reradiated light goes back in the direction from which the light originally came; it is reflected energy.

In describing what happens when light strikes a surface, we shall have occasion to speak of rays and beams of light; rays will be represented by arrows in our diagrams. We must think of a ray as representing the direction in which a wave front is advancing. A bundle of rays is called a beam.

When light is transmitted through space, the waves spread out at equal speed in all directions from the source, as shown in Figure 2. By the time a wave of light from a point on the sun reaches the earth, some 93,000,000 miles away, the part of the wave that strikes a surface such as a mirror or a pavement is only very slightly curved. To all intents and purposes the

wave front is a straight line and the rays of light are parallel to each other (Figure 3).

The light rays that fall upon a surface are called incident rays. (*Incidere* in Latin means "to fall upon.") If the rays strike a surface perpendicularly, they are reflected straight back. Figure 4 shows what happens when they fall obliquely upon the surface. AB is the incident ray, which strikes the surface at B. BC represents the ray after it has bounced back from the surface, as a reflected ray. Let us draw a line, BD, at right angles to the surface at the point B. This perpendicular is called the normal. The angle ABD (or I), formed by the incident ray and the normal, is the angle of incidence. The reflected ray and the normal form another angle, CBD (or R), which is known as the angle of reflection. The angle of incidence is always equal to the angle of reflection; the incident ray, reflected ray and normal, all lie in the same plane. This statement is the law of reflection.

If the entire reflecting surface is very smooth, regular reflection takes place — that is, light is reflected in a regular, orderly way. The surface need not be flat; but if it is curved, the curve must be smooth and gradual. If such a smooth surface, flat or curved, is capable of reflecting most of the light that falls on it, it is called a mirror (Figure 5).

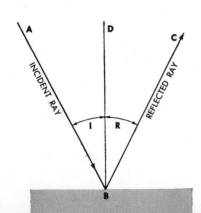

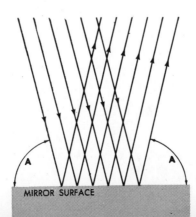

In Fig. 4, extreme left, we see light striking an ordinary surface and reflected from it. Angle I is the angle of incidence; angle R, the angle of reflection. The other drawing (Fig. 5) shows light being reflected from a mirror surface.

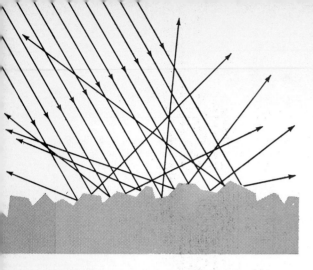

6. When light strikes a rough surface, made up of many tiny smooth surfaces, it is reflected in many directions.

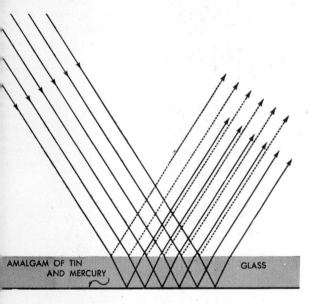

AMALGAM OF TIN AND MERCURY   GLASS

7. Light that strikes an ordinary mirror of plate glass is reflected from the coating on the back of the glass.

8. Light reflected from a precision mirror is reflected from a metal film on the front surface of the mirror.

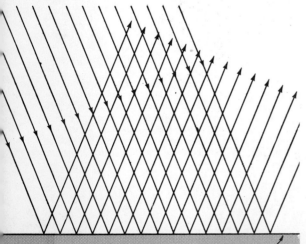

GLASS   SILVER OR ALUMINUM FILM

Mirrors are scarce in nature. Most objects would absorb too much light to be good mirrors, even if they were smooth; and comparatively few are smooth. Dirt, stones, leaves, wood, hair, skin — nearly all common materials have surfaces that are very rough and irregular, from the "atom's-eye view" of a light wave. We may think of such a surface as made up of a great number of very small smooth surfaces (Figure 6). Light is reflected from each small surface element according to the law of reflection, but the rays from the various elements go off in many different directions. The reflected light is diffused, or scattered. This process is called diffuse reflection.

Our visual judgment of the textures of different surfaces depends mainly upon the degree of diffuseness of the light they reflect. Consider, for example, the difference in appearance between a piece of black velvet and one of polished black marble or obsidian, or between a good mirror and a slab of plaster of Paris.

Good mirrors are relatively hard to make. Not only must the surface be smooth to within a few millionths of an inch; it must also absorb no more than a few per cent of the incident light. With a few special exceptions, metals are by far the best mirror surfaces. Why? Metals conduct electricity, and that means that their electrons are not bound to their respective atoms so strongly as in the non-metals. Consequently, the electrons in metals have more freedom to vibrate in unison with the incoming light waves, and to reradiate the energy they absorb. In nonmetals, more of the absorbed energy is trapped as heat. Except for gold, silver and copper, metals are rarely found uncombined with other elements. That is why natural mirrors are scarce.

Ordinary household mirrors are made of plate glass coated on the back with an amalgam of tin foil and mercury, which is painted black to protect it. The metal film itself is the mirror; the glass serves only to shape the film and protect it (Figure 7). Such a mirror is very durable, but it is not satisfactory for a telescope or any other precision optical instrument. For such exacting purposes, a piece of plate glass is so irregular that it may be compared to a plowed field. The front surface of the glass in an ordinary mirror reflects enough light itself to produce an undesirable secondary image.

Glass for a precision mirror must be ground and polished to near-perfect smoothness. A thin layer of metal — usually aluminum — must then be evaporated in a vacuum onto the front surface of the glass. The glass itself merely supports the metal film (Figure 8). Such a mirror reflects up to 95 per cent of the light that falls upon it.

Most substances absorb practically all the light they receive before it penetrates to any appreciable distance. Such substances are said to be opaque. A few natural substances (besides gases) and many more artificial ones are transparent. They permit substantial fractions of light to penetrate through moderate thicknesses of material. All substances absorb some light energy; transparency is only relative. Good optical glass will transmit more than 99 per cent of the light entering a plate one inch thick; but such transparency is uncommon.

The light that strikes transparent substances usually is a mixture of all the different wave lengths that we see as white light. Transparent materials absorb some wave lengths more than others. Since different wave lengths cause different sensations of color to arise in the brain, very few transparent materials appear colorless in transmitted light. Ordinary glass favors the green wave lengths slightly; it is practically opaque in the ultraviolet region. A thin film of silver, such as is used on some mirrors, is quite opaque to visible light, but is transparent in the ultraviolet range.

Some materials allow light to pass through, but they scatter and diffuse it. Such a material is called translucent; its effect can be compared to that of diffuse reflection. Opal glass and certain kinds of paper are very familiar examples of translucent substances.

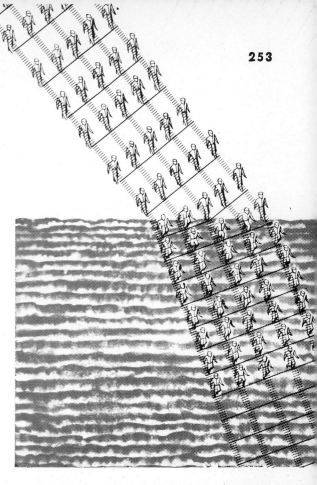

9. A column of marching men would change direction as it passed from a smooth parade ground to a rough field. On the plowed field the men could not march so rapidly.

When an incident ray of light enters a transparent substance at some angle other than the perpendicular, or normal, to the surface, its direction is changed. This effect is called refraction (bending). The ray whose course has been changed in this way is known as a refracted ray. Refraction takes place because the speed of light in a vacuum (or in air, which in this case amounts to nearly the same thing) is greater than in water, say, or glass.

To illustrate the principle involved in refraction, let us suppose that a column of marching men crosses the boundary between a smooth parade ground and a plowed field at an angle, as in Figure 9. They cannot march as rapidly on the rough ground as they can on smooth terrain. In each rank the men who first reach the rough ground slow down while those to their left gain on them. As a result of this difference in marching speed, the direction of march of the column is changed.

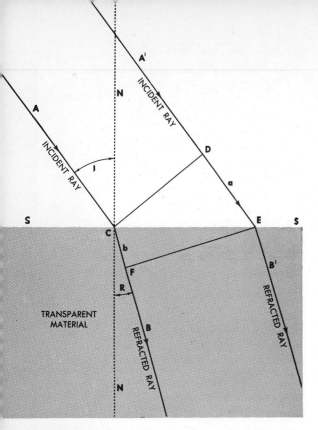

10. This diagram shows how a wave of light is refracted. Details about refraction are given on this page.

In Figure 10 we apply this principle to the refraction of light. The rays AA′ indicate the direction of a train of light waves advancing in air and approaching the surface of a transparent material SS. The speed of light in this material is less than the speed of light in air. As the advancing wave front reaches the position CD, the part near C enters the material and is slowed down. The portions of the front nearer D continue to advance in air at their former speed. As a result, the advancing wave front swings toward the normal, or perpendicular.

By the time the wave front at D has reached the surface at E, traveling the distance a, the light that entered at C has traveled a lesser distance, b, to the point F. Other points of the wave front will have traveled intermediate distances, so that the new wave front in the transparent substance will be EF. Note that the angle between the refracted ray and the normal NN (angle R), called the angle of refraction, is smaller than the angle of incidence (angle I). A ray will be turned toward the normal when its velocity is reduced; it will be turned away from the normal when the velocity is increased. This happens, for example, when light passes from water into air.

The ratio of the speed of light in air *
to the speed of light in a given substance is called the index of refraction of that substance. The index of refraction can be defined more exactly as the ratio of the sine of the angle of incidence to the sine of the angle of refraction. Let us see what this means. In Figure 11 a ray of light strikes the surface of a transparent substance — say, water — so as to form the angle of incidence I. It is then refracted; the angle of refraction is R. We mark off equal distances AC and BC on the incident ray and the refracted ray. Then we draw perpendiculars AD and EB to the normal NN. ADC and ECB are two right-angled triangles, with equal hypotenuses.

The sine of any angle of a right-angled triangle is the ratio of the side opposite that angle to the hypotenuse. The sine of angle I, therefore, is AD/AC; the sine of angle R is EB/CB. Since the index of refraction is the ratio of the sine of the angle of incidence (I) to the sine of the angle of refraction (R), it is

$$\frac{AD/AC}{EB/CB}$$

The hypotenuse AC is equal to the hypotenuse CB; therefore the index of refraction is AD/EB.

The sines for all angles up to 90° are given in trigonometric tables. Once we know the angle of incidence and the angle of refraction as a ray of light strikes and then penetrates a given substance, it is a simple matter to find the sines of the two angles and to determine the index of refraction. The index of refraction of water is about 4/3, or 1.3; that of common glass is about 3/2, or 1.5.

We saw that when light passes obliquely from a transparent substance into air, the rays bend away from the normal. This may result in the effect called total internal reflection. In Figure 12, the ray

* As we have seen, the speed of light in a vacuum is practically the same as its speed in air.

of light A passes through glass and is bent away from the normal as it moves through air (as A'). In this case, the angle of incidence is I^1; the angle of refraction is R^1. If the angle of incidence is increased, the angle of refraction will be increased even more, proportionately, and the refracted ray will be closer to the surface of the glass.

When incident ray B passes through the glass, the angle of refraction (R^2) as it penetrates the air (as B') is 90°; the ray will graze the surface of the glass, as shown. Angle I^2, formed by the incident ray B and the normal, is called the critical angle. If the critical angle is exceeded, the ray does not leave the glass at all but is totally reflected from the surface.

This happens, for example, when ray C passes through the glass, forming the angle of incidence I^3, which is greater than the critical angle I^2. The ray is reflected from the surface (as C'), following the law of reflection.

The critical angle depends on the index of refraction on either side of the boundary surface. It is about 41° for a glass-air boundary, and about 49° for water-air. Total reflection can occur only in the medium that has the higher index of refraction.

The effects we have been describing apply to light of one particular wave length. Light is usually made up of many wave lengths, which are not refracted uniformly. Nearly all transparent materials refract short waves more than long ones. This effect is the basis of the most common type of spectroscope, in which a refracting prism is used to separate the wave lengths that make up light. (See Index, under Spectroscopes.)

Refraction effects may be used to identify small samples of mineral crystals. This is done by immersing them in a succession of liquids, each with a different refractive index. A transparent crystal or other material is visible in air mainly because it refracts light passing through it; it also reflects some of the incident light. It is visible for the same reasons in liquids that have a different index of refraction from its own. Suppose, however, that the crystal is placed in a transparent liquid whose index of refraction is identical with that of the crystal. The light rays will then pass through both liquid and crystal without reflection or refraction and the crystal practically disappears. An imitation diamond of ordinary flint glass vanishes in carbon disulfide, while a real diamond continues to sparkle in the liquid.

How images are formed

The simplest way to form an optical image is to allow light from a glowing or illuminated object to pass through a pinhole and to let it fall on a white screen. Some light will be diffracted from the edge

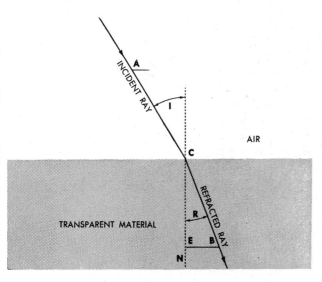

11. How to find the index of reflection of a transparent substance. Details are given on the preceding page.

12. Total internal reflection: ray C is reflected from the surface of the glass as C'. See pages 254 and 255.

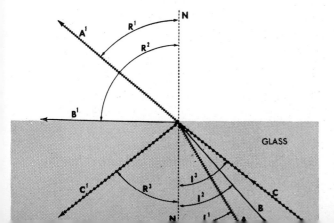

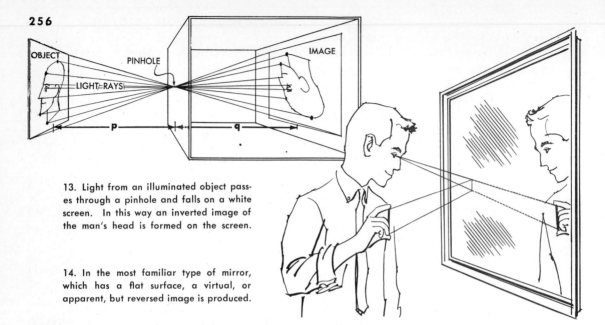

13. Light from an illuminated object passes through a pinhole and falls on a white screen. In this way an inverted image of the man's head is formed on the screen.

14. In the most familiar type of mirror, which has a flat surface, a virtual, or apparent, but reversed image is produced.

of the pinhole. Most of it, however, will pass through the hole and will form an inverted image on the screen. We show how such an image is produced in Figure 13. Light from the bottom of the head passes through the pinhole and forms the top of the inverted image on the screen. Rays from the top are directed by the pinhole to the corresponding part of the image. Of course light is spreading in all directions from the illuminated object, but the pinhole allows only the light traveling in certain definite directions to reach the screen.

An important factor in any image-forming system is its magnification. This term means the ratio of any dimension — height or width, for example — of the image to the corresponding dimension of the object. In a pinhole system, such as is shown in Figure 13, the magnification is the ratio of the image distance q to the object distance p. These distances can be made whatever one likes.

To obtain a brighter image than the pinhole permits, we need a device that will collect a substantial quantity of the light radiating from an object. The mirror offers such a device.

The most familiar type of mirror is the plane mirror, which has a flat, or plane, surface. Figure 14 shows a typical mirror of this kind. Light coming from a series of points on the object is reflected

by the mirror to the eye as shown, creating the illusion of coming from a corresponding series of points behind the mirror. As a result, a virtual, or apparent, reversed image is formed. It is the same size as the object itself and is at the same distance behind the mirror as the object is in front of it. The image is reversed from left to right. The plane mirror is the only optical device that is capable of forming a completely accurate image (in reverse, of course) of an object of appreciable size.

Some mirrors are curved. Such mirrors generally form part of the surface of a sphere. In Figure 15 we show a concave spherical mirror. The center of the sphere of which the mirror forms a part is called the center of curvature. The line passing through the middle of the mirror and the center of curvature is the principal axis.

Most spherical mirrors are relatively flat. This means that the diameter of the mirror is small compared with the radius of curvature — any straight line drawn from the center of curvature to the surface of the mirror. In concave mirrors of this type the rays of light parallel to the principal axis will pass through or very near a common point after they have been reflected from the mirror surface. This point (F in Figure 16) is called the principal focus; it is on the principal axis, half way between the center of curvature and the surface of the mirror. It is known as a real focus because the rays of light actually pass through it. We can show that this is so by placing a piece of paper at the focus of a concave mirror upon which sunlight is falling. The rays converging at the focus will soon cause the paper to burst into flames.

Not all the rays parallel to the axis of a concave spherical mirror meet exactly at the principal focus. This failure of the rays to converge at a single point is called spherical aberration (see Figure 17). Spherical aberration is comparatively small when the mirror is almost flat; it is larger when the mirror is more rounded. This effect can be remedied by making the curve of the mirror a paraboloid instead of a sphere (Figure 18). This figure is commonly used in searchlights, automobile headlights and various other devices. Large mirrors used in telescopes are usually paraboloids.

Some spherical mirrors use the convex side of a spherical surface (Figure 19). In this case, parallel rays diverge from the surface when they are reflected. However, they all seem to come from a single point behind the mirror (F in the diagram). This is called a virtual (apparent) focus, because the rays do not actually converge at this point but only appear to do so. If one exposed a convex mirror to sunlight and placed a piece of paper at the virtual focus, nothing would happen to the paper.

If all the rays that struck the surface of a spherical mirror were parallel to the principal axis, they would not produce an image of appreciable size. The reason is that they would all really meet (concave mirror) or apparently meet (convex mirror) at about the same place. Actually, however, many of the rays are not parallel to the principal axis. As a result, they produce an appreciable image.

The images produced by concave spherical mirrors may be smaller or larger than the objects; they may be virtual or real. As in the case of the plane mirror, an image is virtual if the light rays only seem to form it after they are reflected from the

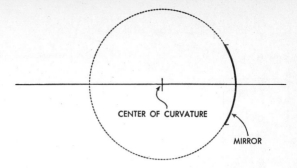

15. A concave spherical mirror, part of the surface of a sphere whose center is called the center of curvature.

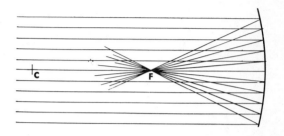

16. The light rays that are reflected from the surface of a concave mirror pass through the principal focus, F.

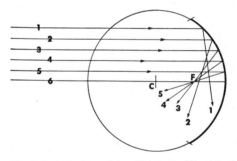

17. Spherical aberration, which is explained in the text.

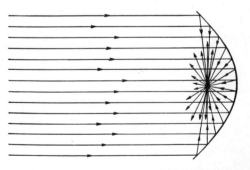

18. Spherical aberration can be remedied by making the curve of a mirror a paraboloid rather than a sphere.

19. Some spherical mirrors use the convex side of a spherical surface. The virtual (apparent) focus is F.

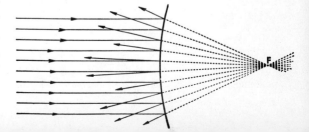

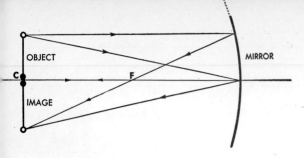

20. If an object is placed at the center of curvature of a mirror that has a concave surface, the image is real, inverted and also of the same size as the object.

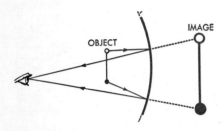

21. When an object is placed between the principal focus and the mirror, a virtual, erect image, larger than the object, will appear — and apparently behind the mirror.

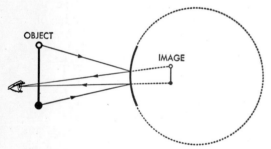

22. The image in a convex spherical mirror is virtual, erect instead of inverted, and smaller than the object.

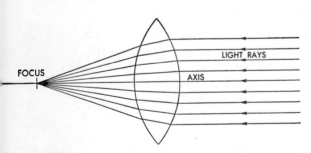

23. A converging lens. Rays parallel to the principal axis are bent so that they meet at the principal focus.

24. A diverging lens. Rays parallel to the principal axis seem to meet at the virtual principal focus, F.

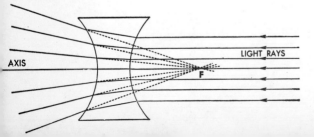

mirror surface. The image is real if it is actually produced by the rays. Both types of images look the same; however, a real image can be thrown on a screen, while a virtual one cannot. A real image is an actual convergence, or concentration, of light; a virtual image is an optical illusion.

If an object is placed at the center of curvature of a concave mirror, the image is real, inverted and the same size as the object (Figure 20). The image is also real and inverted if the object is placed farther away from the mirror than the center of the curvature; in this case, however, the image will be smaller than the object. It will dwindle to a point as the object is moved to a very great distance from the center of curvature.

Suppose we place the object between the center of curvature and the principal focus. The image will still be real and inverted but it will be larger than the object. We could also put the object between the principal focus and the mirror; in that case a virtual, erect image, larger than the object, would appear apparently behind the mirror (Figure 21). This is what happens when one uses a concave shaving mirror. One sees an erect image of one's face larger than life.

Convex spherical mirrors can produce only virtual images, which seem to be located behind the mirror. These images are erect (not inverted) and smaller than the object (Figure 22). Small convex mirrors are sometimes fastened outside a car near the windshield to bring into the driver's view cars that are approaching him from behind.

So far we have been dealing with images formed by mirrors. They can also be produced by lenses. A lens is a piece of transparent material, such as glass, shaped in such a way that it intercepts rays of light from an object and bends them by refraction so that they meet at a common focus. A simple lens has two opposite regular surfaces, which are generally spherical; in some cases, however, one surface may be a plane.

There are two chief kinds of lenses — converging and diverging. A converging

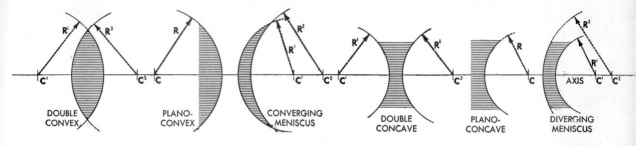

DOUBLE CONVEX   **PLANO-CONVEX**   **CONVERGING MENISCUS**   **DOUBLE CONCAVE**   **PLANO-CONCAVE**   **DIVERGING MENISCUS**

25. Six important types of lenses are illustrated above. The first three — the double convex, the plano-convex and the converging meniscus — are converging lenses. The last three — the double concave, the plano-concave and the diverging meniscus — are diverging lenses. C stands for "center of curvature"; R stands for "radius."

lens is thicker at the center than at the edges. Light rays parallel to the axis converge, or meet, at a point beyond the lens (Figure 23). This point is the principal focus. A diverging lens is thicker at the edges than at the center. When light rays parallel to the principal axis pass through the lens, they are spread apart. In this case the rays seem to come from a virtual focus on the same side of the lens as the light source (Figure 24). This point, F, is the principal focus. In both converging and diverging lenses, the distance from the principal focus to the lens is called the focal length. The focal length depends on both the index of refraction of the material and the curvatures of the two surfaces of the lens.

In Figure 25 we show some common types of converging and diverging lenses.

Note that in each case the principal axis is a straight line drawn through the centers of curvature of the two spherical surfaces. If one surface is plane, the principal axis passes from the center of curvature to the center of the lens.

The relationship between the object and the image formed by a converging lens is much the same as in a concave mirror of the same focal length. The only difference is that since light passes through the lens, real images are formed on the side opposite the incoming light, while virtual images are formed on the side from which the light comes. As in the mirror, the nature of the image will depend upon the distance of the object from the lens (Figure 26). Diverging lenses produce much the same type of image as convex mirrors; the image is virtual, erect and

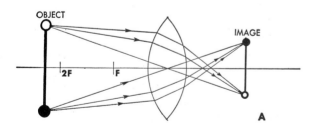

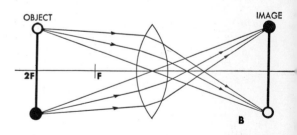

26. Converging lenses. A, object beyond twice focal distance — image inverted, real and smaller than the object. B, object at twice focal distance — image real, inverted and same size as object. C, object at focus — no image. D, object between focus and lens — the image is erect, virtual and larger than the object.

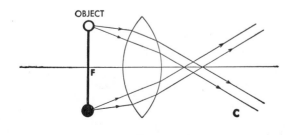

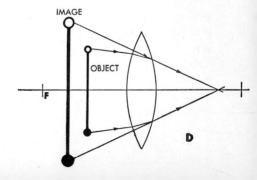

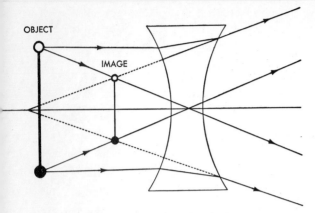

27. Diverging lenses produce very much the same type of images as those produced by convex mirrors. The images are virtual, erect and smaller than the objects. The only difference is that the virtual images are behind the mirror while they are in front of the diverging lens.

smaller than the object. The only difference is that the virtual image is behind the mirror, while it is in front of the lens (Figure 27).

Optical images except those formed by plane mirrors are distorted to a greater or lesser extent as a result of various defects. We have already mentioned spherical aberration in spherical mirrors; this defect occurs also in lenses. Another defect found in mirrors and lenses is astigmatism. In this case rays that come to a lens or mirror from a point far from the principal axis do not meet at a common point of the image. A blurred effect is produced. Astigmatism in lenses can be corrected by using several lens elements. This arrangement is found in the expensive camera lenses called anastigmats.

Chromatic aberration is a familiar type of lens defect. The index of refraction depends upon wave length. Hence any simple single-element lens will focus short-wave light nearer to the lens than it will longer-wave light. This chromatic aberration causes serious blurring and confusion in the image in an instrument of any power. It is usually reduced to a tolerable level by a two- or three-element combination called an achromatic lens. The principle of this correction is somewhat complicated. It depends upon the use of two or three different kinds of glass to bring all

wave lengths approximately to the same focus without lessening the over-all power of the lens combination.

Some familiar optical instruments

A number of devices, called optical instruments, are based on the refraction or reflection of light. Here is a brief survey of a few of the better known ones.

One of the most familiar optical instruments of any complexity is the camera. Basically this familiar device is a converging lens arranged to form a real image on a piece of sensitive photographic film. Usually the image is smaller than the object, but this is not always the case. The lens and the film are mounted in a light-tight box, and a shutter is provided for opening and closing the lens opening, thus governing the amount of exposure the

THREE FAMILIAR OPTICAL INSTRUMENTS

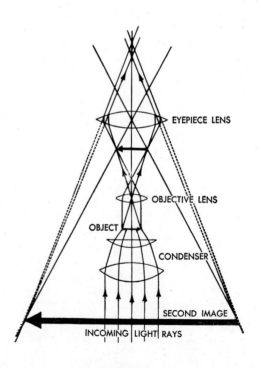

EYEPIECE LENS

OBJECTIVE LENS

OBJECT

CONDENSER

SECOND IMAGE

INCOMING LIGHT RAYS

film receives. In most cameras it is possible to vary the distance from lens to film so as to bring nearby or distant objects into sharp focus in the picture. In most cameras, too, an adjustable mask, called an iris diaphragm, is set in front of the lens; it controls the size of the aperture through which the light passes. We describe cameras in detail in The Versatile Art of Photography, in Volume 9.

A projector is simply a camera in reverse. If a transparent photograph or drawing is placed in the film position in a camera and strongly illuminated, an image will be projected on a screen placed at the normal object position in front of the camera. This is the principle of the photographic enlarger, the slide and motion-picture projector and other such devices. Opaque objects, such as pictures in books, can be projected if suitably illuminated.

The simple magnifier, commonly called a magnifying glass, is a converging lens used to view an object placed nearer to the lens than its principal focus. Under these conditions the eye sees a magnified virtual image, as in Figure 26d. This is a very convenient arrangement because the image is upright. For low-power work, up to magnifications of three or four times, a single lens element works acceptably as a magnifier. For more power, however, it is necessary to use a more complicated lens to correct the chromatic and other aberrations that cannot be avoided in the simpler instrument. The eyepiece of a telescope or microscope is a high-quality simple magnifier, used to view a real image formed by other optical elements — lenses or mirrors or both.

The compound microscope is used to obtain high magnifications — up to sev-

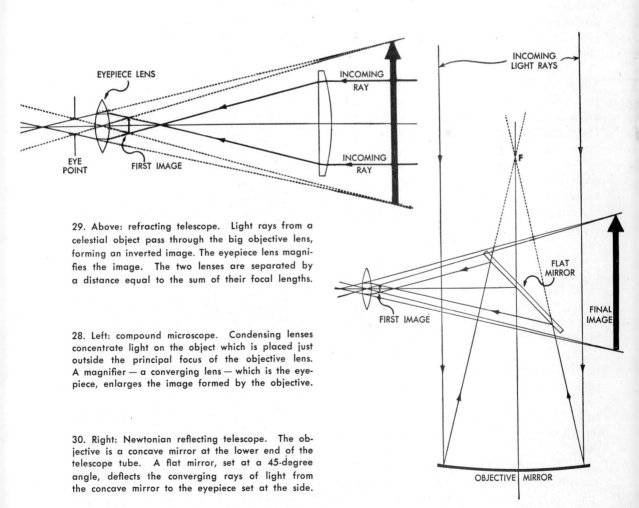

29. Above: refracting telescope. Light rays from a celestial object pass through the big objective lens, forming an inverted image. The eyepiece lens magnifies the image. The two lenses are separated by a distance equal to the sum of their focal lengths.

28. Left: compound microscope. Condensing lenses concentrate light on the object which is placed just outside the principal focus of the objective lens. A magnifier — a converging lens — which is the eyepiece, enlarges the image formed by the objective.

30. Right: Newtonian reflecting telescope. The objective is a concave mirror at the lower end of the telescope tube. A flat mirror, set at a 45-degree angle, deflects the converging rays of light from the concave mirror to the eyepiece set at the side.

eral thousand times — of very small objects. In principle it is very simple, as Figure 28 shows. Condensing lenses concentrate light on the object, which is placed just outside the principal focus of a small converging lens, called the objective. This objective forms a magnified real image of the object a few inches away. Since this image, which is inverted, is still quite small, a magnifier (consisting of another converging lens) is used as an eyepiece to produce a further enlarged virtual image of the first image.

Although the principle of the microscope is simple, its actual construction is one of the most difficult accomplishments in optics. Since it is designed to view very small objects, the objective lens must be on a similarly small scale. The problem of controlling image defects, especially spherical aberration, is a serious one. The result is that high-powered-microscope objectives are quite complicated, consisting in some cases of as many as ten separate lens elements.

The oldest optical instrument of any complexity is the astronomical refracting telescope (Figure 29). As originally devised by Galileo, the instrument used a diverging lens as an eyepiece. However this system, which still serves in cheap opera glasses, has serious defects. The arrangement shown in the diagram is almost universally used in refracting telescopes.

The optical relationships in this instrument are beautifully simple. When properly adjusted for a normal eye, the real image formed by the objective lies at the principal focus of this lens. If the principal focus of the eyepiece is brought to this same point, the observer sees the final virtual image projected to a great distance. It may be noted that the question so often asked of astronomers: "What is the magnifying power of your telescope?" is meaningless. The magnifying power depends on the eyepiece in use at any particular time; it is simply the ratio of the focal length of the objective to that of the eyepiece.

The reflecting telescope (Figure 30) substitutes a concave mirror for the objective lens. This arrangement leads to some difficulties in design, since the light bends back on itself and some means must be found to bring the image out of the incoming beam. This is done in the Newtonian reflector, shown in Figure 30, by means of a flat mirror, set at an angle of 45° to the optical axis of the mirror. Mirrors are free from chromatic aberration, which plagues the objective lenses of all refractors. Another advantage is that they can be made in much larger sizes than lenses. However astigmatism and certain other image defects are more of a problem than in refracting telescopes.

The eye as an image-forming device

The eyes of animals are image-forming devices. The eyes of human beings and other mammals, as well as those of most other highly developed species, are similar in principle to the ordinary camera. They use a small lens that forms an inverted real image on the sensitive film of the retina. The retina then relays information on light intensities to the brain. The eye has an iris diaphragm to control the amount of light entering the organ. Instead of varying the lens-to-film distance in order to focus for various ranges, the eye makes this adjustment by varying the curvature, and consequently the focal length, of the lens itself. Thus it performs a feat that would be quite impractical with artificial optical materials.

At least one living species has long since appropriated the simplest image-forming device — the pinhole. The eye of the pearly nautilus is merely a hollow, enclosed cup with a small hole leading to the exterior. On the inside facing the hole there is a primitive retina. The compound eyes of insects are rather complicated versions of the same principle; they consist of a great number of narrow tubes pointing in different directions. Each tube "sees" a tiny portion of the environment; all the tubes together build up a picture somewhat on the principle of a halftone print or a television screen.

See also Vol. 10, p. 281: "Light and Illumination."

ORGANIC CHEMISTRY

The Study of
the Countless Compounds of Carbon

BY DONALD C. GREGG

WHEN chemistry first became a bona fide science, in the latter part of the eighteenth century, a number of chemical compounds were obtained from plants and animals. Scientists called them organic substances because they were derived either from living or formerly living organisms. The branch of chemistry devoted to their study became known as organic chemistry. The name "inorganic chemistry" was applied to the study of the chemical compounds obtained from the mineral kingdom.

Chemists soon recognized that the organic compounds derived from plants and animals had one thing in common: they all contained carbon atoms. Because of this fact, organic chemistry was also called the chemistry of the compounds of carbon.

263

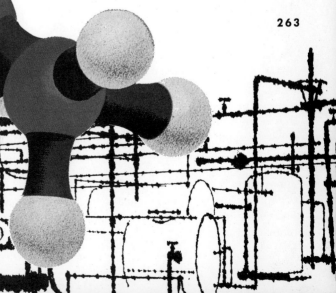

The early chemists spent most of their time studying the acids, bases and salts; they were interested chiefly in the substances obtained from, or related to, the mineral kingdom. It was much easier to work with such inorganic compounds, since most of them were soluble in water or in the common acids. Usually their reactions were quite rapid. It was not too difficult to understand these reactions, or at least to explain them logically by using the theories that were then believed to be correct. Chemists could prepare many inorganic compounds by using very simple methods.

Organic compounds seemed to be quite different in nature and behavior. Many of them reacted with most common inorganic substances very slowly, or not at all. Heat played an important part in the preparation of inorganic compounds, but it usually caused organic compounds to decompose. As we pointed out, chemists could make inorganic compounds in the laboratory; but apparently it was impossible to create an organic compound. Chemists wondered how such compounds were produced in plants and animals.

The Swedish scientist Joens Jakob Berzelius was one of the foremost authorities in chemistry in the nineteenth century. This learned man made no effort to explain the behavior of the organic compounds. He simply stated that since they were derived from plant or animal organisms, their formation was probably due to some "vital force," a force totally lacking in inorganic material. It would be impossible for man, he maintained, to duplicate this "vital force"; hence nobody would ever be able to prepare an organic compound.

In 1828, a young German chemist, Friedrich Woehler, succeeded in doing what Berzelius had declared was impossible. He heated an inorganic compound, ammonium cyanate, and produced urea, an organic compound that occurs in urine. Chemists now began to doubt the existence of Berzelius' "vital force"; they theorized that perhaps other organic compounds might be prepared in the laboratory.

In the years that followed, chemists succeeded in producing many such compounds. More and more researchers devoted themselves to organic chemistry, and

In 1828, the brilliant young German chemist Friedrich Woehler produced the first synthetic, or man-made, organic compound. By heating an inorganic compound, ammonium cyanate, he produced urea, an organic compound found in urine. From Woehler's remarkable achievement stem the innumerable modern applications of synthetic organic compounds. They are used, among other things, for drugs, perfumes, dyes and a host of plastic materials.

the number of synthetic (man-made) compounds mounted steadily. In a few decades they far outnumbered the inorganic compounds. We have traced the history of this development in the chapter Science and Progress I, in Volume 5.

Today we know that the basic laws that govern the formation and behavior of inorganic compounds also apply to the compounds of carbon.* However, we still study organic compounds separately for several reasons. For one thing, carbon atoms have the capacity (which is almost unique) of producing chains and rings of atoms as they combine with each other and with other atoms.** Then too, the organic compounds as a group are much less stable than the inorganic, so that chemists must use different techniques in working with them.

Why do we still call the compounds of carbon "organic" when we realize that

many of them are man-made and not found in any living organism? The chief reason is that the compounds derived from nature and those produced in the laboratory are closely related to one another chemically. It has always seemed best, therefore, to give the name of organic compounds to *all* the compounds of carbon (except the few we have indicated in the first footnote of this article), whether they are of natural origin or produced in the laboratory.

Hundreds of these compounds are synthesized or discovered every year. It would be impossible to overemphasize their importance. Plants and animals are composed mostly of organic compounds; so is most human food, derived directly or indirectly from living things. Many fabrics are produced from organic materials (such as wool, cotton and linen) obtained from plants or animals. Many, too, consist of synthetic organic materials (rayon, nylon, Dacron and the like), developed in the laboratory in an attempt to duplicate natural fibers or to improve upon them. Plastics, rubber, paper and wood are all organic substances. Petroleum, from which we

* Not all carbon compounds are organic. The carbon oxides, carbonates and cyanides are considered to be inorganic.
** Alternating silicon and oxygen atoms also form a network of atoms when attached to various organic groups. The resulting compounds are called silicones (see Index).

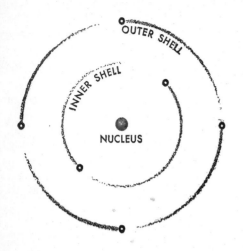

Highly simplified diagram of carbon atom. Its six electrons revolve around the nucleus in two shells. The inner shell has only two electrons; the outer one has four.

obtain fuels and many industrial products, is a mixture of carbon compounds. So is coal tar, which yields drugs, perfumes, flavorings and other useful products.

The structure
of the carbon atom

In other articles * we have pointed out that each atom has a certain number of negatively charged particles, called electrons, revolving around a central core in a number of shells. There are two such shells in the carbon atom; the first one has two electrons and the second has four. Only the electrons in the outer shell are involved in chemical changes.

This shell, with its four electrons, is not filled to capacity; there is room for four more. The carbon atom can share electrons with other atoms in order to fill up the gap — a type of combination called covalence (see Index). When a carbon atom combines with four hydrogen atoms, sharing four pairs of electrons, we say it has a covalence, or combining power, of four.** There are so very few cases in which a carbon atom has a covalence of three that we may leave them out of con-

* Inside the Atom, Volume 1; How Atoms Combine, Volume 3.
** Covalence may be expressed in terms of hydrogen, since this element always has a covalence of one and never more. Other atoms may have different covalences.

sideration for the purposes of this article. We shall assume that in all simple organic compounds, such as we shall take up, each carbon atom has four covalent bonds.

In order to have a clear idea of the carbon atom, we may picture the four valence electrons as being equidistant from each other and from the nucleus of the atom. We can visualize this best if we think of them as located at the corners of a regular tetrahedron,* as shown at the bottom of this column. Of course, it must always be remembered that electrons are in constant movement in space. However, for our purposes the tetrahedral arrangement is a reasonable representation.

The simplest organic compound is methane, in which one atom of carbon (C) is combined with four atoms of hydrogen (H). The formula for this compound is CH_4.** A formula such as this is called compositional; it indicates the kinds and numbers of atoms in a compound.

A space model of CH_4 is shown in the left-hand diagram on the next page. Four electrons, each from a hydrogen atom, which has but one electron, have been added to the outer shell of the carbon atom. This shell, therefore, now has its full quota of electrons — that is, eight.

* A regular tetrahedron is a figure or solid with four faces, consisting of equal triangles — triangles whose sides are all equal.
** A small figure after and near the bottom of a chemical symbol, such as C or H, indicates the number of atoms. If no figure is added, the number 1 is understood.

VALENCE ELECTRON

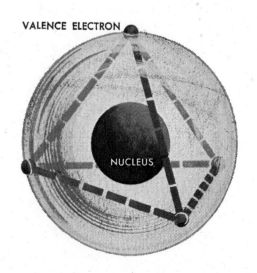

NUCLEUS

At the bottom of column 2, we give a simplified projection of this model; a carbon atom is shown linked to four hydrogen atoms. This is not a true representation, of course; for one thing, it is two-dimensional, while a real molecule has three dimensions. However, chemists represent organic compounds in this simple manner because it is convenient and gives at least some idea of the actual three-dimensional structure. Such pictures of the way in which atoms are joined together are called structural formulas.

The methane, or paraffin, series of hydrocarbons

Methane, CH_4, is called a hydrocarbon because it is made up only of atoms of hydrogen and carbon. There are a great number of other hydrocarbons, all containing only carbon and hydrogen atoms. They occur as gases, liquids or solids. They are found in the gases formed by heating coal or wood in a closed vessel. Petroleum and natural gas are made up almost entirely of hydrocarbons.

Methane is a member of the so-called methane series of hydrocarbons. Each compound in the series has the general formula C_nH_{2n+2}. If there is only one carbon atom in a compound of the methane series, n is equal to 1; if there are two carbon atoms, n is equal to 2 and so on. In the case of methane, for example, since there is only one carbon atom, n equals 1. Substituting 1 for n in the formula C_nH_{2n+2}, we have $C_1H_{2\times1+2}$, or CH_4. In the hydrocarbons of the methane series, the quota of electrons for the outer shells of the carbon atoms is filled; hence these atoms cannot take on any more hydrogen atoms. In other words, they are "saturated" with hydrogen atoms; for this reason they are called saturated hydrocarbons.

The compounds of the methane series are sometimes known as paraffin hydrocarbons. The word paraffin comes from the Latin *parum affinis,* meaning "having very little affinity." The name paraffin hydrocarbons is quite appropriate, since the compounds of the methane series have little affinity with other atoms: that is, little tendency to join them in chemical combinations. The members of the methane series are also called alkanes.

Methane, the first of the methane series, is a gas occurring in natural gas. It is sometimes called marsh gas because it is found in marshes, where it is formed by the decay of vegetable matter. Sometimes it collects in coal mines; miners call it fire damp. When methane is mixed with the oxygen contained in air, it is explosive; many disastrous accidents in mines have been due to the ignition of such a mixture. Methane is a common source of hydrogen; it is also used in the manufacture of carbon black.

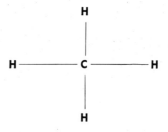

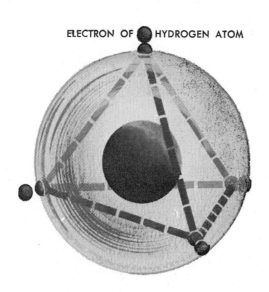

ELECTRON OF HYDROGEN ATOM

On column 2 of the preceding page, we give a space model of a carbon atom. The model shows the nucleus and the outer shell; the inner shell is not indicated, since it does not take part in chemical reactions. Note that the four outer electrons of the carbon atom are represented as equidistant from the nucleus and from each other. In the left-hand diagram on this page, we show a carbon atom combined with four hydrogen atoms to form the compound methane (CH_4). Four electrons, each from a hydrogen atom, have been added to the carbon atom. Right: two-dimensional structural formula for the methane atom.

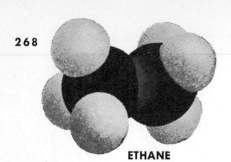

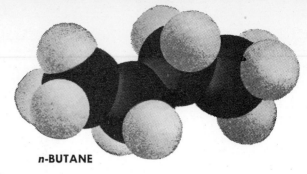

ETHANE   n-BUTANE

In these diagrams of ethane and <i>n</i>-butane molecules, each black sphere represents a carbon atom; each white one, a hydrogen atom.

The second member of the methane series is ethane, which has the compositional formula C_2H_6. It is a colorless, odorless gas, found in coal gas and natural gas. Its structural formula is as follows:

$$\begin{array}{c} \text{H} \quad \text{H} \\ | \quad\; | \\ \text{H}-\text{C}-\text{C}-\text{H} \\ | \quad\; | \\ \text{H} \quad \text{H} \end{array}$$

Each link in the above formula, of course, indicates a covalent bond. Whenever we write a structural formula such as this, we must check it to make sure that the four covalent bonds of the carbon atom are accounted for. Note that in each case a carbon atom is linked to three hydrogen atoms and also to the other carbon atom.

The structural formula shown above can also be written as CH_3CH_3. This is called a simplified structural formula. It indicates that a carbon atom (C), which is connected to three hydrogen atoms (H_3), is linked to another carbon atom (C), also connected to three hydrogen atoms (H_3).

Propane is the third member of the methane series; it has the compositional formula C_3H_8 and the structural formula:

$$\begin{array}{c} \text{H} \quad \text{H} \quad \text{H} \\ | \quad\; | \quad\; | \\ \text{H}-\text{C}-\text{C}-\text{C}-\text{H} \\ | \quad\; | \quad\; | \\ \text{H} \quad \text{H} \quad \text{H} \end{array}$$

The simplified structural formula of propane is $CH_3CH_2CH_3$. This means that a carbon atom (C), connected to three hydrogen atoms (H_3), is linked to another carbon atom (C), joined to two hydrogen atoms (H_2); the second carbon atom is linked in its turn to a third one (C), joined to three hydrogen atoms (H_3). Compressed propane is used as a fuel gas.

The fourth member of the methane series is butane, with the empirical formula C_4H_{10}. Since we have so many building blocks, in the form of atoms, to work with in butane, we might suspect that the different carbon and hydrogen atoms can be combined in more than one way. Actually this is the case. One structure has the normal straight-chain form:

$$\begin{array}{c} \text{H} \quad \text{H} \quad \text{H} \quad \text{H} \\ | \quad\; | \quad\; | \quad\; | \\ \text{H}-\text{C}-\text{C}-\text{C}-\text{C}-\text{H} \\ | \quad\; | \quad\; | \quad\; | \\ \text{H} \quad \text{H} \quad \text{H} \quad \text{H} \end{array}$$

The other has a chain of three carbon atoms, with a carbon attached to the middle carbon of the chain as follows:

$$\begin{array}{c} \text{H} \qquad \text{H} \qquad \text{H} \\ | \qquad\; | \qquad\; | \\ \text{H}-\text{C}\text{------}\text{C}\text{------}\text{C}-\text{H} \\ | \qquad\; | \qquad\; | \\ \text{H} \quad \text{H}-\text{C}-\text{H} \quad \text{H} \\ | \\ \text{H} \end{array}$$

This forms what is called a branched-chain configuration; it shows a chain with a branch leading from one part of it.

The two butane structures we have shown above represent two different compounds, each with the empirical formula C_4H_{10}. The compound with the straight-chain structure is called normal butane or <i>n</i>-butane; the branched-chain compound is known as isobutane. They have different physical properties and chemical behavior. <i>n</i>-butane is a major source of butadiene, the principal chemical in some kinds of synthetic rubber; isobutane is used in the manufacture of certain superior grades of motor fuel.

Two or more compounds with the same compositional formula but with different structural formulas are called isomers. *n*-butane is the straight-chain isomer of butane; isobutane, the branched-chain isomer. There are only two isomers in the case of butane; as we shall see, other compounds have a great many. The existence of isomers is a common and significant phenomenon in chemistry.

In the fifth member of the methane series — pentane — five carbon atoms are combined with twelve hydrogen atoms; the compositional formula is C_5H_{12}. There are three different isomers of pentane, each with a different structural formula: *n*-pentane, isopentane and neopentane. Here are the structural formulas:

a) *n*-pentane

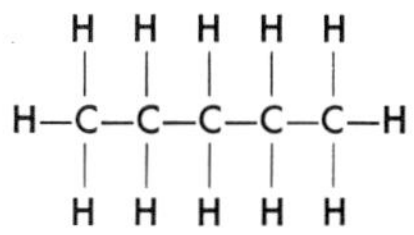

b) Isopentane

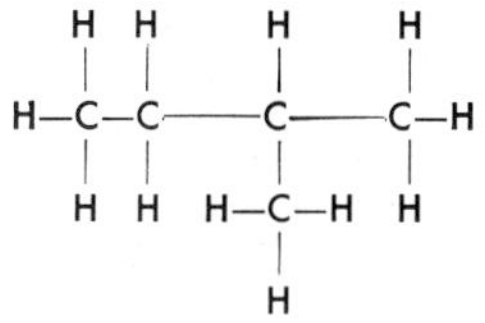

c) Neopentane

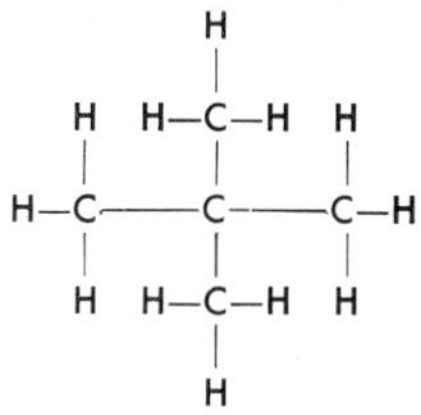

Other members of the methane series include hexane (C_6H_{14}), heptane (C_7H_{16}) and octane (C_8H_{18}). The number of possible isomers increases as the number of carbon atoms in the molecule increases. For $C_{10}H_{22}$ there are at least 75 possible isomers; for $C_{20}H_{42}$, at least 366,319. Only a few of the latter are known. Obviously, because of the existence of isomers, the number of possible organic compounds reaches almost astronomical proportions.

The lighter members of the methane series of hydrocarbons, containing less than five carbons per molecule, are gases. Those having from five to fifteen carbons are usually liquids; those with more than fifteen carbons are solids.

The paraffins are constituents of natural gas and petroleum. Natural gas is mostly methane with some of the other gaseous paraffin hydrocarbons. Crude oil consists of a mixture of many different liquid and solid hydrocarbons. In gasoline, there is also a mixture of hydrocarbons, generally ranging from C_6H_{14} to $C_{12}H_{26}$. The yield of gasoline from a given amount of crude oil may be increased by building up hydrocarbons containing less than six carbons to produce molecules with from six to twelve carbons. It can also be increased by breaking down the hydrocarbons that have more than twelve carbon atoms. The latter process is known as "cracking."

Pure paraffin hydrocarbons are colorless, almost odorless and inflammable. They are very insoluble in water and they do not form ions in water solutions. When they undergo complete combustion, they yield only carbon dioxide (CO_2) and water. It is rather difficult, however, to bring about this type of combustion; hence carbon and carbon monoxide are often produced in addition to carbon dioxide and water. For example, when gasoline is burned in the motor of a car, carbon may be deposited in the cylinders and carbon monoxide (CO), a poisonous gas, may issue from the exhaust pipe.

Certain saturated hydrocarbons have a ring structure. We can form such a structure by joining the ends of a chain together, provided there are at least three carbon atoms. Cyclopropane, with the compositional formula C_3H_6, is a good example of a ring formation. This substance, which is used as an inhalation anesthetic, has the following structural formula:

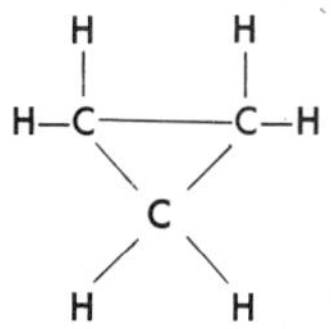

Saturated hydrocarbons with a ring structure are called cycloparaffins. Cyclobutane (C_4H_8), cyclopentane (C_5H_{10}) and cyclohexane (C_6H_{12}) are well-known cycloparaffins.

The unsaturated
hydrocarbons

In all the carbon compounds we have discussed thus far, there is only one bond between any two carbon atoms. There is more than one such bond in certain hydrocarbons. For instance, in the substance called ethylene, C_2H_4, there are two links between the two carbon atoms in its molecule, as the structural formula shows:

$$\begin{array}{ccc} H & & H \\ | & & | \\ H-C & = & C-H \end{array}$$

When there is more than one bond between carbon atoms, the molecule is not saturated since other atoms could be added to it.

For example, instead of the double carbon link in ethylene, we might substitute a pair of hydrogen atoms, enclosed in squares in the following formula:

$$\begin{array}{ccc} H & & H \\ | & & | \\ H-C & - & C-H \\ | & & | \\ \boxed{H} & & \boxed{H} \end{array}$$

This is exactly what happens when ethylene is treated with hydrogen gas and a suitable catalyst. You will note that the new substance that is produced in this reaction is ethane gas (C_2H_6).

Substances such as ethylene, in which carbon atoms are joined by double bonds, are called unsaturated, because they are capable of taking up more hydrogen atoms (and other atoms as well). They are quite reactive.

In some unsaturated hydrocarbons, there are three bonds between carbon atoms. To give a familiar example, acetylene (C_2H_2) has the following structural formula:

$$H-C \equiv C-H$$

Acetylene is widely used for oxyacetylene welding and cutting and in the preparation of many important organic compounds, such as acetic acid and acetone. It is the first member of a series of hydrocarbons with triple bonds between carbon atoms — the acetylene series.

The unsaturated hydrocarbons play an important part in the rubber industry and many others. The basic unit of rubber is isoprene (C_5H_8), an unsaturated hydrocarbon with the following structure

$$\begin{array}{ccccccc} & H & & H & & H & H \\ & | & & | & & | & | \\ -C & - & C & = & C & - & C- \\ & | & & | & & | \\ & H & & H-C-H & & H \\ & & & | & & \\ & & & H & & \end{array}$$

Rubber is a polymer; this means that it is made up of a great number of identical small units. This is how two of the many isoprene units in rubber would be linked:

$$\begin{array}{ccccccccccc} H & & H & H & H & & H & H \\ | & & | & | & | & & | & | \\ -C & - & C = C-C-C & - & C = C-C- \\ | & & | & | & | & & | & | \\ H & & H-C-H & H & H & H-C-H & H \\ & & | & & & | \\ & & H & & & H \end{array}$$

| Isoprene unit | Isoprene unit |

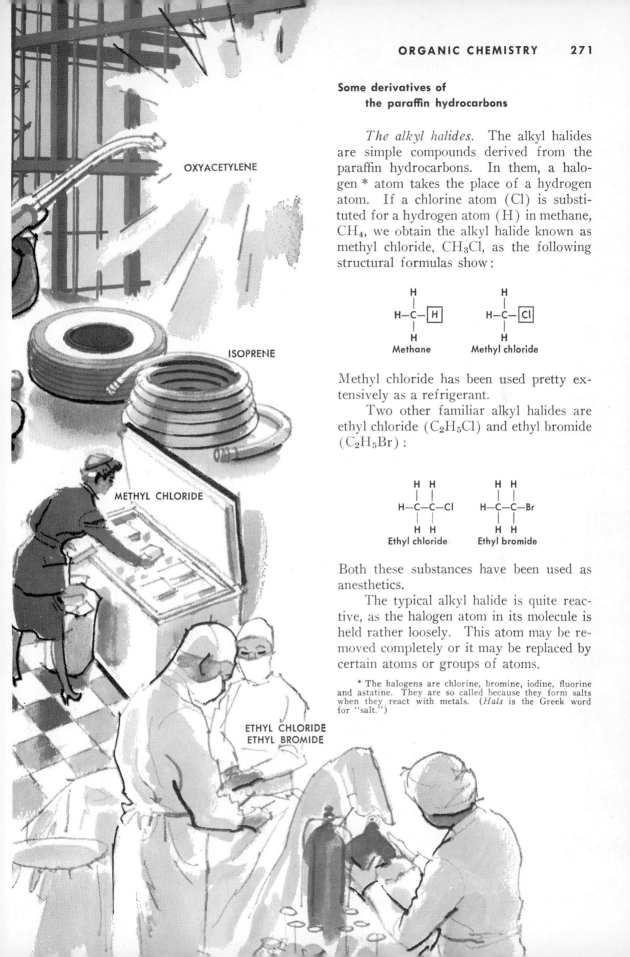

OXYACETYLENE

ISOPRENE

METHYL CHLORIDE

ETHYL CHLORIDE
ETHYL BROMIDE

Some derivatives of the paraffin hydrocarbons

The alkyl halides. The alkyl halides are simple compounds derived from the paraffin hydrocarbons. In them, a halogen * atom takes the place of a hydrogen atom. If a chlorine atom (Cl) is substituted for a hydrogen atom (H) in methane, CH_4, we obtain the alkyl halide known as methyl chloride, CH_3Cl, as the following structural formulas show:

$$H-\underset{\underset{H}{|}}{\overset{\overset{H}{|}}{C}}-\boxed{H} \qquad H-\underset{\underset{H}{|}}{\overset{\overset{H}{|}}{C}}-\boxed{Cl}$$

Methane   Methyl chloride

Methyl chloride has been used pretty extensively as a refrigerant.

Two other familiar alkyl halides are ethyl chloride (C_2H_5Cl) and ethyl bromide (C_2H_5Br):

$$H-\underset{\underset{H}{|}}{\overset{\overset{H}{|}}{C}}-\underset{\underset{H}{|}}{\overset{\overset{H}{|}}{C}}-Cl \qquad H-\underset{\underset{H}{|}}{\overset{\overset{H}{|}}{C}}-\underset{\underset{H}{|}}{\overset{\overset{H}{|}}{C}}-Br$$

Ethyl chloride   Ethyl bromide

Both these substances have been used as anesthetics.

The typical alkyl halide is quite reactive, as the halogen atom in its molecule is held rather loosely. This atom may be removed completely or it may be replaced by certain atoms or groups of atoms.

* The halogens are chlorine, bromine, iodine, fluorine and astatine. They are so called because they form salts when they react with metals. (*Hals* is the Greek word for "salt.")

The alcohols and ethers. In the compounds called alcohols, a hydroxyl group, OH, or more than one such group, is substituted for one of the hydrogen atoms in a paraffin hydrocarbon. When a hydrogen atom of methane (CH_4) is replaced in this way, we have methyl alcohol (CH_3OH):

Methane   Methyl alcohol

Methyl alcohol, often known as wood alcohol, can be made by the destructive distillation of wood; it can also be prepared from water gas, consisting of carbon monoxide and hydrogen ($CO + H_2$). Methyl alcohol is a colorless liquid with an odor suggesting that of wine. When taken internally, it can bring about blindness or cause death. Methyl alcohol is used to denature ethyl alcohol (grain alcohol), making it unfit for human consumption. It is also useful as a solvent in the manufacture of varnishes and shellacs. Quantities are employed in the preparation of formaldehyde, a preservative and disinfectant.

The hydrocarbon ethane (C_2H_6), as we have seen, has the structural formula

Replacing a hydrogen atom in this molecule by a hydroxyl group, OH, we have ethyl alcohol (C_2H_5OH), or common alcohol, with the structural formula

Ethyl alcohol, also called grain alcohol, is produced by the fermentation of sugars, and is an ingredient in alcoholic beverages and many medicinal preparations. It has many important industrial uses. Among other things, it is an important solvent in the preparation of varnishes, shellacs and enamels.

In some alcohols, two or more hydroxyl (OH) groups are attached to carbon atoms. An alcohol with two such groups is called dihydric. Ethylene glycol, $C_2H_4(OH)_2$, is a typical dihydric alcohol; it has the following molecular structure:

It is used as an antifreeze in automobile radiators and as a solvent.

The alcohol glycerol, or glycerine, $C_3H_5(OH)_3$, has three OH groups:

This substance is a useful solvent and serves in the manufacture of nitroglycerin and dynamite.

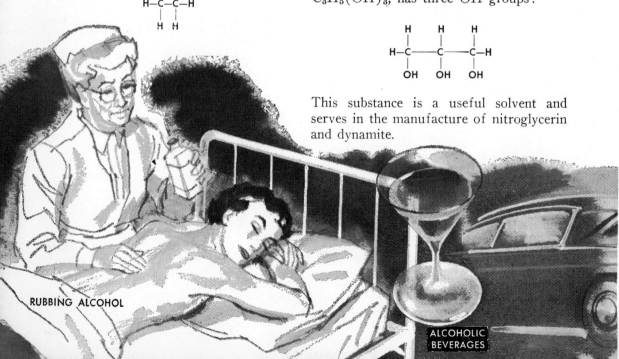

RUBBING ALCOHOL

ALCOHOLIC BEVERAGES

The ethers are obtained by eliminating water from alcohols. The most common ether is diethyl ether, $(C_2H_5)_2O$, with the structural formula

$$\begin{array}{ccccc} H & H & & H & H \\ | & | & & | & | \\ H-C-C & -O- & C-C-H \\ | & | & & | & | \\ H & H & & H & H \end{array}$$

It may be prepared from ethyl alcohol, with concentrated sulfuric acid serving as the dehydrating (water-removing) agent. Diethyl ether serves as an anesthetic.

The aldehydes, ketones and organic acids. When simple alcohols are oxidized (combined with oxygen), aldehydes, ketones or organic acids are produced.

The aldehydes contain the group

$$\begin{array}{c} H \\ | \\ -C=O. \end{array}$$

A simple aldehyde is formaldehyde (HCHO), which can be prepared by passing the vapor of methyl alcohol over a heated metal catalyst. The structural formula of formaldehyde is

$$\begin{array}{c} H \\ | \\ H-C=O \end{array}$$

Formaldehyde, a gas, is used in the manufacture of plastics, dyes and drugs. When it is dissolved in water, this most useful compound serves as a potent disinfectant and antiseptic.

In the ketones, we always find the group

$$\begin{array}{c} O \\ || \\ -C- \end{array}$$.

Dimethyl ketone, or acetone, $(CH_3)_2CO$, is the most important of the ketones; it is widely used as a solvent. Its structural formula is

$$\begin{array}{ccc} H & O & H \\ | & || & | \\ H-C-C-C-H \\ | & & | \\ H & & H \end{array}$$

Organic acids contain the group $O=C-OH$. The simplest of these acids is formic acid (CH_2O_2). The name is derived from the Latin word *formica*, meaning "ant"; the acid was formerly prepared by distilling red ants. A much more effective method of preparation is to oxidize methyl alcohol in the presence of a platinum catalyst. Formic acid is a colorless liquid with an unpleasant odor; it is used, among other things, to dye cloth. The structural formula of the acid is

$$\begin{array}{c} O=C-OH \\ | \\ H \end{array}$$

Other important organic acids are lactic acid, tartaric acid, oleic acid and stearic acid.

The esters. The esters are the most important derivatives of the organic acids. They are prepared by heating an organic acid with an alcohol in the presence of a small quantity of concentrated sulfuric acid. (Esters can also be prepared from inorganic acids.) For example, when acetic acid reacts with ethyl alcohol, it forms an ester called ethyl acetate ($CH_3CO_2C_2H_5$). This substance is used in synthesizing various organic compounds. Its molecular structure is

$$\begin{array}{ccccc} H & O & & H & H \\ | & || & & | & | \\ H-C-C & -O- & C-C-H \\ | & & & | & | \\ H & & & H & H \end{array}$$

ANTIFREEZE

NITROGLYCERIN

BAKELITE

FORMICA

NYLON

The aromatic hydrocarbons

Chemists were able to obtain certain hydrocarbons in large quantities by the distillation of coal tar long before petroleum became available in vast amounts. Since these coal-tar hydrocarbons had rather marked odors, in contrast to the practically odorless paraffins, they received the name of aromatic hydrocarbons. They have since become exceedingly important in industry and medicine. They are used in the manufacture of explosives, synthetic dyes, perfumes, photographic developers, drugs and many other products.

The simplest aromatic hydrocarbon is benzene, C_6H_6. This substance was first obtained by Michael Faraday by the distillation of coal. For a number of years chemists were puzzled about the structure of the benzene molecule; they wondered how six atoms of carbon could be combined with six atoms of hydrogen. The German chemist August Friedrich Kekule solved the problem.* He pointed out that the six carbons formed a ring, with single bonds between carbon atoms alternating with double bonds, as follows:

* It is said that the solution came to him in a dream. We give this interesting story in the article Science and Progress I, in Volume 5.

Since Kekule's structure is actually inadequate, organic chemists generally represent the benzene molecule as a hexagon:

In derivatives of benzene, one or more of the hydrogen atoms are replaced by other atoms or groups. To indicate the structural formulas of such derivatives, a hexagon may be used, with the substituted group at one of the corners. For example, chlorobenzene, C_6H_5Cl, can be represented either by the complete structural formula or by a hexagon, with a chlorine atom (Cl) at a corner, as follows:

or

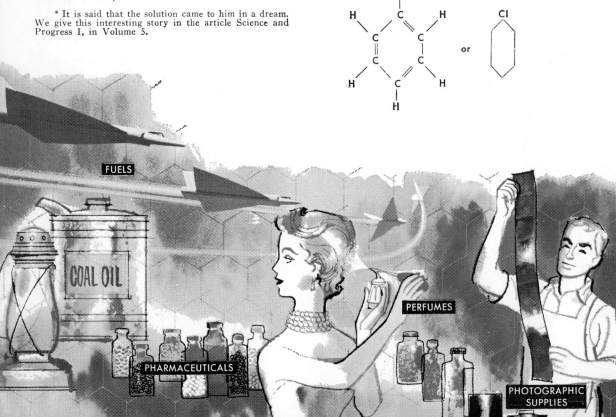

A well-known aromatic carbon is trinitrotoluene, $C_6H_2(CH_3)(NO_2)_3$, popularly known as TNT, and used as an explosive in peace and war. It has the structural formula

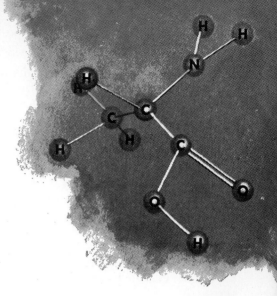

Molecular structure of alanine, one of the amino acids.

CH_3, a radical, or inseparable group, known as methyl, has replaced one of the hydrogens in the benzene ring. Three other hydrogens have been replaced by NO_2, the nitro group.

Some organic compounds found in living matter

The processes of life take place in a jellylike substance, the protoplasm, which is found in all living cells. It is made up mostly of water (85 to 90 per cent). It also contains a small percentage of minerals and 10 per cent or more of the organic substances known as proteins, fats and carbohydrates. Let us examine briefly the chemical structure of these substances.

The proteins. The proteins are among the most complex of all molecules; their molecular weights * sometimes come to many millions. The human body has hundreds of different proteins, with special structures that enable them to perform different tasks. All these proteins contain nitrogen, carbon, hydrogen and oxygen, and often various other elements, including iron, sulfur, copper and phosphorus.

Proteins can be broken down chemically into substances called amino acids. There are a number of different kinds of such acids; all contain an amino group, $-NH_2$, consisting of nitrogen and hydro-

gen, and also a carboxyl group, $-COOH$, made up of carbon, oxygen and hydrogen. The simplest amino acid is glycine, $CH_2(NH_2)COOH$, with the structural formula:

More than twenty amino acids in all have been obtained by breaking down plant and animal proteins. We list them in the article Basic Food Elements, in Volume 4. It is believed that the proteins consist of long chains of these acids, stretched out in zigzag form, or coiled in a spiral or compactly folded together.

The fats. The natural fats are made up of carbon, hydrogen and oxygen atoms. They are esters of the alcohol glycerol and a fatty organic acid.* Among the fatty acids are palmitic and stearic acids, which are saturated, and oleic and linoleic acids, which are not. The saturated fatty acids yield solids (at ordinary temperatures), when they react with glycerol; the unsaturated ones, liquids. Generally the fats occurring in nature are mixtures of solid and liquid fats. For example, the solid fats palmitin (from palmitic acid) and stearin (from stearic acid) make up about two-

---

* The molecular weight of a molecule is the sum of the weights of the atoms of which the molecule consists. The molecular weight of water (H_2O) is 18; of carbon dioxide (CO_2), 44; of glucose ($C_6H_{12}O_6$), 180.

* We have pointed out that esters result from the reaction of an alcohol and an acid.

fifths of hog lard; about three-fifths of it is a liquid fat, olein (from oleic acid).

The carbohydrates. The carbohydrates contain three chemical elements — carbon, hydrogen and oxygen. The general formula for all of them is $C_xH_{2y}O_y$. In the case of the carbohydrates d-glucose, x is equal to 6 and so is y. Hence the compositional formula for this compound is $C_6H_{12}O_6$. Note that in every carbohydrate there are always twice as many hydrogen atoms as there are oxygen atoms, just as in the water molecule (H_2O). The d-glucose molecule has the following structure:

$$
\begin{array}{c}
H \\
| \\
H-C-O-H \\
| \\
H-C-O-H \\
| \\
H-C-O-H \\
| \\
H-C-O-H \\
| \\
H-C-O-H \\
| \\
H-C=O
\end{array}
$$

The carbohydrates are called saccharides (compounds made up of sugars). The simplest are the monosaccharides, which are also known as the simple sugars.

a disaccharide, or double saccharide. The general formula for this type is $C_{12}H_{22}O_{11}$. The three most important disaccharides are sucrose (table sugar), lactose (milk sugar) and maltose (malt sugar).

The more complex carbohydrates are known as polysaccharides; they are made up of combinations of monosaccharides and their molecules are very large. They include starch, glycogen and cellulose, all of which have the general formula ($C_6H_{10}O_5$). The polysaccharide starch is found chiefly in the seeds or tubers of plants and serves as a food reserve. Glycogen occurs principally in the internal organs and blood of animals; it furnishes most of the fuel required for cellular activity. Cellulose forms the cell walls and woody structure of plants. We discuss starch, glycogen and cellulose in detail in several other chapters of THE BOOK OF POPULAR SCIENCE. (See the entries Starches, Glycogen and Cellulose, in the Index.)

In this short article we have been able to give only the briefest sort of introduction to the vast field of organic chemistry. However, you have learned something about the significance of molecular structure and the intimate relationships existing

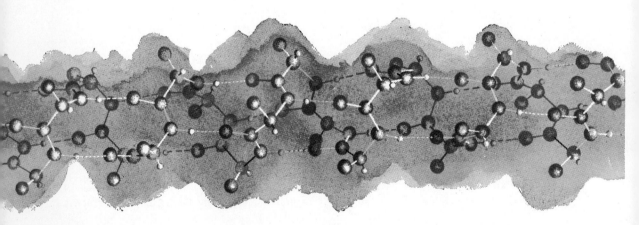

It is believed that proteins consist of long chains of amino acids.

Glucose, mentioned above, is a monosaccharide; so are fructose and galactose, which have the same compositional formula as glucose — $C_6H_{12}O_6$. When two monosaccharides combine chemically, they form

between organic compounds. If you would like to know more about this fascinating subject, you will find ample material in the books we list in Volume 10, page 280, under the heading "Organic Chemistry."

THE WORLD OF COLOR

Effects Produced by Different Wave Lengths of Light

BY J. H. RUSH

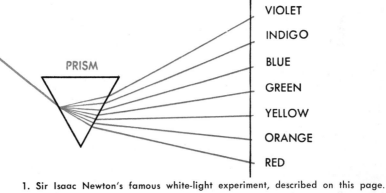

SUNLIGHT

PRISM

VIOLET
INDIGO
BLUE
GREEN
YELLOW
ORANGE
RED

1. Sir Isaac Newton's famous white-light experiment, described on this page.

THE perception of color adds greatly to our knowledge of the world around us, and is also a never ending source of pleasure. It causes us to appreciate fully the beauty of a sunset, the striking hues of autumn foliage or the manifold colors of the rainbow. Men have been keenly aware of color since the early times, as countless writings attest. Yet its nature remained a complete mystery until the seventeenth century. Sir Isaac Newton's classic experiments in 1666 yielded the first clues to its physical basis.

Newton admitted a narrow beam of sunlight into a darkened room through a very small hole in a window shutter. When he placed a triangular glass prism in the beam, the light falling on a normally white screen appeared as a series of colors (Figure 1). This is because light is refracted in passing through a prism; in

other words, its path is bent away from the original direction. This effect had long been known, but Newton's experiments showed that different colors are bent through different angles. Red is bent least; then, in order, orange, yellow, green, blue, indigo and violet. To Newton it seemed evident that these colors must be derived from the white sunlight. To prove this point, he used a second prism to bend the colored rays back to their original path. When the rays were recombined in this way, they produced white light.

Much later, when Thomas Young and other investigators established that light is a wave disturbance, they found that color is related to the wave length of light. Waves of light are not directly visible, as are the crests and troughs of waves on water; but they can be demonstrated by indirect methods.

Light is one of many different kinds of electromagnetic radiations — disturbances that travel through the universe in the form of waves.* (See the article Electromagnetic Radiation, in Volume 7.) These radiations all travel at a fixed speed of about 186,300 miles per second in free space. They show an indefinitely long range of wave lengths. Lightning generates waves that may be several miles long; the waves of gamma rays emitted from radioactive materials may be less than $\frac{2}{100,000,000,000}$ of an inch in length.

Scientists express very small distances in units called angstroms. An angstrom is equal to $\frac{1}{100,000,000}$ of a centimeter, or about $\frac{1}{254,000,000}$ of an inch. The wave lengths of visible light range from about 7,500 angstroms (about $\frac{30}{1,000,000}$ of an inch) in the red end to 4,000 angstroms (about $\frac{16}{1,000,000}$ of an inch) in the violet end. Electromagnetic waves in this range excite certain nerve endings in the retina of the eye. The impulses transmitted to the brain by these nerves give rise to sensations of light and color. We shall consider this process later in more detail.

A wave length of light is not a color; rather, it is related to a sensation of color. Different wave lengths excite different color sensations. The sensations can be produced without the agency of light. If you have reasonably good visual imagination, you can "see" colors at will by closing your eyes and imagining them. You can generate the sensation of color by pressing gently on your closed eyes. A blow on the forehead may cause you to "see stars."

How colors are isolated by refraction

We pointed out that a ray of light is refracted, or bent, as it passes through a prism. It is always refracted whenever it crosses the boundary between air (or a vacuum) and a transparent substance, except when it strikes the substance at right angles.** The paths of the different wave lengths of light will diverge within the

* Light and the other kinds of radiation behave like particles in certain respects.
** We discuss refraction in the article An Introduction to Optics, in this volume.

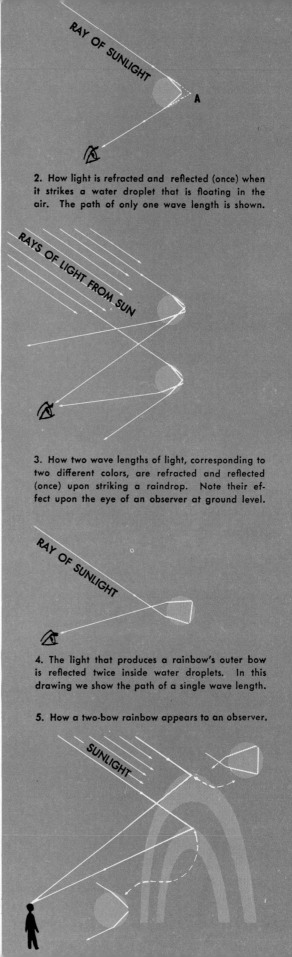

2. How light is refracted and reflected (once) when it strikes a water droplet that is floating in the air. The path of only one wave length is shown.

3. How two wave lengths of light, corresponding to two different colors, are refracted and reflected (once) upon striking a raindrop. Note their effect upon the eye of an observer at ground level.

4. The light that produces a rainbow's outer bow is reflected twice inside water droplets. In this drawing we show the path of a single wave length.

5. How a two-bow rainbow appears to an observer.

substance, because each wave length is bent at a different angle. This divergence is the basis of many natural phenomena and is applied by man in various ways.

A familiar example of color dispersion by refraction (and also by reflection) is the rainbow. This phenomenon is produced as sunlight strikes water droplets floating freely in the air. The droplets are nearly spherical. The sunlight that strikes them is refracted and reflected in such a way that the observer sees a huge arc extending above the horizon and showing the full range of color. A rainbow generally shows a bright primary, or inner, bow and a fainter secondary, or outer, bow.

The rays of light that produce the inner bow are refracted in water droplets, reflected once at the internal surface and then refracted again on emerging (Figure 2). Since the different wave lengths of sunlight are refracted differently, they follow different paths within a given drop and also as they emerge from it. Figure 3 shows how two wave lengths of light — those corresponding to violet and red — are refracted from a drop.

An observer can see color only from certain raindrops. In each case a given wave length of light, refracted and reflected from a drop, must strike the observer's eye. This happens only if the ray striking the drop and the emerging wave length that strikes the eye form the appropriate angle (A in Figure 2). The angle in question ranges from about 42° for red to 40° for violet. The observer will obtain the sensation of color from a vast number of drops, which from his viewpoint form an arc in the sky. The different colors of the inner bow follow one another in order from the outer part of the arc to the inner part, red being outermost.

The secondary, or outer, bow of the rainbow is formed by light that has been reflected twice inside each droplet (Figure 4). The order of colors is reversed from that of the primary bow and the colors are fainter. The two bows will appear to an observer as in Figure 5. Theoretically a third bow can be produced by

three internal reflections. However the third bow is rarely, if ever, intense enough to be visible.

The effects called halos and sundogs are produced quite similarly by the refraction of sunlight or moonlight in ice crystals high in the atmosphere. These colored rings or spots occur at definite angles from the sun or moon; they result from somewhat more complex optical processes than those that produce the rainbow.

The dispersion of wave lengths by a prism of glass (or other material) is the basis of one of the most useful instruments of science — the spectroscope. A familiar form of prism spectroscope is shown in Figure 6. Basically it is simply a prism of the type used by Newton, plus various accessories. The first section of the instrument is the collimator. This consists of a tube with a slit at one end and a lens at the other. The slit limits the width of the entering beam, so that the final images in different wave lengths will be narrow and will not overlap. The lens of the collimator bends the rays of light so that the rays from any point on the slit are all parallel as they enter the prism.

As the prism disperses the light, wave lengths are sent off in different directions. The lens on the other side of the prism brings each beam of parallel rays to focus again, forming a sharp image of the slit in each wave length. These slit images are viewed through a magnifying eyepiece; the second lens and the eyepiece form a small telescope. Figure 7 shows a side view of this type of spectroscope. The telescope, which can be swung around on a vertical axis, can be focused on different wave lengths of light.

There are various other kinds of spectroscopes. In the instrument called the spectrograph, the spectral images are focused on film and are photographed; the photographic records they produce are known as spectrograms. The prism is not the only device used to disperse the different wave lengths of light. The diffraction grating, described later, serves the same purpose; it is superior in certain respects to the prism spectroscope.

Characteristics of sources of light

The band of colors, or wave lengths, produced by a spectroscope is called a spectrum. All wave lengths are present in white light. Its spectrum forms a continuous, unbroken spread of color from the longest visible waves of red light to the shortest visible waves of violet. This is called a continuous emission spectrum (see page 284a). Incandescent solid or liquid substances, or even gases under unusual conditions, radiate white light and show a spectrum of this kind. Sunlight is essentially white; but its spectrum is interrupted by thousands of narrow dark lines, representing missing wave lengths (page 284a). As we shall see, these wave lengths have been absorbed by atoms in the sun's atmosphere. The sun's spectrum is called an absorption spectrum.

When a gas at low pressure is heated or excited by electrical discharge, it shows a different kind of spectrum. The molecules of the gas will always radiate certain sharply defined wave lengths, just as each string of a musical instrument "radiates" (that is, produces by vibration) its characteristic tone. This kind of radiation appears in the spectroscope as a bright-line spectrum — a series of isolated sharp images of the slit (page 284a). Each of the images represents a different wave length.

Even iron and other metals, normally solids, emit bright-line spectra in the vapor state. The vapors of the chemical elements and of many compounds can be identified by the characteristic wave lengths they radiate. Sodium vapor, for example, emits only two wave lengths in the visible spectrum: these are in the yellow-orange region. They are only six angstroms apart, and are easily mistaken for a single line in a small instrument.

The wave lengths of the bright sodium lines match those of two strong dark lines in the spectrum of the sun (page 284a). This is why. Any relatively cool vapor absorbs exactly the same wave lengths of light that it radiates when it is hotter. The region outside the brilliant white photosphere of the sun is cool, compared to the photosphere itself. Therefore the sodium vapor in this outer region absorbs its characteristic wave lengths out of the continuous spectrum radiated from the sun's dense interior. Instead of bright lines of sodium the sun's spectrum shows dark lines of this element. Other dark lines in the solar spectrum similarly reveal the presence of other elements in the sun.

Many substances radiate visible light even when cool and in the solid or liquid state if they are excited by other radiation. Such substances are said to be fluorescent. (Materials that store the exciting energy and reradiate it visibly later are said to be phosphorescent.) Zinc sulfide glows yellow under blue or ultraviolet light or X rays; barium platinocyanide glows blue. Mix these two materials, and you get a fluorescence that approaches white light.

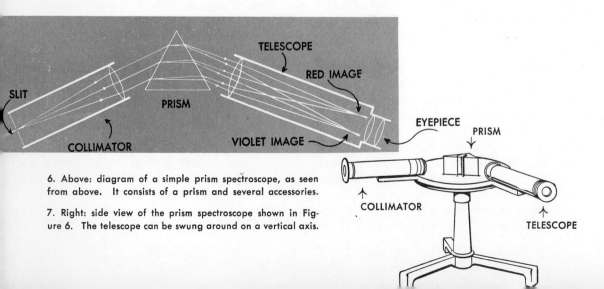

6. Above: diagram of a simple prism spectroscope, as seen from above. It consists of a prism and several accessories.

7. Right: side view of the prism spectroscope shown in Figure 6. The telescope can be swung around on a vertical axis.

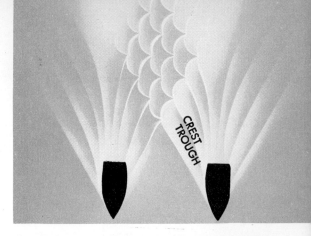

CREST
TROUGH

8. Waves produced by the boats show an interference pattern. The water is choppy where crests or troughs intersect each other; quiet where crests intersect troughs.

This principle is used in fluorescent lamps. A glass tube is coated inside with a mixture of substances known as phosphors that emit a great number of wave lengths when they are excited. The total effect of these wave lengths on the eye is similar to that of white light. The tube is filled with mercury vapor at low pressure. When the vapor is excited by a high-voltage discharge, it radiates strongly in the violet and ultraviolet regions. These radiations bring about visible fluorescence in the coating on the glass.

Effects of interference, diffraction and polarization

Bubbles of nearly colorless soap solution develop brilliant colors before they break. Similar colors appear in cracked glass or ice, or on the surface of very old glass, or in insect wings, or in oil films on water, or in light viewed through a window screen or through one's eyelashes. All of these color effects are caused by the interference of light waves and by a related effect called diffraction.

Interference can be illustrated by a familiar example, illustrated in Figure 8. When a motor boat cuts through calm water, it throws out a series of waves that run off at an angle from the bow. These waves move across the water at a definite speed. If two boats cruise on parallel courses some distance apart, their bow waves meet and cross each other. As they do so, they interfere. If crests on both waves pass a certain point at the same time, the water at this point rises to twice the height of one wave. This effect is called construction interference, or reinforcement. At an adjacent point, a crest of one set of waves arrives simultaneously with the trough of another set, and the two cancel each other. This is destruction interference. If the two wave motions interfere with one another in this way, the water at the said point will hardly be disturbed. As a result of reinforcement and destructive interference, the water between the two boats will show a checkerboard pattern — smooth at some points and with up-and-down motion at others.

Such interference effects are commonplace in the case of light waves. Let us see what happens when light is reflected from a thin film, such as the wall of a soap bubble (Figure 9). When light encounters a surface of transparent material, such as this, most of it goes on through; but a small fraction of its energy is reflected. We see a soap bubble, therefore, partly by light that is reflected back to us from the outside surface of the watery film. But some of the light is reflected from the second, inside surface also, and most of this light also comes back to our eyes. What we then see depends on the degree of interference that takes place between these two reflected beams of light. It will depend on the wave length of the light and the thickness of the bubble film.

In certain instances the two reflected wave trains for a given wave length will be in phase: that is, their positive crests will coincide and so will their negative crests * (Figure 10). The film will then appear bright in the color corresponding to that wave length because of what is known as constructive interference. The reflected wave trains for another wave length will be out of step, and will cancel one another as a result of destructive interference (Figure 11). The film will then appear dark as far as this particular wave length of light is concerned.

Usually we see a soap bubble film in white light, which presents the entire visible range from red to violet. Since the

* A wave of light can be represented as having a positive and negative crest, corresponding to the crest and trough of a wave of water.

wave lengths differ, the film cannot cancel or reinforce all colors at once. If the thickness of the film is such that one wave length will be canceled, another wave length will be partially reinforced. The colors that are reinforced and that are therefore visible will change continually as the film grows thinner by evaporation.

An air space between two surfaces of glass, ice, mica or other transparent material behaves much the same as a water film in air. If two such surfaces meet at an angle, the thickness of the air gap between them will vary along the wedge that is formed (Figure 12). If light passes through, one color and then another is reinforced. The colors of the spectrum are repeated along the wedge like a series of rainbows. Exceedingly small distances can be measured by instruments that depend on slight changes in the width of the air gap and the consequent shifts of the colored fringes.

The phenomenon called diffraction is basically an interference effect. A wide, uninterrupted wave front moves forward without appreciable change of form. If the wave is interrupted, however, by an opaque edge or a narrow obstacle, the light bends around the obstacle into the shadow zone. Light and dark fringes then develop outside the edge of the shadow. When white light is diffracted in this way, differences in wave lengths lead to color effects as in other cases of interference.

Diffraction effects ordinarily become noticeable only when light passes through very small openings or around very narrow obstacles. A very common and beautiful demonstration is the appearance of the moon through a door screen or window screen. Four bright streaks of light appear to radiate from the round globe, parallel to the horizontal and vertical wires of the screen. These streaks appear nearly white when seen with the naked eye. But a binocular or a small telescope will reveal that the streaks consist of a series of white-light spectra, in which the sequence from violet to red is repeated over and over again. This principle is applied in the diffraction-grating spectroscope (Figure 13). The diffraction grating used in this instrument consists of a surface of glass on which many thousands of fine grooves have been ruled; there are generally several thousand evenly spaced grooves to the inch. When white light strikes the diffraction grating, a series of continuous spectra is produced.

Striking color effects can be produced when certain objects are viewed in plane-polarized light. To understand what this kind of light is, it must be pointed out that light does not vibrate in the direction in which it travels, but at right angles to that direction. It vibrates in all possible directions perpendicular to the ray — up and down, sideways, obliquely (Figure 14).

Suppose we let a beam of white light pass through a thin plate cut from a crystal of the mineral called tourmaline. If the beam is directed to a screen after it passes through the tourmaline plate, the

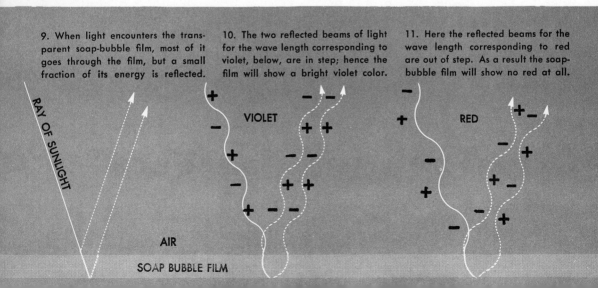

9. When light encounters the transparent soap-bubble film, most of it goes through the film, but a small fraction of its energy is reflected.

10. The two reflected beams of light for the wave length corresponding to violet, below, are in step; hence the film will show a bright violet color.

11. Here the reflected beams for the wave length corresponding to red are out of step. As a result the soap-bubble film will show no red at all.

RAY OF SUNLIGHT

VIOLET

RED

AIR

SOAP BUBBLE FILM

light will appear practically unaltered. Yet a definite change in its character has taken place. The tourmaline plate has permitted only the vibrations in a single plane to pass through (Figure 15). Let us now put a second tourmaline plate behind the first. This second plate can be set in such a way that light vibrating in one plane will pass through both plates and will appear on the screen (Figure 16A). If we now turn the second plate 90°, the light vibrations in the single plane can no longer pass through and no light at all will fall on the screen (Figure 16B).

When vibrations of light can vibrate only in a single plane as in Figures 15 and 16A, we say that the light has been plane-polarized. The plate that polarizes the light is called the polarizer; the second plate, which serves to analyze the light, is called the analyzer.

If we put a flake of mica between the plates, a series of beautiful spectral colors will appear on the screen. This is due to variations in the thickness of the crystal and in the wave lengths of light that pass through it. Various other transparent crystalline substances will show similar effects in plane-polarized light. If glass or transparent plastic is stretched or compressed, the stresses set up in this way alter its structure. When plane-polarized light is passed through it, colored fringes will appear.

Why is the sky blue? Why is the setting sun red? Why is deep water blue or green? The cause in each case is the scattering of light — a diffraction effect. Dust particles and even the molecules of atmospheric gases interrupt the wave fronts of sunlight advancing into the atmosphere. The light that was advancing in a straight line from the sun to the earth is dispersed; tiny new wave fronts develop in all directions. In a word, the light is scattered.

Like all diffraction and interference effects, the resulting color patterns depend on wave length. A given obstacle "looks bigger" to a short wave of violet light than it does to the longer red; therefore, the effect of the obstacle on the violet wave will be more drastic. Consequently,

violet light is scattered most, blue somewhat less and green, yellow and orange still less, in that order. Red is affected least of all.

The sky generally looks blue because the short blue waves are scattered more than the longer waves of red light. It is true that the violet waves are dispersed even more than the blue. However, the sky does not appear violet because the sun is relatively weak in violet light. Besides, the eye is less sensitive to violet light than to blue. Deep water appears blue for the same reason. However, impurities in the water often absorb the blue, and green becomes the predominant tint. The sky looks red near the horizon because at that angle, the path through the atmosphere is long and traverses much low-flying dust. As a result the bluer light is effectively scattered out of the direct beam before it can be observed by the viewer.

Twilight is also the result of the light scattered by particles and molecules in the atmosphere. Were it not for this scattering, we would be in darkness as soon as the sun went below the horizon.

The selective absorption
of wave lengths of light

A white dress, red roses, green leaves, brown paint, blue enamel, a black diamond — each of these terms implies that a definite color is associated with a particular kind of stuff? How do such colors arise?

First, let us see what we mean by "black" and "white." We have already

12. A wedge-shaped air space has been left between two glass surfaces. A given color will be reinforced at one place and canceled at another.

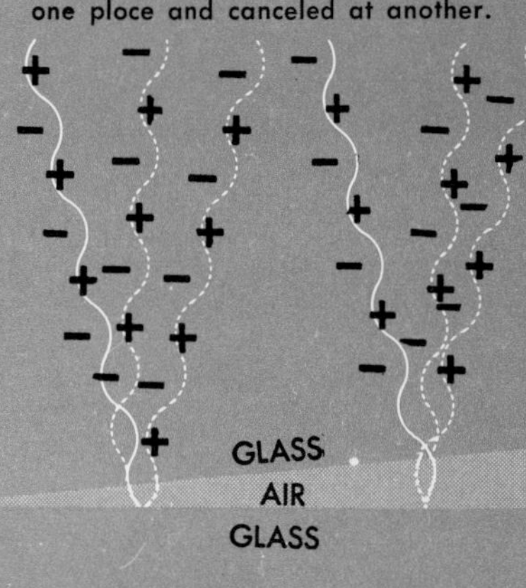

GLASS
AIR
GLASS

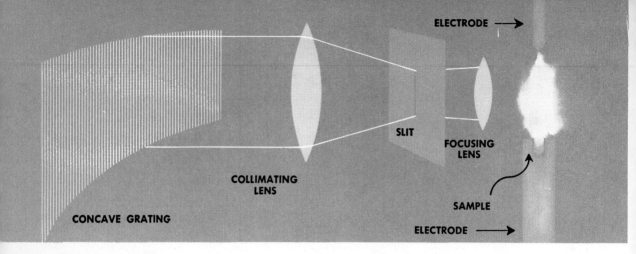

Labels in figure: ELECTRODE →, COLLIMATING LENS, SLIT, FOCUSING LENS, SAMPLE, ELECTRODE →, CONCAVE GRATING

13. The chemical composition of a given sample can be analyzed by a diffraction-grating spectroscope.

pointed out that white light is a mixture of all visible wave lengths of light. Obviously, when an opaque object * reflects all wave lengths equally well, it will appear white in white light. Surfaces that reflect all wave lengths weakly will look gray; extreme gray merges into black. A truly black surface would be one that absorbed all the light falling upon it, reflecting none whatsoever. Actually there is no absolutely black surface.

If an opaque object absorbs certain wave lengths and reflects others, we see a blend of the colors that it reflects. A red rose absorbs violet, blue, green and yellow wave lengths much more strongly than it does red. Blue enamel absorbs red and other longer waves and reflects blue and perhaps some green. If an object absorbs wave lengths at both ends of the spectrum — violet and red — and reflects the middle lengths, we see yellow-green.

The apparent color of an opaque object depends upon the quality of the light that falls upon it. Suppose the molecular structure of a substance is such that it can reflect only blue light. If it is illuminated by white light, it will absorb all wave lengths except that of blue. It will reflect blue light and will appear to be blue. If a blue beam of light falls upon the object, the beam will be reflected; the object will still look blue. Suppose, however, that we direct a red light upon the object. The red light will be absorbed. The object will

reflect no color at all; therefore, it will appear to be black.

It is a sensible precaution to take a piece of fabric to the front door of a store that does not have daylight illumination to see how the fabric will look in white light. Any amateur photographer knows that daylight and the light of a tungsten lamp differ greatly in color values they give objects. A special filter or special film must be used to equalize the rendering of color in these two kinds of illumination.

Visual judgments of color are very rough. They tell only the predominant behavior at the surface of an opaque object. Only the spectroscope can reveal just what the object really does with the light that falls upon it. Frequently the actual color pattern is quite complex.

What happens when light passes through a colored transparent or translucent substance? Some wave lengths are absorbed as they pass through the material; those that emerge in perceptible strength determine the color (Figure 17). The absorption pattern is complex; it varies according to the material. For example, if cellophane in a certain shade of green is folded and refolded and viewed against a strong white light (such as that from a photoflood lamp), the green color gradually darkens as the number of folds increases. After the cellophane has been folded fifteen or twenty times, the light that comes through will no longer be green but deep red.

The spectroscope would show why this mystifying color change occurs. The dye in the cellophane is only relatively

* A substance is opaque if no light can pass through it. A transparent substance transmits light freely so that objects on the other side can be clearly seen. A translucent substance also lets light pass through it; but it diffuses the light in such a way that objects on the other side cannot be seen clearly.

Spectrum of incandescent light, produced by a 4-by-4-inch Bausch and Lomb diffraction grating.

the world of
COLOR

Light is a form of energy, traveling through the universe in waves. The wave lengths of visible light range from rather less than 4,000 angstroms to more than 7,000. (An angstrom is 1/100,000,000 of a centimeter.) The waves of light in this range excite certain nerve endings in the retina of the eye. Impulses are transmitted by these nerves to the brain; here they give rise to sensations of color. Here we show some interesting color phenomena.

284-a

When light from a given light source is broken up by the instrument called the spectroscope into different wave lengths, the resulting color combination is called a spectrum. Below are some spectra.

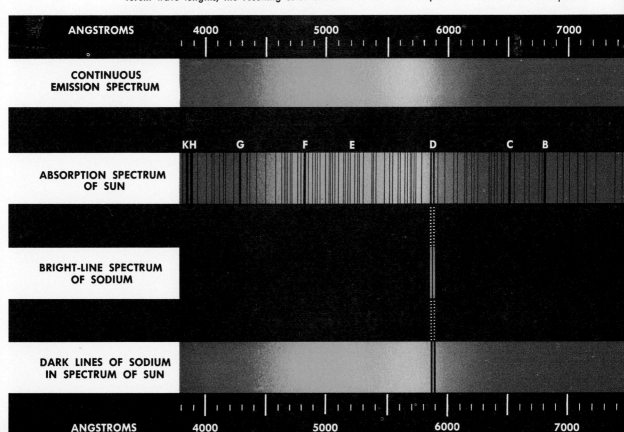

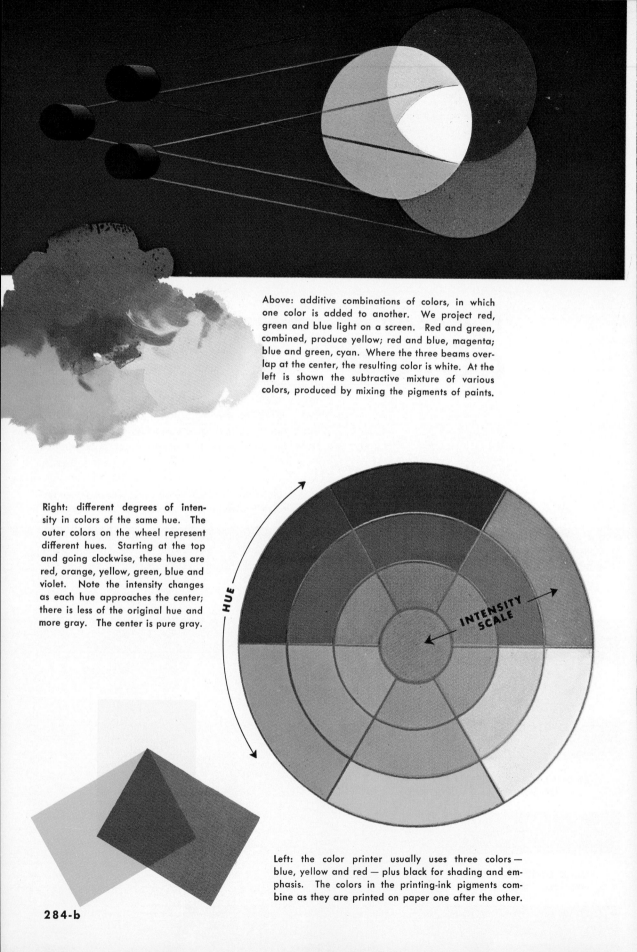

Above: additive combinations of colors, in which one color is added to another. We project red, green and blue light on a screen. Red and green, combined, produce yellow; red and blue, magenta; blue and green, cyan. Where the three beams overlap at the center, the resulting color is white. At the left is shown the subtractive mixture of various colors, produced by mixing the pigments of paints.

Right: different degrees of intensity in colors of the same hue. The outer colors on the wheel represent different hues. Starting at the top and going clockwise, these hues are red, orange, yellow, green, blue and violet. Note the intensity changes as each hue approaches the center; there is less of the original hue and more gray. The center is pure gray.

HUE

INTENSITY SCALE

Left: the color printer usually uses three colors — blue, yellow and red — plus black for shading and emphasis. The colors in the printing-ink pigments combine as they are printed on paper one after the other.

284-b

How white light is transformed to produce a series of striking color patterns. (1) Upper left: a full-scale replica of a bolt has been made out of a plastic called photoelectric Fosterite. The bolt has been heated and subjected to tension; the stress pattern that has been produced is preserved as the plastic cools. The bolt is sliced; when it is viewed in polarized light, the stresses that have been brought about show up as a definite color pattern. Note the vivid colors at the points of greatest tension. (2) Upper right: the closely spaced ridges on a phonograph record convert white light into colors through diffraction. (3) Lower left: every color of the spectrum is produced as light strikes cut-glass carvings, acting as prisms. (4) Lower right: color patterns brought about by interference as white light strikes an oil slick that has been spread on a black asphalt pavement.

#1 — Albert Fenn.   #2, 3 and 4 — Andreas Feininger.   LIFE © Time, Inc.

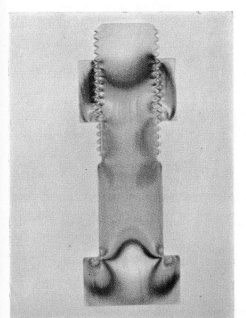

1 2

3 4

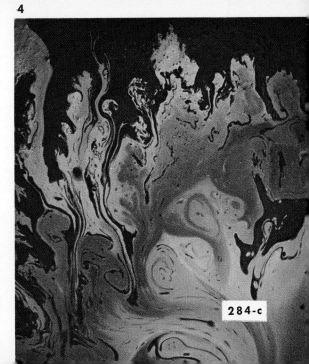

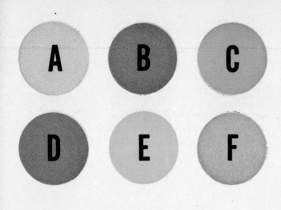

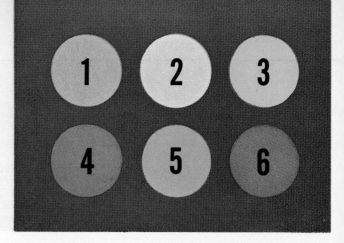

THE
PSYCHOLOGY
OF COLOR

Identical colors may appear to be quite different when they are seen against different backgrounds. Here is a good way to test the keenness of your color perception. The six lettered circles shown at the left are also given, in different order, at the right; here they are indicated by numbers. Try to match the lettered colors on the left with the numbered colors on the right. Then check your answers against the correct pairings, which are printed here upside down:

A-2; B-6; C-5; D-4; E-3; F-1.

CONTRAST CONTRAST

The colors used for the lettering in the two panels are identical. Note how faint the lettering in the left-hand panel appears, because the background is so light in color. The lettering in the other panel is far more legible, because it contrasts sharply with the background.

The five circles, above, are apparently different in color. As a matter of fact, these "different" colors are identical in hue, value and intensity. The apparent differences between them are due entirely to the fact that we view them against five different backgrounds.

Under very strong illumination, fix your gaze on the lowest right-hand star in the flag for fifteen seconds. Then look at a blank piece of paper. The flag will now appear in its proper colors. This effect is due to retinal fatigue, discussed in considerable detail in this article.

transparent to green. It is even more transparent to deep red than to green — that is, it permits a greater percentage of the red light to pass through. But the eye is not very sensitive to deep red, and so it responds to the green instead of to the red. When enough thicknesses of cellophane have been provided, the absorption of the green light becomes so much greater than the absorption of the red that the eye at last perceives red as the predominant wave length.

Color mixing:
lights and pigments

We know that when wave lengths of roughly equal intensities from all regions of the spectrum are combined, they give the color sensation of white light. Curiously enough, however, by mixing certain color pairs, we can deceive the eye into believing that the entire spectrum is present. Blue and yellow are one such pair; red and cyan (bluish green) are another. The colors in a pair of this kind may consist of light from very narrow portions of the white-light spectrum. Yet, if the wave lengths and intensities are properly balanced, the total impression they give is the same that we would get from viewing the whole spectrum of light.

Obviously there is more to the perception of color than the simple matter of wave lengths in the spectrum. As a matter of fact, this very complexity makes the task of the artist or dyer easier than it would be if every color had to correspond to a definite wave length. Two colors can match perfectly, as far as the eye can see, and yet appear altogether different in the spectroscope. Because of the flexibility of color sensation, it has been difficult to develop a science of *color,* as distinguished from a purely physical science of *wave length.*

The most familiar approach to the analysis of color is based on a set of three primary colors. Many different three-color combinations can produce the sensation of white light when they are superimposed. Red, green and blue are often chosen; they can be used as the basis of

a color wheel, which will show the relationship between the primary and secondary colors (Figure 18). When two primary colors, such as red and green, are combined, they give a secondary color. Red and green light produce yellow; blue and green give cyan; blue and red yield magenta, a shade of reddish purple. Each secondary color produced by the mixing of two primaries lies half way between them on the color wheel. For example, we saw

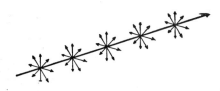

14. Ray of unpolarized light. The light vibrates in all possible directions in a plane at right angles to the ray.

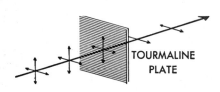

15. The plate permits only light vibrations in a single plane to pass through; the light has been plane-polarized.

16. In A, below, plane-polarized light passes through both plates and appears on the screen. In B; plane-polarized light passes through the first plate but is stopped by the second; it does not appear on the screen.

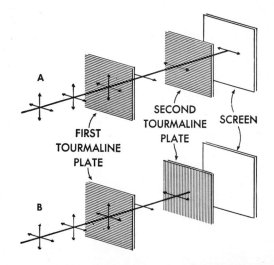

that red and green produce yellow; therefore, yellow lies half way between red and green on the wheel. There is an infinite variety of intermediate hues between adjacent primary and secondary colors.

It is interesting to note that the secondary color magenta is not a spectral color: that is, it does not correspond to any wave length in the white-light spectrum. It is produced when we bend the spectrum back on itself, so to speak, allowing the red and violet ends to overlap.

Each primary color on the wheel is opposite the secondary formed by the mixture of the other two primaries. The colors in such a primary-secondary pair are called complementary; when they are combined, they produce white. Green and magenta are complementary; so are red and cyan; so are blue and yellow.

The color relations we have been discussing apply only to *additive* mixtures of colored lights, in which one color is *added* to another. This kind of color mixing must be distinguished from the process of *subtractive* combination. In this process, blended colors *subtract* from white light the colors that they cannot reflect.

Let us examine the difference between the two processes. To illustrate the additive process we project yellow and blue light onto the same area of a white screen. We add one color (yellow) to the other (blue); both colors are reflected and the eye registers the combination as white light. When we mix yellow and blue paints, we are using the subtractive process. The resulting mixture, as any child who has dabbled in water colors knows, is green, not white.

To understand why subtractive combination produces this result, we must point out that paint contains coloring ingredients called pigments. These are finely divided particles of a substance that absorbs certain wave lengths and reflects others. Pigments do not absorb or reflect sharply defined regions of the spectrum. Yellow pigment absorbs violet and blue light almost completely; but it reflects some green and red; together with the yellow that it reflects most strongly. Similarly, blue absorbs practically all the red and yellow but reflects some green and violet, as well as blue. When yellow pigment and blue pigment are mixed and exposed to white light, each is free to absorb any light not absorbed by the other. The only region of the spectrum that is not fully absorbed by either pigment in this case is green. Therefore the mixture of yellow and blue pigments gives green.

Subtractive color mixing is the basis of color printing. The color printer applies his colors one at a time. Usually he uses three colors — blue, yellow and red — plus black for shading and emphasis. Each of four printing plates carries those details of the picture that appear in one of these colors; the plate is rolled with ink of that color and printed. When the four colors are printed on the paper one after the other, in exact register, the colors of the pigments combine subtractively and reproduce approximately the colors of the original.

We discuss the subtractive method of mixing colors, as applied to photography, in the article called The Versatile Art of Photography, in Volume 9.

The classification and measurement of color

Everyone recognizes that a word such as "yellow," or "green" or "blue" is not enough by itself to describe a color impression adequately. We commonly use the terms "strong," "rich," "pale," "weak," "dark" and "light" in our efforts to describe variations of colors. It is possible to analyze color impressions more accurately by giving three qualities or characteristics: hue, saturation and brightness.

Hue is the spectral characteristic of a color; in other words, it corresponds to a definite wave length of the spectrum. What of colors, such as magenta, that do not appear in the spectrum? In such cases, the hue represents the value obtained by the additive mixing of two spectral colors, representing two definite wave lengths of light. When we say that something is red, or green or yellow, we refer to its hue. No amount of shading or thinning can

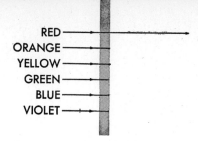

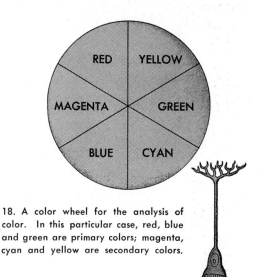

17. All wave lengths of light except that for red are absorbed in the transparent object above. The wave length for red passes through; the object appears red.

18. A color wheel for the analysis of color. In this particular case, red, blue and green are primary colors; magenta, cyan and yellow are secondary colors.

19. A cone of the retina. Cones are cells responsible for the appreciation of color and the perception of fine details.

turn red into green. Hue is really the fundamental quality of color sensation.

Saturation (sometimes called intensity) expresses the quality that we commonly call richness or strength of color. Suppose we project a beam of red light from the spectrum onto a white screen. The color we see is saturated. All of the light that comes to the eye from the screen is capable of exciting the sensation of red. Now, suppose we project a beam of white light onto the same spot as the red. The color is now diluted, paler; it is a combination of red sensation and the general sense of illumination that we call white. By varying the intensities of the white and red beams, we can get any degree of saturation. In handling pigments, adding white or gray to a hue is equivalent to adding white light. This results in a decrease in saturation.

Brightness (lightness, brilliance, intensity or value) is a measure of the effectiveness with which an object reflects light. A brightly colored object is one that reflects or transmits a large fraction of the light falling on it, so that it appears brilliant, or luminous. We can demonstrate brightness by projecting a spectral color onto a series of screens of various shades of gray. The brightness of the resulting color will vary according to the reflecting qualities of these gray shades. On a white screen, which reflects the greatest possible amount of light, the brightness of the color would be at its maximum. A black screen would not reflect any light; the brightness would be zero.

To provide a basis for comparing and cataloguing color samples, several systems of classification have been developed. All depend upon the fact that a color can be uniquely classified, provided its hue, saturation and brightness are specified. One of the systems is the Munsell color atlas, giving colored samples of different hues and various degrees of saturation and of brightness for each hue. The intervals between the successive hues, saturations and brightnesses are small enough so that they form a practically continuous series.

If a textile merchant orders a fabric of a color corresponding to a particular Munsell specification, he will have a good idea of what the material will look like. A system such as the Munsell color atlas cannot, however, be absolutely precise. In the first place, the same kind of light source must be used to illuminate both the Munsell standard and the sample that is to be compared with it. This requirement is very important, as we have already noted. Daylight and tungsten lamps, for example, give radically different impressions of some color samples, particularly mixtures that involve green or blue. Differences in texture of the colored surfaces also affect the color sensation one obtains from them. Perhaps the most troublesome factor in color standardization is the human element. Few people would agree exactly on a match of two colors. At least one person in twenty would disagree violently with the majority, and he would not necessarily be color-blind.

The Munsell atlas and other catalogue systems depend upon the matching and identification of colors in terms of the color sensation they arouse, and not in terms of their wave lengths. For many purposes, a more accurate standard is desirable. Instruments called colorimeters have been developed to measure color properties according to wave lengths of light. One type of colorimeter uses an ingenious approach, based on the hue, saturation and brightness of the sample. A spectroscope selects the wave length that matches the sample hue. An auxiliary beam of white light is then added to the hue and its intensity is adjusted to match the saturation of the sample. Finally the brightness of the combination is adjusted to match the brightness of the sample. In this way the three factors of color are related to the white-light spectrum.

Of course, no measurement means anything unless the standard to which it relates is accurate and consistent. It is useless to refer a color sample to a white-light spectrum if the reference light is not really white, according to an agreed standard. Much of the work of colorimetry, or color measurement, has gone into specifying and developing stable, reproducible illumination sources.

A theory of
color vision

Many attempts have been made to explain the perception of colors. A number of puzzling details must be accounted for; that is why, even today, no theory of color vision is completely satisfactory. Most scientists agree, however, upon one basic approach to this problem. This approach was suggested by Thomas Young and developed by the distinguished German physicist and physiologist Hermann von Helmholtz a century ago.

According to the Young-Helmholtz theory, the cones * in the retina of the eye are of three types. One type is most sensitive to the red region of the spectrum (that is, the wave lengths corresponding to the red region). Another is particularly sensitive to the green region, a third to the blue-violet (see Figure 20). Each type of cone is sensitive to a wide range of other wave lengths. For this reason, as the diagram shows, the sensitivity curves overlap. The red-sensitive cones respond not only to red, but also, though less strongly, to orange, yellow and green. The green-sensitive elements respond also in some degree throughout the range from red to blue; the blue-sensitive cones respond also to violet and green.

Light of almost any single wave length must excite at least two sets of cone-elements to some extent, because of their overlapping ranges of sensitivity. We pointed out that light from the red and green regions of the spectrum, received

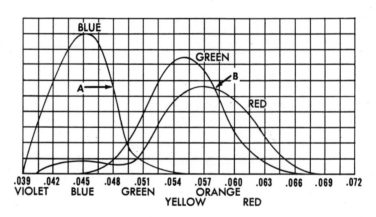

20. Chart illustrating the Young-Helmholtz theory of color vision, described on this page. According to this theory, the cones in the retina are of three types. The first is most sensitive to the red region of the spectrum; the second, to the green region; the third, to the blue. Each type is also sensitive to a wide range of other wave lengths. The figures in the chart show the wave lengths in microns. (A micron is a thousandth of a millimeter.) Increasing distance from the base line indicates increasing intensity of sensation.

.039   .042   .045   .048   .051   .054   .057   .060   .063   .066   .069   .072
VIOLET   BLUE   GREEN   ORANGE
YELLOW   RED

* Cones are curiously shaped cells (see Figure 19) that are responsible for the appreciation of color and also for the perception of fine details in the image that is focused on the retina. See Index, under Rods and cones of eye.

as a single impression, is perceived as yellow. If we accept the three-color theory, yellow is a composite sensation that can be produced by stimulation of both the red-sensitive and green-sensitive cones by wave lengths near their points of maximum sensitivity. We also know that a single wave length in the spectrum between the red and green also produces the sensation of yellow. In the diagram we note that the yellow region of the spectrum is situated where the red and green response curves cross.

If all three sets of retinal cones are equally stimulated, the resulting sensation is that of white light. Two or more isolated wave lengths can produce the white-light sensory response if they are so located that they cross the sensitivity curves at equal heights. Suppose color sensation

21. The Benham disk (above) is mounted on a spindle piercing the center. If the disk is spun, colors appear.

is produced by two wave lengths of light. One of these, say, is yellow, which intersects the red and green curves at B (in Figure 20). The other wave length is a shade of blue that intersects the blue curve at about the same height (at A). These two wave lengths will combine to produce the sensation of white light.

There is nothing in the structure of the retina to support or refute the three-color theory of color vision. All the color-sensitive cones look alike under the microscope. The theory has not successfully explained some details of color sensation; however, it is supported by many bits of indirect evidence. One of the most interesting of these is retinal fatigue.

If you stare fixedly for at least fifteen seconds at a colored pattern, so that the colored image is held on a definite region of the retina, and then close your eyes or look steadily at a white surface, you will see the pattern in its complementary color. This seems to be due to the fact that exposure to a red pattern, for example, produces an effect similar to fatigue in the retina or the brain and reduces the sensitivity to red. Then, when you expose the retina to white light, green- and blue-green-sensitive cones in the fatigued area respond more strongly than the recently exposed red-sensitive cones. The resulting color sensation is that of the white-light spectrum with the red region removed — that is, cyan, the color that is complementary to red.

Color blindness affords further evidence for the three-color theory. A person with green color blindness, for example, responds to various light stimuli as if the green-sensitive curve in Figure 20 were missing. Other types of color blindness are just as consistent. It is as if one or more of the response curves shown in Figure 20 were lacking.

Oculists' charts for detecting color blindness make use of this principle. A chart of this kind is composed of a field of many small colored spots. The pigments in these spots all look alike to a person with normal vision; but the spectral colors they reflect are so chosen that certain spots look different from the others to one who is color-blind. These "peculiar" spots are arranged to form numerals or other symbols, and the person being tested is asked to read what he sees. From his response, the oculist knows immediately what elements of color vision he lacks.*

The device known as the Benham disk (Figure 21) demonstrates the color-response characteristics of the retina. It consists of a black semicircle and a white one with four sets of black arcs marked on it. The disk can be made of cardboard and mounted on a toothpick piercing the center of the disk. When the latter is spun

* Editor's note: For a different theory of color perception, see The Land Theory of Color Vision, in Volume 8.

like a top, colors appear. They are different for the four sets of arcs shown in the diagram, and they change with the direction or speed of rotation. These color sensations are not fully understood. Evidently they depend upon differences in the promptness with which the three sets of color-sensitive retinal cones respond to light stimuli.

The psychology of color perception

The visual effects we have been discussing depend upon stimuli affecting the retina and other structures of the eye. Many subtler peculiarities of color vision involve the complex processes by which visual sensations are interpreted in the brain. Among them are harmony or discord of colors, the effects of adjacent colored areas on each other and the influence of colors upon emotional moods.

Most people are aware that certain combinations of colors are pleasing, while others are distinctly displeasing or even painful to one of artistic sensibilities. The harmonious blending of colors is a fine art, which is based on certain definite principles and practices.

As in any kind of artistic composition, colors must give an effect of clear relationship and unity of purpose. Whatever colors are used, one color must clearly dominate. The others must relate themselves to the dominant color, but must not compete with it. The simplest harmonious effects depend upon the colors that have something in common. They may be different shades or tints of the same hue, or they may be neighboring hues on the color wheel. Red and yellow might clash by too much contrast. However, an intermediate orange will blend with either, or even unite all three colors harmoniously.

One of the strongest color discords occurs when highly saturated complementary colors compete with each other. We have already pointed out that the afterimage of any color in the retina is its complementary color. This fact may have something to do with the peculiarly unpleasant effect when complementaries are combined. Each is in a sense the negative of the other. However, if one complementary is subdued by reducing its saturation or brightness, it serves to enhance the other complementary, rather than to compete with it.

Subdued colors nearly always harmonize because no single color stands out strongly enough to dominate or rival the others. Compositions of this type are monotonous, however. The recent trend in clothing and decoration has been toward the more difficult but much more positive harmonious blends of bright hues.

Colors influence each other strongly by contrast, especially when they are adjacent to one another. Such contrasts usually follow the principle of black-and-white contrast effects. For instance, a small gray circle on a white background appears distinctly darker than the same circle on a background of black. In much the same way, a color generally appears lighter in contrast to a dark shade than to a light tint, regardless of the contrasting hue. Such effects are obviously important in clothing design and interior decoration, as well as in painting.

Color can deceive the eyes in many ways. A white wall seems nearer than the same wall painted a dark shade. Red seems nearer than blue; this effect seems to depend upon the way in which the eye accommodates to radically different wave lengths.

Certain more or less definite emotional effects are produced by colors. Green is restful; red exciting and disturbing. These two colors have long served for traffic signals. Green indicates that an automobile driver can safely proceed; red, that it would be dangerous to go forward. Our emotional reactions to these two colors may be based on the reassuring green of grass and trees and the alarming color of arterial blood; but this is mere guesswork. Blue is a "cool" color; do we think of blue water in this connection? It is favorable to sleep; but it can also be depressing. We say that we are "feeling blue" when we are unhappy; "blue songs" are sad.

See also Vol. 10, p. 280: "Color."

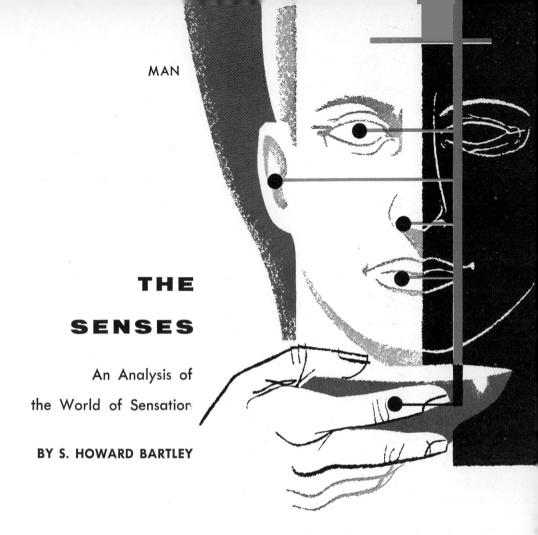

THE
SENSES

An Analysis of
the World of Sensation

BY S. HOWARD BARTLEY

FOR centuries it was accepted as self evident that there are five senses: vision, hearing, touch, taste and smell. We now realize that this list is quite inadequate. Besides the five senses it includes, there are senses of pressure, pain, heat and cold; there is a sense of balance and a muscle sense. They all function through certain specialized organs called sense organs. Some of them are more or less modified nerve endings. Others — especially the eyes and ears — are comparatively large and very complex structures.

When scientists want to find out about the senses, their experiments follow a pretty definite course. First, they try to find out the tiniest stimulus that will affect a sense organ so as to result in a conscious experience, or sensation. The stimulus that will just suffice to do this is called the threshold amount, or simply threshold. There are many kinds of thresholds. For example, there are thresholds for red, green and other color sensations, a threshold for touch and a threshold for pain.

The researcher is also concerned with the variations in the strengths of stimuli that will bring about just-noticeable differences in sensations. These just-noticeable differences, also called J.N.D.'s, represent the number of perceptible steps into which a range of sensations can be divided. If there are 60 J.N.D.'s in a band of gray tapering from white to black, it means that to the person looking at the band, there are 60 perceptible grays. It is also important, in studying the senses, to find out how the strength, amount and duration of a stimulus compare with the sensation that the stimulus produces.

The sense
of vision

The sense of vision is one of the most versatile of all. It supplies us with information about objects at very close range,

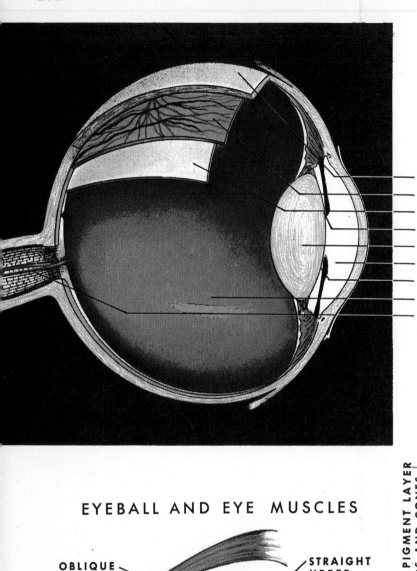

CUTAWAY DIAGRAM OF EYE

SCLEROTIC COAT
CHOROID COAT
RETINA
IRIS
LENS
AQUEOUS HUMOR
IRIS
VITREOUS HUMOR
OPTIC NERVE

LAYERS OF RETINA

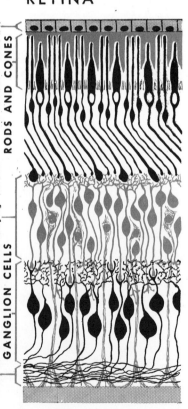

PIGMENT LAYER
RODS AND CONES
BIPOLAR CELLS
GANGLION CELLS
NERVE FIBER LAYER

EYEBALL AND EYE MUSCLES

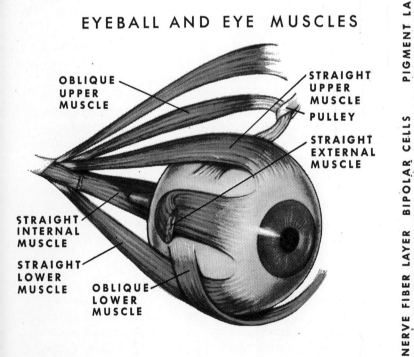

OBLIQUE UPPER MUSCLE

STRAIGHT UPPER MUSCLE
PULLEY
STRAIGHT EXTERNAL MUSCLE

STRAIGHT INTERNAL MUSCLE

STRAIGHT LOWER MUSCLE

OBLIQUE LOWER MUSCLE

at a great distance and at an infinite number of distances between these two extremes. We become aware of these objects through the sense organ known as the eye. Rays of light pass through a transparent part of the eyeball, penetrate through the inner part of the eye, and fall upon a layer of sensitive tissue — the retina. As a result impulses arise, and these impulses are carried by way of nerve fibers to the brain. Here the impulses give rise to the sensation of sight.

The eye, which is enclosed in a bony cavity called the socket, is more or less spherical in shape and is about an inch in diameter. It can be turned in different directions by six muscles, attached at one end to the eyeball and at the other end to the wall of the socket. The eyeball wall, which encloses the liquid and semiliquid interior sections, is made up of three layers. The outermost one is the tough sclerotic coat, which protects the delicate inner structures of the eye. We can see part of this coat; it is what we call the white of the eye. In front there is a circular, transparent area, called the cornea. The outer surface of the sclerotic coat is lubricated by tear fluid, secreted by the lachrymal gland. From the eye surface the fluid passes into small tubes — the lachrymal ducts — and from these into the nose. Protection for the outer surface of the eye is provided by the eyelids. These useful structures close as a result of reflex action when something is about to touch the eye.

The middle layer of the eye wall is known as the choroid coat. It carries a multitude of small blood vessels, which extend throughout the inner part of the eyeball except in front, where there is a hole, called the pupil of the eye. Surrounding the pupil is a circular band, called the iris; it gives the eye its color. The third and innermost layer — the retina — lines most of the eyeball interior (but not the front part). This tissue, as we have observed, converts the energy of light waves into nerve impulses, which are sent to the brain.

The retina consists of several layers, of which the most important are the rods and cones, the bipolar cells and the ganglion cells. The rods and cones are curiously shaped nerve cells (see diagram) that are sensitive to light. Impulses arise in them when they are struck by light rays. As a result other impulses are set up in the bipolar cells. These impulses are then transmitted to the brain by way of the fibers of the ganglion cells. The fibers make up a cablelike bundle, called the optic nerve, which issues from the back of the eye.

The rods are sensitive to far smaller amounts of light than the cones; they are responsible for vision in dim light. Cones are particularly concerned with the perception of fine detail and of color. The rods are sensitive to light because of a chemical reaction brought about by the light rays that enter the eye. The outer part of the rods contains a red pigment called visual purple, or rhodopsin. When acted upon by light, it decomposes into a substance called retinene and then into vitamin A and a protein. The vitamin A and the protein are later reconverted into visual purple. A certain amount of vitamin A is lost in the process. This loss must be made good by fresh stores of vitamin A, supplied through the circulation. If a person's diet is lacking in this vitamin, he will not be able to see well in dim light — a condition called night blindness. It is believed that much the same kind of chemical reaction takes place in the cones.

Rods and cones are not uniformly distributed throughout the retina. A small region in the center, called the *fovea centralis,* contains only cones; this region is concerned with acute vision. The cones are distributed more and more sparsely toward the edge of the retina, while the relative number of rods increases. We see clearest in daylight by using the center of the eye — the *fovea centralis;* at night, by using a region somewhat to the side. We are blind at all times to objects that cast their images only on the part of the retina where the optic nerve leaves the eye. There are no rods and cones in that part of the retina, which is called the optic disc. This results in a blind spot in the field of vision.

We can demonstrate the existence of the blind spot by a simple experiment, illustrated in the diagram on this page. Close your left eye and look at the cross in the diagram, holding the page close to your eye. Gradually move the page away from the eye. At a distance of about ten inches, the black spot beside the cross will become invisible. The reason is that the image of the spot will be formed on the part of the retina where there are no rods or cones — the optic disc.

Thus far we have described the wall of the eye, with its three layers — the sclerotic coat, the choroid coat and the retina. Let us now examine the internal structure of the organ. A short distance behind the cornea is the elastic crystalline lens, which focuses the light rays entering the eye. The lens, which is supported by a hammock-like ligament, is transparent in the case of a healthy eye. Sometimes it becomes clouded — a condition called cataract. In the commonest type of cataract — senile cataract, generally found in persons over fifty years of age — vision is gradually lost as the condition develops. To restore sight, the lens is removed and special glasses are provided to take its place. In other types of cataract the cloudiness may disappear more or less completely in time. This is particularly true in the case of cataract caused by perforation of the lens, due to injury.

The interior of the eyeball is divided by the crystalline lens and its supporting structures into two chambers. The one in front of the lens is filled with a watery solution called aqueous humor. The other chamber, occupying the space between the lens and the retina, is much larger. It contains a jellylike substance called vitreous humor.

In the article An Introduction to Optics, in this volume, we pointed out that light comes to us in a series of waves. Different wave lengths excite different sensations of color when they strike the retina. The eye is not equally sensitive to light rays of all wave lengths. Certain rays will not result in any sensation of vision; these are the ultraviolet and infrared rays.

We see objects because waves of light are reflected from them and enter the eye. The organ of vision has often been compared to a camera. In the camera, light reflected from an object passes through a lens, is focused on a film at the back of the camera and forms a small image. In the eye, reflected light passes through the crystalline lens and is brought to a focus on the light-sensitive retina. An adjustable diaphragm in the camera increases or decreases the amount of light that reaches the film. In the eye this sort of adjustment is made by the iris. Its opening can be made wider or narrower by means of muscles radiating from the inner rim of the iris.

Rays of light from a distant point are brought to a focus closer to the camera lens than rays reflected from nearby objects. To cause these dissimilar rays to be focused sharply on the film, the camera lens must

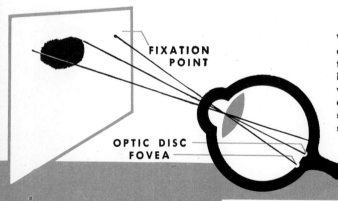

FIXATION POINT

OPTIC DISC
FOVEA

We see most clearly in daylight by focusing on a small pit, or fovea, in the retina. At all times we are blind to objects that cast their images on the optic disc, the part of the retina where the optic nerve leaves the eye. The existence of the blind spot can be demonstrated by a simple experiment, utilizing the cross and circle shown below and described in detail on this page.

be moved nearer to the film or farther away from it, as the case may be. In the case of the eye the proper adjustment is made by changing the shape of the lens, which is an elastic body. This is done by means of tiny muscles within the eyeball. In the diagram we show how the lens adjusts for far and near vision. The process of changing the shape of the lens in this way is called accommodation.

The lens of the eye produces an inverted image in the retina. (See the diagram on this page.) This image is interpreted in the brain in such a way that it appears erect. For example, when the lower part of the retina is stimulated by light rays reflected from a given object, it seems to us that the object is in the upper part of the field of vision. If we put on glasses which focus the incoming rays in such a way that upright images appear in the retina, everything seems topsy-turvy at first. In time, however, we learn to make the correct interpretation. If we remove the glasses and let inverted images be formed on the retina, we must learn all over again how to interpret the images.

The eye is able to adapt to the various levels of light that one meets in everyday life, except to such extreme intensities as those of the sun or of arc lights. There is a striking example of this faculty of adaptation when we enter a movie theater from broad daylight. At first, little is visible; soon, however, darkness gives way to dim illumination, which enables us to find our way to a seat.

In the article The World of Color, in this volume, we tell how the different sensations of color are interpreted in the brain. We also deal with the interesting phenomenon called color blindness.

It is important for our sense of vision to tell us where objects are. This is called localization. When we try to decide where an object is, our decision will depend upon *what* we think the object is. For instance, if you saw an illuminated white rectangular area in a dark room, you might consider it to be a movie screen; in that case, it would seem to be a number of feet away. Suppose you thought of it as a blank filing

card; the bright rectangular area might seem to be only inches from your eyes. In a case such as this, the observer does not have to analyze the situation at length in order to place an object. He makes an instant judgment as to what the object is, and he sees it at what he considers the appropriate distance.

If one object is partly covered by another, the covered one will appear to be farther away. This is called overlay. Elevation is another factor. Objects lying higher in the field of sight generally appear

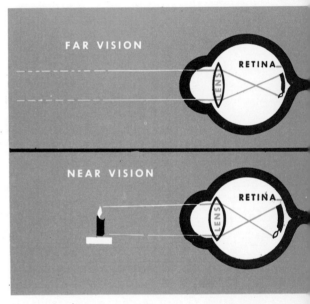

How the lens, which is an elastic body, accommodates — that is, changes its shape — for far and near vision.

farther away than lower-lying ones. This effect is sometimes reversed; in that case the higher object seems nearer. Relative size also helps to localize objects; smaller ones seem farther away. Relative brightness is still another factor in determining apparent location. Usually the brighter of two otherwise similar objects seems to be nearer. An object appears to move toward the observer when it is made brighter, and to recede when it is dimmed.

Still another important factor in the localization of objects is parallax. This effect is obtained as one moves past objects or simply moves one's head. Certain objects will seem to stand still; others will

Rays from distant objects come to a focus in front of the retina.

seem to move in the direction opposite the observer's motion. The "still" objects will appear to lie beyond those that "move." For example, suppose the train on which one is riding moves past a telephone pole and a building. The pole will apparently move in a direction opposite that in which the train is moving, but the building will stand relatively still. The building will seem to be farther away.

The fact that we have binocular vision — that is, vision through two eyes — helps us to perceive depth. Each eye gets a somewhat different view of a given object, particularly at close range. When the two retinal images are interpreted together in the brain, we have a clear three-dimensional view. The depth of an object is not so clearly perceived when we look at it through one eye.

The interpretation of objects in the brain may not correspond to their actual appearance. In that case there will be a series of optical illusions. A number of such illusions are shown on the next page.

The vision of certain persons is limited or defective. The crystalline lens may refract rays of light effectively enough, but the retina is either farther away from the lens than in a normal eye or nearer to the lens. In such cases special artificial lenses, spherically curved, must be prepared in order to provide normal vision. These lenses may be set in spectacle frames or may be applied directly to the eyeball in the form of contact lenses.

In the condition called myopia, or nearsightedness, the retina is closer to the lens than in a normal eye. When light rays are reflected from objects near to the eye, they focus properly on the retina. Distant rays, however, come to a focus in front of the retina. To correct this defect in vision, a concave (minus) artificial lens is placed in front of the eye. It diverges rays

and makes even parallel rays from far-off objects reach the lens of the eye as if they came from nearby objects. Hence the lens focuses them on the retina and not in front of it.

In farsightedness, or hyperopia, the retina is farther away from the lens than is normal. Rays reflected from far-off objects are focused properly on the retina; rays from nearby objects are focused behind it. A convex (plus) lens is used to remedy this defect.

Certain corrective lenses are in the form of prisms; they bend incoming rays of light so that the eyes need to converge ("turn in") either more or less than without glasses to obtain the same visual results. These lenses prevent one from seeing double. Other lenses are intended to remedy the condition called astigmatism, in which light rays are brought to focus on the retina not in the form of sharp points, but as short lines. Astigmatism is due to the fact that the crystalline lens has raised or depressed areas on its surface. A person suffering from this condition must wear lenses with raised or depressed areas to compensate for corresponding defects in the crystalline lens.

In fitting glasses, sharpness of vision, or visual acuity, is an important factor. There are several standard tests for visual acuity. One of the most common is the Snellen Chart. It consists of several lines of letters of different sizes. A person with normal vision will be able to read letters farther away than the person with poor visual acuity. A person is said to have 20/20 vision if at twenty feet (the standard distance) he can read letters that a person with normal vision can read at the same distance. If at twenty feet, a person can just read letters that are so big that the normal person can read them at forty feet, the person is said to have 20/40 vision.

Rays from objects near at hand come to a focus behind the retina.

OPTICAL ILLUSIONS

Things are not always what they seem when we use our senses to obtain information about the outer world. On this page we show how the sense of vision sometimes plays us false. For example, in the figure within the circle, at the right, the outer vertical lines will sometimes seem nearer to the eye than the central vertical line; sometimes they will seem farther away. In the next figure, the two diagonals are of the same length.

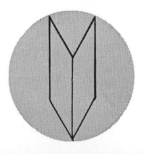

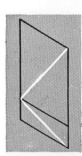

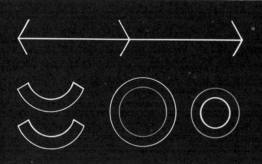

The two horizontal lines forming part of the two arrows at the left are equal in length. The two curved figures under the arrows (left) are equal. In the case of the circles within circles, the inner circle of one is equal to the outer circle of the other.

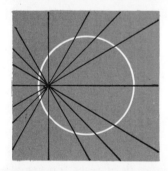

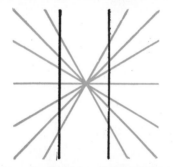

The curved figure set within the radiating lines is a true circle. The two heavy vertical lines in the next figure are straight and are parallel to each other. They certainly do not appear to be so.

If the eye of an observer were at the point where the four lines converge, he would not be able to distinguish between objects A, B, C, D and E. To one standing beside the observer, the objects would seem quite different.

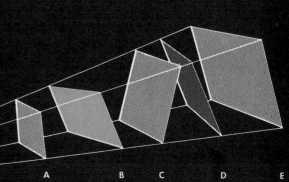

A B C D E

The sense
of hearing

When we hear sounds, we are really becoming aware of waves of air that strike our ears. These waves arise as the result of vibrations caused by some disturbance or other, such as the ringing of a bell or the tooting of a horn or the drawing of a bow across the strings of a violin. The vibrations are sent through the air in a series of alternating compressions and rarefactions. We describe the production and transmission of sound waves in the article on sound (see Index, under Sound).

All the air waves in our vicinity strike our ears, but not all produce the sensation of sound. The human ear registers only the vibrations that range from about 16 a second to between 20,000 and 30,000 per second. We could also say that the vibrations have a frequency of from 16 to between 20,000 and 30,000 cycles per second.

The ear, in which we receive the vibrations, is divided into three compartments — the outer ear, the middle ear and the inner ear. The outer ear is made up of two parts. One is a structure of cartilage and skin, called the auricle, or pinna. The other is a short canal — the auditory canal — which penetrates the bone of the skull and is closed at the inner end by a delicate membrane — the eardrum. The auricle is the part of the ear that protrudes from the side of the head; it serves to direct sound into the ear.* The sound waves entering the auditory canal set the eardrum, at the end of the canal, into motion. It vibrates in time with the air waves.

The middle ear is a small chamber, hollowed out in the skull bone. In this chamber, there are three small bones, called

* Animals, such as horses or cats, can turn their ears in the direction of a sound. Man cannot. It is true that a few persons can move (wiggle) their ears somewhat. The muscles that enable them to do so are rudimentary vestiges of the muscles that once enabled man to move his ears freely.

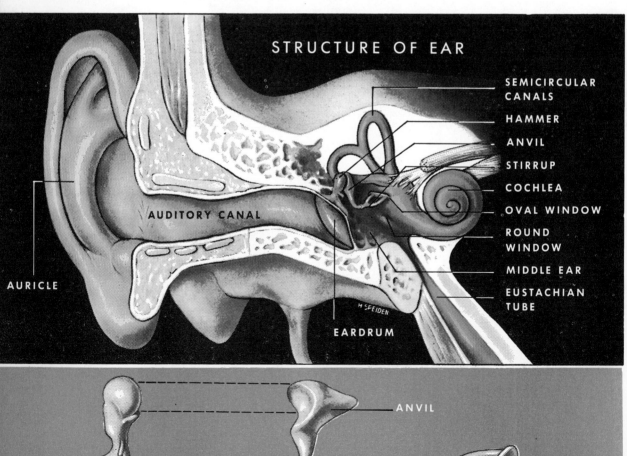

STRUCTURE OF EAR

SEMICIRCULAR CANALS

HAMMER

ANVIL

STIRRUP

COCHLEA

OVAL WINDOW

ROUND WINDOW

MIDDLE EAR

EUSTACHIAN TUBE

AUDITORY CANAL

AURICLE

H SFEIDEN

EARDRUM

ANVIL

HAMMER

STIRRUP

ossicles, which are joined together; they form a connecting link between the eardrum and the inner ear. The ossicles act as a lever system, transmitting to the inner ear the vibrations that strike the eardrum. The outermost of the three bones is attached to the eardrum; because of its shape this bone is known as the hammer. The middle bone vaguely suggests an anvil; that is what it is called. The name stirrup, applied to the innermost bone, is very apt; the resemblance to a stirrup is quite striking. These bones are sometimes given the

The outer and middle ears serve only to transmit vibrations to the inner ear, where we find the basic organ of hearing. This is contained in a spiral chamber, called the cochlea ("snail shell," in Latin), which is imbedded in bone. The cochlea makes about two and a half turns. The inner ear also contains three semicircular canals and two small sacs; these structures have to do with the sense of balance, which we shall discuss later, and not with the sense of hearing. Only the cochlea is concerned with the latter sense.

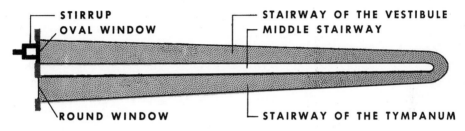

How the three tubes of the cochlea would look if they were removed from it and uncoiled.

Latin names malleus (meaning "hammer"), incus ("anvil") and stapes ("stirrup"). The footplate of the stirrup fits into an oval window, beyond which lies the inner ear; this window has a membrane stretched across it. A little below the oval window is a round window, also closed by a membrane.

A small tube, called the Eustachian tube, runs from the middle ear to the back of the nose. The middle ear connects by means of this tube with the outer air, whenever the tube is open. It opens only in the act of swallowing; whenever this takes place, air passes up the tube into the middle ear and causes the air pressure on both sides of the eardrum to be equalized. If the Eustachian tube is blocked because of a cold in the head, part of the air in the middle chamber is absorbed into the blood. There is then more pressure on the outside of the eardrum than on the inside, and the eardrum cannot vibrate freely. Sounds are imperfectly transmitted, and the person becomes slightly deaf. Infections of the middle ear may occur when organisms in the mouth or nose make their way into the middle ear by way of the Eustachian tube.

Within the cochlea is a tube which follows the windings of the "snail shell"; it is separated from the walls of the cochlea by a liquid called perilymph. This coiled tube is really made up of three adjacent tubes, separated from one another by membranes. Each tube is called a stairway, or scala ("stairway," in Latin), because it suggests somewhat a spiral staircase. The first tube is called the stairway of the vestibule (scala vestibuli). At its base is the membrane of the oval window, to which the stirrup of the middle ear is attached. Next to the stairway of the vestibule is the middle stairway (scala media). Finally there is the stairway of the tympanum (scala tympani), which has the round window at its base. All three tubes are filled with a fluid, called endolymph. Since the two outer tubes are connected with each other at the tip of the cochlea, endolymph extends in a single channel from the oval window to the round window. In the diagram on this page we show how the three tubes would look if they were removed from the cochlea and uncoiled.

The membrane separating the middle tube from the stairway of the tympanum is

ORGAN OF CORTI

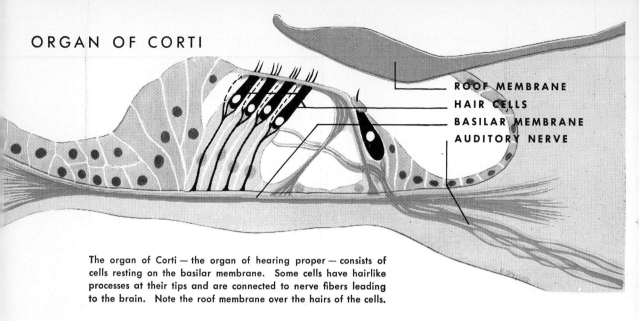

ROOF MEMBRANE
HAIR CELLS
BASILAR MEMBRANE
AUDITORY NERVE

The organ of Corti — the organ of hearing proper — consists of cells resting on the basilar membrane. Some cells have hairlike processes at their tips and are connected to nerve fibers leading to the brain. Note the roof membrane over the hairs of the cells.

known as the basilar membrane. It is made up of a great number of fine cross fibers. Upon it rest the cells that make up the organ of hearing proper — the so-called organ of Corti. Some of the cells, called hair cells, have hairlike processes at their tips. These cells are connected to nerve fibers that lead to the brain. Extending over the hairs of the cells like the eaves of a roof is a delicate membrane — the roof membrane, which can move freely.

When vibrations are transmitted to the oval window by way of the stirrup of the middle ear, the fluid in the stairway of the vestibule is alternately pushed inward and then outward again. Since the outer and inner tubes are connected at their tips, the movement of the fluid is transmitted to the round window. When the footplate of the stirrup pushes the oval window membrane inward, the membrane of the round window is moved outward. When the stirrup moves back toward the middle chamber, the movement is reversed. As a result of the to-and-fro movement, different parts of the basilar membrane are set to vibrating in turn. The hair cells resting on the vibrating part of the basilar membrane also move in unison, and impulses are generated in the appropriate nerve fibers. These impulses are transmitted to the brain.

Why does only a part of the basilar membrane vibrate at any one time? The cross fibers of the membrane increase in length; the longer ones are near the top of the spiral, the shorter ones near the bottom. Their rates of vibration depend upon their length. When a vibration of a given frequency — say, 1,500 cycles a second — is transmitted, only that part of the basilar membrane that has the same vibration rate will respond. This is called sympathetic resonance. (See Index, under Resonance.)

The sounds that are transmitted by the ear may be divided into two classes — tones and noises. Any sound having a fairly well-marked pitch is classified as a tone. Pitch is the quality of sound that enables us to classify it as higher or lower than another sound. Only if differences in pitch are clearly indicated, do we refer to sounds as tones rather than as mere noise. It is true that one type of noise — a rumble, say — may be lower than another — a rattle — but the difference between their sounds is not nearly so well defined as the difference between a C and D on a musical instrument, such as a violin or piano. Sounds vary not only in pitch but also in loudness, tone quality and duration. We analyze the different qualities of sounds in the article Musical Sounds, in Volume 5.

Sounds have the property of direction; they seem to come from somewhere. In certain cases they appear to come from the direction in which the vibrations are actually set up. But in many everyday situations the source of the sound may be a baffling mystery to the listener. A common example is the difficulty one has in locating a squeak or other noise that develops in an automobile. Many times the source of the noise cannot be discovered by listening. It may be necessary to determine the frequency range of the sound and then to test all possible sources of the squeak to see which one produces vibrations within this frequency range.

Often sounds appear to come from a plausible source (that is, from the place where we could expect them to arise), even though actually they arise elsewhere. This is one of the factors in ventriloquism. If the ventriloquist's dummy moves its lips while the lips of the ventriloquist remain motionless, the plausible source of sound would be the dummy. Therefore to us it seems that the dummy is talking and the ventriloquist is silent, even though the actual source of the sound is obvious enough if we give the matter any thought.

Sound movies and public-address systems involve ventriloquistic principles. When a movie is shown on a screen, the sound source may be located just behind the screen. Yet the sounds seem to come from a variety of plausible sources. We are not conscious of the fact that the faint sound of a trumpet in the distance originates in the same place as a "near-up" whisper uttered in a close-up. Of course this is also true of sound effects on the radio and TV.

When loud-speakers are distributed throughout a hall to transmit the voice of a speaker on a platform, the sounds you hear may come mainly from a place beside you or even behind you. This may prove to be annoying at first. In time, however, the sounds will appear to come from the plausible source — that is, from the speaker himself.

The sense
of taste

The organs of taste are found chiefly on the tongue, though there are a few in other places, such as the soft palate and the epiglottis. The receptive structures for taste are called taste buds. They consist of bundles of cells, which open upon the surface of the tongue (or other area).

The cells have tufts of tiny hairs at these openings. The other ends of the cells are connected to nerve fibers leading to the brain. Substances in solution stimulate the cells through the hairs. It is important to note that unless a substance is dissolved in the mouth, it cannot be tasted. If you put a piece of lump sugar on your tongue, you will taste it only when it starts to dissolve. Ordinarily the saliva in the mouth dissolves foods sufficiently so that they can arouse the sense of taste.

There are at least four fundamental taste qualities — sweet, sour, bitter and salty. Some authorities add a fifth quality — alkaline. All other tastes result from a blending of these four (or five) and also from combinations of taste qualities with other types of sensations.

The basic taste sensations are not equally aroused in every part of the tongue's surface. The front part is particularly sensitive to sweet and salty substances. Bitter substances are tasted at the back of the tongue and also in the throat. Certain substances seem to affect more than one class of taste buds. For example, some substances may taste sweet when applied to the tip of the tongue, and bitter when applied to the back.

Taste sensations to which the different surfaces of the tongue are sensitive. A taste bud is shown at the left.

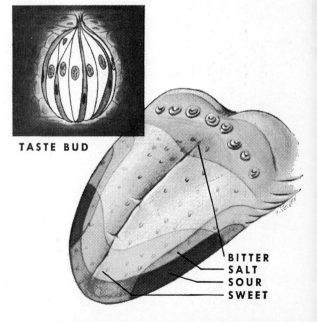

TASTE BUD

BITTER
SALT
SOUR
SWEET

The tongue is sensitive to touch, cold, heat and pain, and these sensations affect the sense of taste. Lumpy mashed potatoes do not taste quite the same as smooth mashed potatoes. It makes a difference, as far as taste is concerned, whether a food is dry or moist, soft or hard. A soda or other soft drink tastes sweeter when it is warm than when it is cold. Thresholds for saltiness, bitterness, sourness and sweetness vary with temperature, but not equally so.

Sometimes one hardly knows whether one is tasting or smelling a substance. Since the mouth and nose cavities are open to each other, particles derived from food in the mouth may ascend to the nose and stimulate the sense receptors there. This stimulation may be much more vigorous than the stimulation of the taste buds on the tongue. In such a case, taste sensations are mostly sensations of smell.

The senses of taste and smell differ greatly in the level of their thresholds. A solution of ethyl alcohol must be 24,000 times more concentrated to reach the taste threshold than to reach the threshold of smell. Only minute quantities of the gases given off from foods are needed to excite the sense of smell. That is why this sense is so often involved when we taste things. If one has a cold, and the mucous membrane of the nose is inflamed, the sense receptors of the nose will not be stimulated. The result will be that the food we eat at such a time may seem tasteless, because we cannot smell it as we chew it.

The sense of smell

The sense of smell is far more important for the lower animals than it is for man. It provides the members of many animal species with invaluable information concerning the outside world. It leads them to food, warns them of danger and helps them to select their mates. In man the sense of smell is generally quite rudimentary. There are exceptions, however. Certain primitive peoples can track down their quarry by the sense of smell. It is said that when one of the other senses is lost, the sense of smell may be developed so that it can partly compensate for the lost sense.

The cells involved in the sense of smell are in the mucous membrane of the topmost part of the nasal passages.

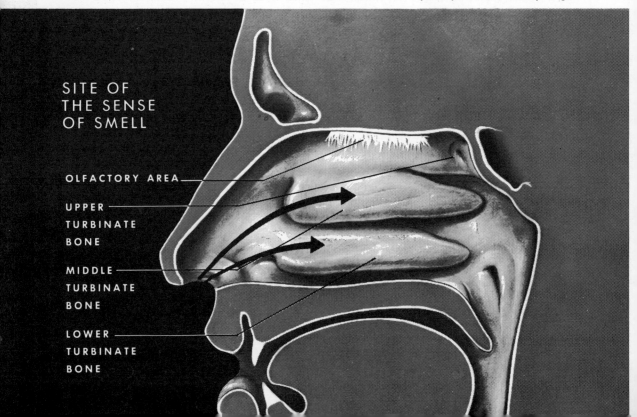

SITE OF
THE SENSE
OF SMELL

OLFACTORY AREA

UPPER
TURBINATE
BONE

MIDDLE
TURBINATE
BONE

LOWER
TURBINATE
BONE

We smell a substance because molecules are constantly thrown off from it in the form of vapor or dust and make their way up the nose. Very small amounts of such vapor or dust can be smelled. If only $\frac{1}{460,000,000}$ of a milligram of mercaptan, a substance smelling like garlic, is contained in a whiff of air, it can be detected. This is a remarkably small amount when we consider that a milligram is only $\frac{35}{1,000,000}$ of an ounce. However, in terms of molecules, the amount is not so small, since it contains some 21,000,000,000 molecules.

If a substance throws off a great many molecules in a short time, it will have a strong smell. This happens, for example, when gasoline fumes arise. On the other

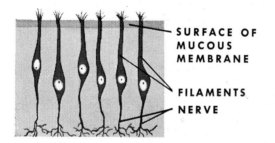

SURFACE OF
MUCOUS
MEMBRANE

FILAMENTS

NERVE

Close-up of the cells involved in smell. Note the little hairs projecting above the mucous-membrane surface.

hand, a substance will have practically no odor if few molecules are emitted. This is true of most metals. However, when metals are heated to a high temperature, molecules are thrown off in quantity, and there will be a strong odor. The strength of the smell sensation does not always depend on the number of molecules a substance releases. Certain substances, such as musk, may give off a strong fragrance for years and yet remain practically unchanged in size and weight.

The organs of smell are located in the mucous membrane of the uppermost part of the nasal passages, away from the direct path of the stream of air drawn in through the nostrils. The cells involved in smell — the olfactory cells — are embedded in the mucous membrane. A long filament of protoplasm leads from each cell to the surface of the mucous membrane; tiny hairs project above this surface from the fila-

ment. A nerve fiber leads from the other end of the cell through a tiny aperture in the floor of the skull to the olfactory lobe of the brain. The surface of the mucous membrane in the olfactory area is bathed in fluid. The particles that are to be smelled must be dissolved in this fluid before the olfactory cells can be stimulated.

We pointed out that the currents of air drawn in from the nostrils do not pass directly through the olfactory area. However, particles of the odor-releasing substance are carried upward to the area by eddies that arise as the currents of cold air from the outside meet the warm air already in the nasal passages. If we want to sharpen the sense of smell as we become aware of a faint odor, we sniff vigorously — that is, we draw in short, vigorous breaths. This causes a reinforced stream of cold air to penetrate the nasal passages. The upward currents become stronger, and more molecules of the odorous substance come in contact with the olfactory area.

The sense of smell is a quickly adapting process. This means that when one is exposed to an odoriferous (odor-producing) substance for a short time, he finds the smell much less noticeable than at first or he may not even notice it any more. That is why people who have been in a room for some time are surprised when a newcomer complains, say, of the odor of stale tobacco smoke. They have become so adapted to the odor that for them it does not exist.

Various attempts have been made to classify odors, but they have not been particularly successful. About all we can do is to say that one substance smells like roses, another like garlic, and a third like almonds.

Touch and the other skin senses

When the skin is stimulated, several different sensations can be aroused — sensations of touch, pressure, heat, cold and pain. All of these, except the sensation of pain, depend on the presence of specialized sense organs at the end of nerve fibers. The nerve fibers responsible for sensations of pain have no special structures at their terminal points in the skin. The different kinds of skin receptors are shown in the diagram on this page. The receptors responsible for touch, pressure, heat and cold each respond to only one type of stimulus. The nerve endings that register sensations of pain can react to different kinds of stimulation.

The various kinds of sense receptors are not evenly distributed on the skin. It is possible to map out various spots where different types of stimuli take effect in a particular skin area. There are touch spots, heat spots, cold spots, pressure spots. Certain areas of the skin are much more sensitive than others to a given type of stimulation. For example, the finger tips are very sensitive to touch, much more so than the back of the hand. Pain fibers are found in every area; they form an extensive network.

The sense of touch. We can show the different reactions to the sense of touch in different skin areas by applying two points point threshold is quite low for the tip of the tongue; when we let this part of the tongue pass over the teeth, a tiny cavity will be felt as a huge pit.

The skin adapts readily to the touch of an object that makes contact with it, to such an extent that the feeling of contact will soon disappear entirely unless the object is kept moving over the skin. We also need change in stimulation when we examine surfaces by touch. We must rub our fingers constantly over a substance if we are to have a good idea of its surface texture. If we put our finger tip on a piece of sandpaper and let it remain there, it will be impossible to tell whether the finger is resting on rough sandpaper or smooth glass.

The sense of pressure. If we lightly touch the skin with a rigid object, such as a matchstick, the sensation of touch will be aroused. If we press more firmly, we will experience a sensation of pressure. This is due to the stimulation of certain sense receptors, called Pacinian capsules, which

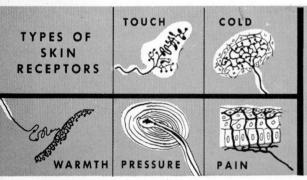

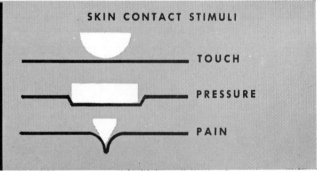

Left: skin receptors for touch, cold, warmth, pressure and pain. Only the receptor that is responsible for pain has no special structure. Right: how different sensations can be aroused by different types of skin contact.

simultaneously to the skin. When the two are very close together, they feel like a single point. As they are gradually placed farther apart, a stage will be reached at which they will just barely be felt as two points. This is called the two-point threshold. The more sensitive to touch a given area of the body surface is, the smaller the two-point threshold will be. It is much smaller on the arm than on the back, and still smaller on the finger tips. The two-

are found in the underlying tissues of the skin. As we increase the pressure, the sensation becomes duller and more pervasive. Finally, if the pressure is increased still more, there will be a sensation of pain.

The senses of heat and cold. If we are to become aware of warmth or cold, something that is at a higher or lower temperature than normal skin temperature — about 78° F. — must be applied to the skin. The sensitivity to warmth will extend to

SEMICIRCULAR CANALS

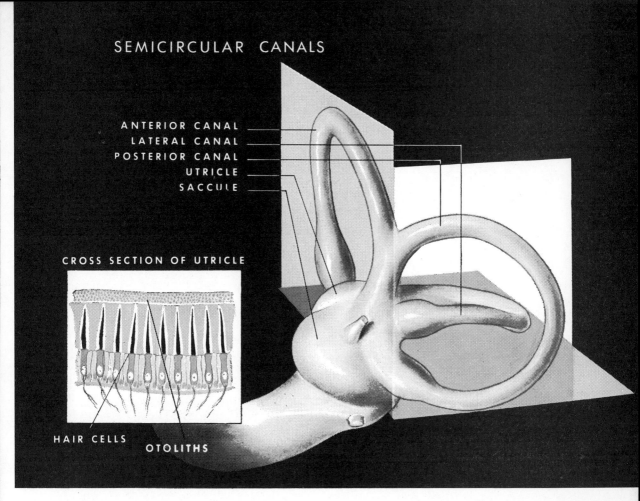

ANTERIOR CANAL

LATERAL CANAL

POSTERIOR CANAL

UTRICLE

SACCULE

CROSS SECTION OF UTRICLE

HAIR CELLS

OTOLITHS

The semicircular canals, utricle and saccule, in the ear, have to do with balance. The semicircular canals occupy different planes. One is horizontal; the second, vertical; the third vertical, at right angles to the second.

a level about 35° above 78°, or 113° F.; sensitivity to cold will extend to about 35° below 78°, or 43° F. The sensations aroused by temperatures above 113° F. and below 43° F. are identical; it is no longer possible to distinguish between heat and cold at such temperatures.

The sense of vision has quite a good deal to do with our experiences of warmth and cold. Some things look warm, others cold. Again and again we are genuinely surprised when we touch an object and find that the sense of temperature does not confirm the impression provided by the sense of sight.

The sense of pain. Pain may be caused by a rigid or sharp object pressed firmly into the skin. However, we pointed out that the nerve endings responsible for sensations of pain can react to different kinds of stimuli. Pain may be aroused by the application of an electric current, by con-

tact with chemicals, by mechanical effects, such as the cutting or tearing of tissue. It may also result when heat or cold is applied. The sensation of pain is a protective device. It warns us of possible danger to the body.

The "control senses" — balance and kinesthesia

Certain senses help us control the reactions of the body to various external stimuli. Among these "control senses" is the sense of balance or equilibrium. It is centered in certain structures of the ear — the semicircular canals and the two sacs known as the utricle and the saccule.*

The three semicircular canals are membranous tubes, filled with fluid. Each leads from the utricle, describes a half circle and then returns to the utricle. The canals

* As we shall see, it is not certain that the saccule is concerned with balance.

occupy different planes. One is horizontal; another is vertical; the third is also vertical but at right angles to the second. One end of each canal forms a swelling, called an ampulla; this contains receptor structures, from which nerve fibers lead.

The nerve endings in the ampulla are stimulated by movements of fluid in the canal. These movements arise when there is a quick turning motion of the head or of the body as a whole — as, for example, when a dancer pirouettes on one foot. Each canal is stimulated most strongly when the turning motion is in the same plane as the canal itself. The movement causes a flow of fluid into the ampulla and stimulates the receptor cells there. It produces consciousness of the rotation; it also brings about reflexes that keep the body in proper equilibrium. If the semicircular canals are diseased or destroyed, serious disturbances of balance arise.

The sac called the utricle contains tiny bodies, called otoliths, which are attached to the hairs of sensitive hair cells. When the head is tilted, the otoliths, responding to the force of gravity, pull upon the hairs of certain hair cells. As the cells are stimulated, they cause righting reflexes — that is, reflex muscular movements that tend to bring the head and body to a normal position. Seasickness and other kinds of motion sickness are probably caused by repeated stimulation of the hair cells in the utricle.

We know little about the functions of the other sac — the saccule, which is considerably smaller than the utricle. Some believe that the saccule is also concerned with the sense of balance; others think that it has to do with the sense of hearing.

Another "control sense" is kinesthesia, or the muscle sense; it has to do with the control of the muscles. Kinesthesia ("movement sensitivity," in Greek) is based on receptors in the voluntary muscles, tendons and joints. These receptors register muscular contractions or elongations; they supply information about the state of tension of the muscles and the position of the limbs. The muscle sense helps us control reflex movements. They cause the re-

flex not to "trigger off" too readily; they help sustain a reflex movement.

The muscle sense also helps us coordinate voluntary movements of the muscles. When kinesthetic impulses are lacking, muscular movements are apt to go wide of the mark. A person will often reach in the wrong direction or will bump into things or will go past objects he wishes to grasp. These ill effects vary in extent from one part of a limb to another. They are most marked at the extremities; the fingers are affected more than the wrists, and the hand more than the forearm. In all these cases, there is nothing wrong with the muscles themselves or with the nerve tracts that supply them.

The muscle sense helps us judge the weight of objects. This is because kinesthetic impulses relay to the brain the amount of effort that the muscles require to heft an object. Usually the senses of touch and sight are also involved.

Other kinds of sensations

Certain sensations supply us with information about conditions in the interior of the body. There are sensations of hunger, thirst, nausea and fullness of the rectum, all of which are described in the article The Digestive Tract, in Volume 4. The sense of pain may also be experienced from within the body.

There may be acute pain when the large intestine is distended because of gas formation or when a stone obstructs the bile duct or when circulation in the heart muscle is interrupted. The inner surface of the abdominal wall is sensitive to various types of stimuli that bring about the sensation of pain. Yet apparently the nerve fibers that carry this sensation are not so widely distributed in the inner part of the body as on its outer surface. Almost any one of the internal organs can be cut or crushed or even burned without causing pain. We feel no pain when the gray matter of the cerebrum (the largest part of the brain) is cut or even removed.

For further reading see Vol. 10, p. 276:
"General Works"; "Physiology."

The Bear and His Cousins

Bears, Pandas, Raccoons and Others

THE bears, which include the largest modern land-dwelling carnivores, or flesh eaters, number certain quite diminutive animals among their kin. At first glance, the huge, lumbering grizzly bear, say, would seem to have little in common with the small, tree-dwelling raccoon. But a brief survey of the evolution of the carnivores will show conclusively that these two animals are really cousins.

Many of the earliest mammals were insect eaters; some of these forms merely needed to increase their size and modify their teeth a bit to become true carnivores. Consequently, as the hoofed, vegetation-eating mammals evolved, there developed a group of flesh eaters to prey upon them. These carnivores had piercing canine teeth, as well as shearing teeth, or carnassials, for cutting through tough sinews and cracking bone. Since flesh is relatively simple to digest and does not need to be vigorously chewed, the grinding molar teeth were much reduced or were completely lost in the course of time. These primitive carnivores — called creodonts by scientists — were rather slow-moving and were decidedly stupid. However, they seem to have had no trouble in coping with their equally clumsy and stupid prey.

In time the early hoofed mammals were supplanted by faster and more intelligent beasts. A carnivore that hoped to make a meal of these creatures would have to be clever enough to stalk them and speedy enough to overtake them. Of the creodonts, only one line, called the miacids, developed these qualities. The earliest miacids were about as large as weasels; they were largely tree dwellers. From the miacids developed the two modern groups of land-living carnivores. The civets, hyenas and cats comprised one group; the weasels, dogs, raccoons and bears, the other.

The dogs retained the use of some of their grinding teeth, for they were not so restricted to a diet of flesh as were some of the other carnivores, such as the cats. The raccoons, giant pandas and true bears — all descendants of primitive members of the dog family — departed still further from the carnivorous way of life.

The raccoons and their kin gradually abandoned the practice of using certain teeth to shear food, but they developed excellent grinding teeth. They added to their diet of frogs, shellfish and small mammals such food as birds' eggs, fruits, nuts and corn. The bears supplemented their meat diet with grasses, berries, fruits, seaweed (in the case of the polar bears) and roots. The grinding teeth of the bears were enlarged for chewing this vegetable matter; the shearing function of the carnassials was gradually lost.

The bears, which are found in every continent, except Australia, are divided into several distinct groups. The black bears of the United States and Canada are closely related to the great brown bears and the grizzlies. The semiaquatic polar bear, too, is closely akin to these members of the family. The honey bear of India and the Malayan sun bear, however, are more distantly related, as is the spectacled bear of South America.

All bears have heavy, long fur and stubby tails. The skeleton is massive; the limbs are strong and are furnished with powerful claws for digging and for fighting. The bear has a shuffling, flat-footed gait, called plantigrade; the sole of its foot is placed upon the ground, leaving a footprint much like that of a man.

Bears are normally slow-moving animals; yet they can break into a lumbering gallop. The black bear can run at a speed of up to twenty-five miles an hour. The big brown bear has a peculiar pacing gait. In this, first the two legs of one side are raised and brought forward together, and then the two legs of the opposite side. The brown bear will sometimes gallop for a mile or more straight up a steep slope without stopping to rest.

Usually only young grizzly and black bears climb, but older individuals, if not prevented by excess weight or large size, occasionally take to a tree. Bears climb by literally hugging the trunk or a branch with all four limbs and hitching their way upward. They carefully back down the same way. Bears will swim across swiftly flowing rivers and navigate lakes of at least five miles in width. Polar bears are excellent divers and swimmers; they have often been sighted, in polar waters, miles from the nearest land.

Bears live from sea level to high mountain country, inhabiting swamps, forests, tundras, snow fields and upland meadows. Black bears, grizzlies and brown bears confine their wanderings to a fairly restricted home range. In this area they forage for food, both by day and night, seek mates during the breeding season and spend the winter in some sheltering spot.

These bears make trails through their home ranges — trails that are followed by the same individuals for years and often used by other bears that happen to be in that particular area. Bear trails along salmon streams and in heavy forests are like any game trail that leads through dense vegetation. On the tundra, big brown bears often establish trails that consist of two parallel ruts, like the ruts of a wagon trail, but, of course, much closer together. Brown bears sometimes walk their range by stepping in the same footprints they made in the past. The result is a trail formed by a series of deep pits laid out in a zigzag pattern.

Bears let their presence be known by so-called bear trees. Often such trees stand conspicuously beside a bear trail. They are marked by smooth places where a bear has rubbed itself or by jagged wounds where the bear has clawed at the bark. Strands of hair adhere to the bark and stick to the pitch that oozes from the claw marks. Sometimes bears strip the bark from trees and scrape off the pulp that covers the wood, leaving tooth marks where the pulp has been removed.

Feeding signs also reveal the presence of bears. Ground-squirrel burrows that have been dug into and ant hills that have been scooped out show where a bear has fed and on what it has preyed. Overturned and smashed logs and rolled-over rocks are signs that a bear has been searching for ants and beetles; a patch of uprooted vegetation may mean a bear has dug for roots. Bee trees are ripped apart for their contents of both honey and bee brood. Since bears cover leftover portions of food with debris, the discovery of such a cache signifies that a bear is active in the vicinity.

Food is fairly hard to come by when snows cover the land; consequently, most bears sleep the winter through. They bed down in a sheltering cavity among rocks or in a hollow under an upturned log; if need be, a den is excavated in the side of a hill. The winter's sleep is not really hibernation, for the bear sleeps less profoundly than do true hibernators, and its body temperature remains close to normal. During this period, she-bears give birth to their tiny, undeveloped cubs. Two is the usual number, but three or four cubs are not uncommon.

The polar bear — a powerful carnivore of the Far North

Polar bears frequent ice floes, islands and coastal areas of the arctic regions. They are large, magnificent animals, the males measuring 7 to 8 feet in length and weighing 600 to 1,100 pounds; females are somewhat smaller. Polar bears are all white or yellowish white, except for their black eyes, nose pads and foot pads. Unlike other bears, only pregnant female polar bears den up during the winter. The males and barren females prowl the sea ice and open water in search of prey.

Mirrorpic

A famous polar-bear family. Brumas, an eleven-weeks-old cub, tempted by mild weather and bright sun, makes her first appearance at the London Zoo. She is held in the firm embrace of mother-bear Ivy.

Throughout the long winter, then, the female lies beneath the snow, and there brings forth her young. She must have within her system a sufficient reserve not only to support her own life during this long period, but to afford milk to her cubs as well.

These usually number two. They are born naked, blind, and helpless, unable even to wriggle a few inches to their dam's side should they be disturbed. Thus hidden in the close confinement of their snowy lair there is not much danger of their being lost in this way, of course.

The lair, dug by the female polar bear before she has her cubs, is a trim burrow in the ice or packed snow. One of these dens, which was found dug at the foot of an iceberg, tunneled downward through the ice for a short distance at an angle of something like forty-five degrees; then it angled upward for a certain space and finally terminated in an oval chamber. Lairs such as this one are temporary shelters, however, for the mother and her cubs abandon them by late February or March.

The mother polar bear
is devoted to her cubs

The female polar bear has proven to be one of the most devoted of mammal mothers. She stands over her little cubs to shield them from fierce blasts of icy winds. She helps them cross open stretches of water between ice masses by towing them as they hang on to her body; sometimes she carries them on her back. There are many travelers' tales telling of the heroism displayed by mother polar bears in defense of their young against human enemies. Though mortally wounded, the mother bear will stay by her cubs and protect them to the last. If they are wounded, she will lick their wounds and try to urge them away or guard them from further harm.

The polar bear is slender of body with a long neck and a small tapering head. Its streamlined shape enables the animal to knife through the water most gracefully. It uses its front legs for swimming, while the hind limbs trail behind to function as a rudder. Oil glands in the skin oil the fur effectively so that it sheds water and keeps the body dry. A fatty layer below the skin protects against the cold, and the thick fur holds little air spaces that keep heat loss from the body to a minimum. The polar bear has broad feet, the soles of which are heavily haired to prevent slipping as the animal travels swiftly and tirelessly over the ice.

This bear shows great skill in stalking its prey. When it sees a seal basking on an ice floe close to the water's edge, the bear paddles in the water, carefully approaching the quarry. Only the bear's head appears above the surface and it resembles a bit of floating ice. If the seal rears up to look about, the hunter remains motionless, drifting. Finally the bear is near enough! It quietly sinks below the surface, swims under water the remaining distance and comes up in a thunderous thrust to knock down the prey with a heavy blow of the paw. The bear uses similar tactics to capture eider ducks and scoters that are resting on the surface of the Arctic waters. With only its eyes and nose visible, the bear swims quietly among a flock of these birds. When a bird dives in alarm, the polar bear quickly overtakes it under the water.

Sometimes the polar bear waits patiently at a hole in the ice; when a seal pokes its head up to breathe, the hunter quickly makes the kill. Water animals are not its only prey; it often tracks down caribou and the Arctic fox. It generally refrains from attacking big male walruses when they are in the water; these animals would be dangerous foes. The polar bear often feasts on the decaying remains of a beached whale, sharing the repast with others of its kind and even with foxes. The animal's sense of smell is so keen that it is said that it can scent a stranded whale, even if the carcass is covered with snow, at a distance of twenty miles.

Polar bears usually hunt alone, though the females are accompanied by their young. From May to July is the mating season, when the adult male and female travel together. Eskimos seem to be their only enemies. Polar bears have been known to live to be thirty-four years of age.

Big brown bears range
over the Northern Hemisphere

The big brown bears have an extensive range over the mountainous and forested areas of the Northern Hemisphere; consequently, there are many local races of these bears, which differ from one another in color and minor characteristics. There are brown bears in the Pyrenees, the eastern Italian Alps, the Balkan Peninsula, northern Scandinavia and Finland. They occur in northern and central Asia.

The Syrian bear is a light-colored brown bear of the mountains from Turkey to Iran; the isabelline bear is native to the Himalayas; a darker race lives on Hokkaido Island, Japan, and on the Kuriles and Sakhalin Island in the Sea of Okhotsk. The big brown bears of North America inhabit the coastal regions of Alaska and adjacent islands, frequenting especially those rivers running with salmon. The closely related grizzly bears live in the Rockies through British Columbia and Alaska.

Grizzlies and brown bears are large animals that often cannot be easily told apart. They both have humps at their shoulders and their heads are massive, though the brown bear's head is larger. The claws of the brown bear are of moderate length and deeply curved; those of the grizzly are long and straight. Usually the big brown bear, which may stand almost five feet at the shoulder, has a uniformly yellowish to dark brown coat. The grizzly has a more frosted appearance due to the fact that the hairs, especially those on the back, are tipped with white. This bear's shoulder height does not often exceed three and a half feet.

Though brown bears and grizzlies eat many kinds of foods, this diet often varies with the season. A survey of what grizzlies eat shows that in the spring they dig up roots; in June and July they feed on reed grass and horsetail and, when op-portunity arises, on rabbits, young caribou and mountain sheep. In August and September they thrive on berries. Ground squirrels are eaten at any time, as are the remains of animals left by wolves.

Usually grizzlies and brown bears restrict their hunting activities to a definite territory, or home range. The home range of the Alaskan grizzly covers about sixty-five square miles. The grizzly is said to mark the boundaries of its range, as a sign to other bears, by making teeth and claw marks on trees as high as it can reach. The brown bear of Russia roams over an area of ten to thirteen square miles. Brown bears leave "scent signals" to indicate their territory by rubbing the back against a tree and leaving urine at its base.

Brown bear young, born in January or February, are only rat-size at birth. The mother transports each infant by carrying the cub with its head held in her mouth. The young develop slowly, and they are not mature, that is able to reproduce their kind, until about six years of age. These bears hear well but they have a better nose for detecting game. Since their flesh is not particularly good, their only enemy seems to be the so-called sportsman who is out to get a trophy. These animals adopt a defensive attitude toward humans. A brown bear will not attack a man unless he blocks its path or molests a she bear that is accompanied by its cubs.

N. Y. Zool. Soc.

A magnificent example of the Alaskan brown bear. The big brown bears are the world's largest carnivores.

During the winter the brown bear hibernates, seeking the shelter of a tree or cave. It may scrape a hole beneath a fallen tree or it may enlarge a burrow begun by another animal. When it goes to rest, it has a large store of fat on which to live during the winter months. Instinct tells the animal when it has accumulated enough to last it through the winter. If food has been scarce during the preceding season, the bear may have to delay its winter sleep. During hibernation, only the bear's essential life functions remain in operation — respiration and circulation. All others are suspended.

In January or February, the female bear gives birth to one to four bear cubs; the number is generally two. The cubs are blind at birth and weigh about one and a half pounds. They first see the light of day in April or May when the mother bear emerges from her winter home. The cubs remain with her for the first two years of their lives.

When the bear emerges from its hibernation in the spring, it is lean and hungry after supporting itself and perhaps several cubs on the fat of its own body during the winter. It feeds voraciously at this time on both vegetation and flesh. It may even raid cattle and sheep herds.

The great brown bears of the north

In the northern parts of North America, Europe and Asia, there are a number of brown bears of great size. Many of them are local subraces, each occurring in only a very small range; they may even be restricted to a single island. For example, the famous Kadiak (or Kodiak) bears are found only in Kadiak (or Kodiak) Island off Alaska. These giant beasts may reach a length of ten feet and weigh more than 1,500 pounds. They are the largest land-living carnivores in existence. They eat both vegetation and animal food.

Another interesting local race is the Kamchatka bear of northeastern Asia. This Asian giant lives entirely on salmon that it catches while wading in a river. It does not knock the fish out of the water with its paw as many people believe. Actually the animal spears its prey on its claws and then leaves the river to eat the fish on land. The Kamchatka bear will never feed while standing in the water.

The Sitka bear is a smaller (though still very large) local race; the Yakutat and Kidder's bears on the mainland of Alaska are huge animals. All the Alaskan bears are reported to be less fierce than the grizzly, though far exceeding the latter in size. Nevertheless, the big-game hunter must be on his guard against them.

The powerful grizzly — the most dangerous of all

At one time, the grizzly bear (*Ursus horribilis*) was the scourge of North America from the badlands of Dakota to the coast of California. As time went on, large numbers of these bears were poisoned and hunted down by ranchers whose flocks they were raiding. As a result, the animal is now quite scarce. It usually reaches about six and a half feet in length and may weigh up to 1,400 pounds. Although not so large as its northern cousin, this bear is far more savage. *Ursus horribilis* is considered by many hunters to be the most dangerous of all North American wild animals. Its terrible claws are highly developed and are wielded by the grizzly with great effectiveness. Grizzlies have been known to kill full-grown bull bisons with a single blow on the neck with a paw. It is said that these bears can drag away a male wapiti weighing a thousand pounds.

The various types of American bears

There is a confusing variety in both size and coloring in the bears of North America. Consequently, naturalists differ widely in their classifications of the different species. One view holds that there are only three true species of bears in North America: the black bears; the grizzlies and brown bears; and the polar bears. According to this view, all of the other varieties are subspecies. Others describe as many as ten different species of North American bears.

The American black bear is the smallest of North American bears. It grows to a length of five or six feet and stands two to three feet at the shoulder. Its claws are comparatively short and curved. The encroachment of human civilization at one time threatened the existence of this animal over much of its territory; but laws for its protection and the bear's own resourcefulness have insured its survival. A forest and swamp dweller, the black bear inhabits much of Canada and Alaska, the eastern mountains from Maine to Florida, the Rockies and the mountains of the West Coast. In the east, the bear usually wears a coat of glossy black; in the west, however, many black bears are reddish brown and are known as cinnamon bears. There is even a creamy white variety of black bear, called Kermode's bear, living on Gribbell Island, British Columbia.

An extremely powerful animal for its size, this carnivore usually hunts, day or night, for the smaller animals such as rodents, frogs and insects. It also eats berries and roots and supplements its diet with an odd assortment of foods, including pine cones, skunk cabbage, river-clam shells, briars and wasps' nests! The nests of bees and wasps are robbed of honey and the young insects.

Black bears are ordinarily solitary animals living in a well-defined home territory. Trees with clawed and chewed bark indicate that a bear lives in the area. Sometimes several bears congregate at a feeding ground, such as a berry patch or garbage dump. Relations among them may be peaceable but often fights take place.

The sexually mature males and females associate during the breeding season, which lasts approximately one month in June or July. After mating, the bears go their separate ways, the male having nothing to do with the raising of the family. Like other bears, black bears den up in rocky cavities, hollow trees or beds of dense foliage for a prolonged winter sleep. In late January or February, cubs are born — usually one cub to a female having her first offspring and two or three young to an older female.

The newborn cubs have a thin coat of fine hair and are blind and toothless. They are about nine inches long and weigh less than ten ounces. In March or April, when the cubs weigh four pounds, they leave the winter's den with their mother. She continues to nurse them for some time, but they also eat other foods. The cubs box and wrestle each other and engage in

Giant, or short-tailed panda, dwelling in bamboo forests in eastern Asia. It is rather smaller than a black bear.

N. Y. Zool. Soc.

Northern Pacific Railway

The American black bear. This animal is found in a number of forested regions of North America, north of Mexico. It climbs trees easily, travels about a great deal and is often captured and tamed. A kindred species, the Himalayan black bear, ranges from Assam, in India, to the frontier of Iran.

games of hide-and-seek. When in danger, they scurry up a tree if they cannot reach their mother. The young remain with the mother through the following winter. In the wilds, black bears probably live from twelve to fifteen years. The females are sexually mature at three years; they mate about once every two years.

In Asia, there are three species of black bears. The most widespread is the Himalayan black bear, which ranges from the eastern portion of Iran through Afghanistan, Tibet and Thailand into China and Manchuria. It is also found on the islands of Hainan and Formosa. Somewhat smaller than the American black bear, it is further distinguished by a light-colored, inverted crescent or Y-shaped patch on its chest. The hair on the neck is long, giving the animal a sort of ruff. This bear is a forest creature, living on grain, honey, roots, fruits and grasses.

The Malayan sun bear, or bruang, is a slightly smaller species restricted to the Malay peninsula and the islands east to Borneo. It has a crescent of white or orange on its chest similar to that of the Himalayan black bear. The animal is an excellent tree-climber. It has a very flexible tongue and lips, which it uses in feeding on insects and honey.

A more specialized insectivorous bear is the sloth bear of India. This creature differs from the rest of the bear family in having two less teeth than the normal quota, a long snout and a baggy, hanging lower lip. The shaggy beast lives on fruits, flowers, honey and insects; it is particularly fond of white ants, or termites, which are its natural prey. Its immensely powerful claws enable it to tear open the hills of these insects. The sloth bear's extraordinary powers of suction then come into play as it draws ants from deep within the nest by a violent intake of breath.

The spectacled bear of the Andes, in South America, is a close relative of the sun bear. It is a small, black species, ranging in the mountains from Colombia to Chile. The name is derived from the whitish, spectaclelike rings about the animal's eyes. It is the only bear in South America and the only member of the bear family found in the Southern Hemisphere. It lives primarily on fruits.

The panda is related to the raccoon

When the giant panda was first discovered in 1869 by the French missionary Armand David it was believed to be a bear with abnormal coloring because it had so many bearlike characteristics. It was classified with the bears for a long time; but as zoologists came to know more about the animal, they realized that it is more closely related to the raccoon. The giant panda reaches a length of about six feet and weighs over two hundred pounds. Like the members of the bear family, it has plantigrade feet, strong claws and a short, stumpy tail; however, it has two less teeth than the ordinary bear. In coloring, the panda is white or cream-colored, except for areas of black around its eyes, ears, legs and shoulders. The animal lives in Tibet and southwestern China at elevations of from 6,000 to 14,000 feet above sea level. Its diet is made up almost exclusively of bamboo shoots. Very few of these animals have been exhibited in zoos.

The true panda, or lesser panda, as it is more commonly known, is a bushy-tailed animal about the size of a large cat. Its back and head may be brown, red or chestnut. Black is the color of its underside, legs, feet and ears; the ears are trimmed with white. The face is white below the eyes and there are white spots above them. On the ground the true panda is slow-moving, but it is an excellent tree-climber. Like the giant panda, it lives mainly on bamboo shoots, but in captivity it will eat bread and milk, eggs and small birds. It hisses like a cat, growls like a bear and laps, at times, like a dog. Generally, however, it plunges its muzzle into water and drinks by suction, after the manner of a bear. The true panda can partially retract its claws, like a cat. The animal is a native of Asia, but formerly ranged over Europe to the British Isles, at a time when this region had a subtropical climate and was abundantly forested.

The raccoons —
"washing-bears"

The German name for the raccoon is *Waschbaer,* which means "washing bear." It is derived from the fact that this little bearlike mammal washes its food before eating it whenever water is available. The raccoon is quite closely related to the bear; for one thing, like the bear, it has plantigrade feet. Raccoons have the same number of teeth as sloth bears and pandas. Internally, too, there are anatomical similarities among these animals.

The raccoon is particularly noted, perhaps, for its food-washing habits. Although not an aquatic animal, it usually lives near water. When it captures food, the raccoon carries it to the nearest water, and holding it squirrel-fashion in its front paws, dabbles it again and again in the water, before putting it in its mouth. It is interesting to note that in this operation the raccoon makes remarkable use of its fore paws as "hands"; it uses them as dexterously as most monkeys.

The common raccoon of the United States and Canada is about as big as a domestic cat, but the length of its legs and the bushiness of its coat make it appear larger. The back is highest over the hindquarters. The head is triangular; patches of black over the eyes give the face a "banditlike" appearance.

The varied diet
of the raccoon

The raccoon has a varied diet. It is fond of frogs, crayfish, oysters and other shellfish; it also eats birds' eggs, poultry, mice, nuts, fruits and corn. The raccoon's appetite for ears of corn when they are young and tender arouses the particular animosity of farmers. They often hunt it down with dogs when it makes raids on their fields or chickens. The dogs drive the animal into a tree where it can easily be caught.

Raccoon attacks on chicken farms and corn fields are particularly bothersome to the farmer because these animals are so clever and so destructive. It is particularly difficult for the farmer to catch them because they seek their food only at night. In the daytime the raccoon sleeps, generally in the hollow of a tree. When it attacks chickens, the animal usually kills far more birds than it can possibly eat. Corn-field raids are equally destructive, because the raccoon not only greedily devours many ears of corn but also knocks down a number of stalks.

Raccoons find most of their food on the ground, but they generally live in hollows in trees because it is so much safer for them. They hibernate in these places of refuge in the colder parts of their range; it is in the trees that their young are born. The raccoon is quite wary, although it has relatively few enemies other than man. It is adept in climbing trees and often uses this way of escape when danger threatens. Despite man's persecution of the animal, its existence has not been threatened; in fact it is increasing in number in some areas.

Raccoons make
interesting pets

When taken into captivity, young raccoons make interesting, although somewhat troublesome pets, because of their restlessness and insatiable curiosity. "Nothing," writes Godman, "can possibly exceed the domesticated raccoon in restless and mischievous curiosity if suffered to go about the house. Every chink is ransacked, every article of furniture explored, and the neglect of servants to close closet doors is sure to be followed by extensive mischief." As we have seen, the animal soaks its food in water; it also likes to play with water. A tubful should always be provided for a pet raccoon. This is one of the most playful of animals. Referring to his two raccoon pets, Godman writes: "They are frequently seen sitting on the edge of [a] tub, very busily engaged in playing with a piece of broken china, glass or a small cake of ice. When they have any substance which sinks, they both paddle with their forefeet with great eagerness until it is caught, and then it is held by one, with both paws, and rubbed between them; or a struggle ensues for possession of it. When it is dropped the same sport is renewed."

Raccoons choose as a home a hollow in a tree, which is made smooth within but which is not lined with straw or other materials. Only in the colder parts of the United States are the raccoons confined to their dens in winter. Moderate cold does not trouble them nor interfere greatly with their hunting, which in winter, in the northern states, is chiefly for mice. Three or four cubs are born in May, and by late autumn they are quite large enough to fend for themselves.

Several other animals are members of the raccoon family — the cacomistle, which dwells in well-watered woods; the kinkajou, which is quite monkeylike in appearance; and the coati, which is to be found in Central and South America.

The kinkajou, as the only member of the family that has developed a prehensile tail, is perhaps more completely at home in the trees than any of the others. The natives regard it as a type of monkey, and it has, indeed, a lemurlike appearance. But the paws are paws, not hands, and they are armed with powerful curved claws, not nails. Furthermore the dentition is unmistakably that of the raccoon family. The kinkajou, which is about as large as a small cat, eats small mammals and birds, fruit and honey. Kinkajous collect in troops. When they move through the trees, they take long leaps from bough to bough.

The cacomistle, or ring-tailed cat, ranges from Central America and Mexico to Alabama, Texas, Colorado and southern Oregon. It resembles a raccoon, but is more slender; it has a long, bushy tail, ringed in black and white. It is a dweller in the woods, making a moss-lined nest in a hollow tree. The cacomistle is a nocturnal animal; it feeds on small mammals and birds, insects and occasionally fruit.

The coati, or coatimundi, is represented by two species: the reddish brown ring-tailed coati, found in South America, and the white-nosed coati, which has white markings around its eyes and on its snout and which ranges from Mexico and Central America south to Peru. The German equivalent for coati is "proboscis bear" — an apt name for an animal with a long snout, short and sturdy limbs and a bear-like gait. (It walks on the sole of the foot, with the heel touching the ground.)

The coatis are good climbers and chase lizards among the trees. The reptiles thus pursued drop to the ground, only to find a second detachment of coatis, obviously co-operating with the hunters overhead. The coati can be tamed easily and makes a docile and interesting pet. All the members of the raccoon family, with the exception of the pandas, are to be found in North and South America.

See also Vol. 10, p. 275: "Mammals."

The eastern raccoon
(*Procyon lotor lotor*).

1. A striking result of wind action — a sand dune in a desolate region of Saudi Arabia.

THE WINDS AT WORK

How Wind-blown Sand and Dust Erode the Land

by H. T. U. SMITH

WINDS rank high among the important forces that shape the surface of our planet. They erode rock materials and soil, transport the eroded material over vast areas and deposit them upon the earth. In this way they level the land here, build it up there, scoop out hollows and thus create the distinctive types of landscapes that make the earth's surface so richly varied.

One might think that the stronger the wind, the more effective it would be as an agent of erosion, transportation and deposition. This is not the case. Violent winds, such as tornadoes and hurricanes, are not particularly important as agents of geologic change. They do not come often enough, or stay long enough or cover a wide enough path to be effective. The more moderate winds that blow for hours at a stretch, time and time again through countless centuries, are far more important than are the more violent types of winds in the never-ending task of altering the surface of the earth.

As the winds erode the land, they work with two types of material — sand and dust. When the sand is carried by the wind, it drifts along close to the ground until it accumulates in the distinctive hills and ridges known as dunes. After the dunes have been formed, they too are moved by the winds. The effects of wind upon dust particles are quite different. The particles are blown high in the air and are transported rapidly and to considerable distances, sometimes traveling more than a thousand miles; finally they are deposited upon the ground. Dust is spread much thinner than wind-blown sand; it covers a much larger area.

Effects of erosion by the wind

It may be difficult to point to definite traces of erosion by the wind. Frequently the layer that has been removed is so thin and the area of removal so extensive that the noticeable effects at any one spot are slight. There are some exceptions, however. In certain desert basins in the western United States, the ground level in various limited areas has been lowered as much as fifteen feet by the blowing away of dust. In North Africa some fairly large desert basins, forming oases, have been carved out, at least partly, by the blowing away of sand and dust. In cases like these, erosion by the wind is limited to loose sand, silt and clay; in other words, the wind picks up only what other processes of erosion have made ready for it.

A special variety of wind erosion, known as sandblasting, is produced by blown sand before it is trapped to form a dune. The grains of sand act like tiny chisels, scratching and scraping whatever surface they strike. Each grain cuts a little deeper than the one before; in this way, even hard materials can be gradually ground down. Bottles and pieces of glass lying in areas of drifting sand soon lose their brightness and take on a dull, frosted appearance. If automobiles are driven in sandstorms, their paint may be worn off and their windshields frosted in a fraction of an hour. In some places where sandblasting is especially strong, wooden telephone poles are gradually worn away. They must either be protected (Figure 2) or replaced by new poles.

The effect on rocks, which naturally have been exposed to sandblasting for a longer period of time, is much greater. The larger blocks and boulders are grooved and pitted in ways not duplicated by other processes of erosion (Figure 4). If different parts of the rock are of unequal hardness, the softer parts are worn down more rapidly, leaving the harder parts standing out. Pebbles are sometimes worn down to Brazil-nut shape; one or more surfaces may become flattened. Such sandblasted rocks are called ventifacts (objects made by wind). Figure 3 shows a typical group of ventifacts. They are found in most of the areas where sand is drifting actively today, and in many other regions where wind action was once vigorous but has since been checked.

In most places, the effects of sandblasting are confined to scattered blocks

2. Protecting a telephone pole from the effects of sandblasting by means of buffer poles and a sheath of metal.

All photos by the author unless otherwise indicated

and pebbles. In a few regions, however, the bedrock itself has been carved into distinctive ridges, troughs and unconnected hollows, all on a relatively small scale. This happens only where sandblasting is unusually strong, and where there is practically no rain to permit erosion by running water — a condition found in parts of southwest Africa.

Sand dune areas of the world

The sand dunes of the world represent one of the most striking results of wind action. In the United States, dunes

3. Ventifacts, or wind-cut stones, in a striking variety of shapes. They are found in many areas of the world.

are found in a great many widely separated places. Altogether, these American dunes cover more than 40,000 square miles, or about 1½ per cent of the entire area of the country. They are far more widespread in other regions of the world; some of the dune areas in the deserts of Asia, Africa and Australia are larger than many of the American states. Smaller formations are found in many countries that do not have deserts, such as France and Germany.

There are various kinds of dunes. In some regions, such as the desert area of southern California shown in Figure 5, the dunes look like great sharp-crested waves in a vast sea of sand. Formations of this kind are called transverse dunes, because they lie athwart the direction in which the wind is blowing. They are more or less curved; each wave formation is

4. Boulders showing the remarkable grooving and pitting effects that are produced by continuous sandblasting.

generally broken up by gaps into several sections. Every dune has a steep side and a gentler slope; the steep side always faces away from the wind. The sand is blown up the more moderate slope and is dropped at the crest.

As we walk over a transverse dune, we find that the sand is loose and easily moved; our feet sink in and each step is an effort. We can climb up the gentler slope quite easily; but it is quite a different matter to scale the steep side. If the wind happens to be blowing hard, sand showers down upon us, getting into our eyes and ears and clothing. Suppose that we beat a hasty retreat to the gentle slope of the next dune. Plenty of sand is blown along here too, but it stays fairly close to the ground; only when the wind is exceptionally strong does it rise to face level.

Wind-blown sand moves in two different ways. Some grains, especially the larger ones, roll and bounce along, often forming ripples that seem to creep over the surface of the sand. Other grains, however, are swept up into the air and move along in a more or less steady stream. The harder the wind blows, the faster and higher the stream moves, and the more sand is carried. The stream flows smoothly up to the top of the dune. As soon as it passes the crest it loses momentum, drops down out of reach of the wind and is trapped behind the dunes.

As the wind blows, sand is taken from one side of the dune and added to the other; therefore, the dune as a whole advances with the wind. The advance continues as long as the wind blows in the same direction. A change in direction will cause a corresponding change in the direction in which the dune will move. If the wind entirely reverses its direction, the dune will do likewise. What was formerly the steep side will become the gentler slope, and vice versa.

Transverse dunes are found in many parts of western North America and also in the deserts of Africa and Asia (including Arabia). In many of these places, the sand waves are lofty and have long valleys, or troughs, between them. Such dunes may be hundreds of feet high, and occasionally may tower to a height of a thousand feet or more.

So far we have been dealing only with the larger transverse dunes on a continuous sea of sand. If we look at the lower right-hand corner of Figure 5, we see a series of smaller, more regular dunes quite distinctly separated from one another. These dunes are more or less crescent-shaped, with stubby horns pointing in the direction toward which the wind is blowing. Dunes such as these are called barchans (pronounced *bar'-kans*). In the photograph, some of the barchans have joined together, forming "Siamese twins" and "Siamese triplets." The more sym-

5. Sand dunes in southeastern California. The small dunes shown in the lower right-hand corner are barchans.

metrical forms are found only where there is no crowding. Barchans are moved by the wind in much the same way as the larger sand waves; the difference is that some sand is blown around the sides as well as over the top, thus accounting for the horns.

In some places the principal forms of dunes are long, low ridges of sand that lie approximately at right angles to the trend of the sand waves, or parallel to the direction of the wind. These large formations are called longitudinal dunes. Some of the ridges are practically straight; others are slightly wavy. They range up to about thirty feet in height and a hundred feet in width; many are more than a mile long.

Both sides of these dunes have practically the same slope. The dunes are more or less covered with grass and bushes; only the crest is bare. On some dunes, even the crest is covered with vegetation; where this occurs, it means that wind action has stopped on this particular dune. There are many varieties of longitudinal dunes; they occur in true deserts and also in regions where there is scanty vegetation. In some parts of the Sahara Desert they represent the main type of dune; they are also widespread in the deserts of Australia and India. There are interesting longitudinal dune formations in the semidesert plateau region of northeastern Arizona.

There is a fourth type of dune, made up of U-shaped ridges, largely covered with bushy vegetation. In some places bare patches, across which sand is blowing, may be seen at the bend in the U. On these so-called active dunes, there is a steep slope away from the wind, as in barchan dunes; but the horns point in a direction opposite to that of the barchans. This type of dune is found only where vegetation is present; the sides of the U are always at least partly covered with vegetation. In some places there are variations of the U shape; individual dunes have been crowded together, producing forklike or rakelike formations. Here and there U-shaped dunes are gradually being buried by oncoming sand waves, as shown in Figure 6. This means that a second series of dunes is being built up upon the older dunes. More than two generations of dunes may be formed in this manner.

U-shaped dunes and similar types are common along sandy shores and other sandy areas, where grass, bushes or trees are growing. They occur at many places in the Great Plains, east of the Rocky Mountains, where winds are strong,

effects of present-day wind action. The typical blowout (Figure 7) is an elongated, troughlike zone, culminating in a broad apron of sand. The trough is formed by wind erosion — the blowing away of sand. At the end of the trough that is away from the wind, the sand generally forms a steep, high front; as more sand is blown over the top, this front gradually moves forward, burying trees and whatever other obstacles may lie before it. Blowouts gradually bring about a breakdown of older dune masses that have become static.

Irregular, grass-covered sand hills, such as those we have described, are called dead dunes; they represent masses of sand that were once shaped by the wind and were then anchored in place by the overgrowth of vegetation. Where blowouts occur, the conflict between grass and drifting sand can be seen today, although on a much smaller scale than before. In some places, the sand drifts fast enough to bury the grass and destroy it. In other areas, certain hardy kinds of grass keep growing fast enough to escape burial, and they stop the drifting sand.

droughts occur frequently and the ground is very sandy. They are never found in true deserts.

In some dune areas, it is difficult to recognize any distinctive types of dune. This is especially true where the formations are largely covered with trees, as along the southern shore of Lake Michigan. From a high vantage point one sees only an irregular pattern of hills and ridges, crossed here and there by long, narrow patches of bare, drifting sand, called blowouts. Only the blowouts have a distinctive appearance; they alone show the

Transverse dunes, barchans, longitudinal dunes and **U**-shaped dunes are all referred to as live, as long as they undergo the action of the wind. They can all become dead if they are invaded by vegeta-

6. Dunes along the coast of southern California. The **U**-shaped dunes on the landward side are being gradually overridden by the waves of sand that advance from the seaward side.

Spence Air Photos

7. A blowout. Winds blowing from the lake have cut through the older dunes.

tion. As dead dunes, they will no longer be subjected to the effects of wind action. If the vegetation is destroyed, wind action may be revived, leading to various changes in the shape of the dune.

All live dunes have the property of movement, either of the entire dune or part of it. The rate of movement depends on the strength and frequency of wind-storms, the constancy of wind direction and the size of the dune; the larger the dune, the slower it moves. The rates of migration (movement) range from a few feet to several hundred feet per year; the average is probably less than a hundred feet. Where strong winds blow regularly in the same direction, large dunes can override and bury whatever lies in their path; farm land, forests, roads, buildings, even entire villages may be overwhelmed. Sooner or later, however, most migrating dunes are anchored by vegetation.

Dust storms
and dust deposits

Dust storms are a vivid memory to those who lived in the Great Plains area of the United States in 1933 and in the years following. While these storms were raging, the air was filled with choking dust. Some-times it swept forward, for all the world

like a great, high cloud with a steep front; those overtaken by such a storm would suddenly find themselves gasping amid swirling clouds of dust particles. During severe dust storms, day became as dark as night. Automobile headlights could be seen only at a distance of a few feet. Breathing was difficult; practically all out-door activity came to a halt.

After a storm had passed, dust was left everywhere. It was a fine, powdery substance that penetrated even into tightly closed buildings. If the storm had been mild, the dust residue was only a thin, gritty film. If there had been an unusually severe storm, it left behind layers of dirt that could be removed by the shovelful.

The air would often be hazy with dust hundreds of miles from the starting point of a storm. Sometimes a thin layer of brownish dust would be deposited more than a thousand miles away from the point of origin; the layer would be particularly noticeable if it fell on snow. A typical storm, in November 1933, was traced from the Great Plains to northern New York, a distance of 1,300 miles. It trav-eled at an average speed of 43 miles per hour.

The amount of dust carried and de-posited by a single storm was enormous. At moderate distances from the starting point, up to forty tons of dust per square mile were deposited. At greater distances, the deposits were not so heavy, of course; yet they amounted to three tons per square mile in some instances. Since the dust storms raged over tens of thousands of square miles and extended over a period of several years, the total amount of dust laid down came to millions of tons.

The dust carried in this way had been blown from farms in which the soil had become dry, powdery and bare. For a long time there had been too little rain; crops had failed to grow; the ground had been left unprotected from the wind. In many places, in a desperate attempt to start new crops, the dry ground had been broken up by farming implements into a crumbly mass. As strong winds swept over this bare, loose earth, the finer material was stirred by gusts and eddies and carried away. The coarser material was blown along the ground, scraping loose more dust for the wind to pick up. Once in the air, the dust was kept continually stirred up by swirling air currents. Only when the wind died down, or when it encountered obstacles such as buildings, was this stirring process checked; the dust then settled down gradually.

The dust storms that we have described resulted from temporary, man-made deserts. The conditions under which they occurred have been duplicated again and again in widely separated regions of the world, and have led to the devastation of immense areas.

The dust storms that have a particularly important bearing upon geologic change occur particularly in two types of environment: in true deserts and in the vicinity of glaciers.

In true deserts, there is no vegetation to protect the ground. The wind sweeps unchecked over bare earth and stirs up any fine material that is present. Severe dust storms are common; frequently the dust is blown long distances beyond the borders of the desert. Showers of reddish dust from the Sahara have long been familiar in southern Europe; they have penetrated as far north as England. One of these dust falls, in 1901, left from 3 to 31 tons of dust per square mile over an area of at least 300,000 square miles, making a total

deposit of some 2,000,000 tons in Europe alone. It is estimated that an average thickness of about 5½ inches has been added to the soil of Europe in the last 3,000 years by dust storms originating in the deserts of Africa. A similar story could be told of other deserts.

In the vicinity of glaciers, too, conditions favor the rise of dust storms. Glaciers carry much ground-up rock material scoured from their rocky channels. During the melting season, running water transports this drift along broad outwash channels away from the glacier. After the melting season, the channels become dry; they then contain large amounts of rock flour, or ground-up rock. This powdery material is easily picked up by winds blowing out from the ice, and local dust storms result. As the winds slow down away from the ice, the dust settles out, collects around vegetation and in time becomes a part of the soil. It may gradually collect to a thickness of many inches or even feet, covering the ground like a blanket. Fairly rapid deposition of this type takes place near the glaciers of Alaska and Greenland.

If a deposit of wind-blown dust is thick enough and distinctive enough to be easily recognized, it is known as loess (pronounced *loh'-ess* or *luhss*). This formation, illustrated in Figure 8, generally shows no horizontal layering or bedding; it occurs in a single vast layer. The thickness of the deposits ranges from a few feet

8. Loess formation exposed in a road cut near St. Louis, Missouri. Loess represents a thick deposit of wind-blown dust.

to more than a hundred feet; Chinese deposits are several hundred feet high in some areas. Loess sometimes forms high bluffs along the sides of valleys. When these deposits are eroded by running water, a distinctive landscape is formed; in this, nearly vertical walls alternate with flat-bottomed valleys and gulleys, forming an intricate pattern.

The dust particles of which loess is composed are well sorted; they consist of grains between $\frac{1}{16}$ and $\frac{1}{32}$ of a millimeter in diameter. Fossils found in loess are mostly the shells of land snails and the bones of land animals.

Loess deposits, dating back to the glacial period are widespread in the central and northwestern parts of the United States. Extensive loess deposits also occur in South America, Asia and Europe. In some parts of China, thousands of people live in cavelike dwellings dug into bluffs of loess. Some Chinese loess deposits show the layering that is generally associated with deposits laid down by water, thus suggesting that running water finished what the wind began.

The depositing
of volcanic ash

The eruptions of active volcanoes provide a special source of dust for the wind to move. This material, which is called volcanic ash, is blasted violently into the air. It reaches far greater heights and is present in far greater quantities than ordinary dust; it is carried farther and is deposited in thicker layers. When Katmai Volcano, in Alaska, erupted in 1912, volcanic ash fell like snow at the village of Kodiak, a hundred miles from the volcano; it covered the ground to a depth of nearly a foot. The clouds of ash were so dense that they brought almost continuous darkness for more than two days. After the ash had settled down, it was drifted like snow by the wind, and was stirred into minor dust storms.

Other regions were similarly affected. It is estimated that a mantle of ash a foot or more in thickness covered an area of 2,000 square miles; the ash mantle was a quarter of an inch or more thick over an area of more than 20,000 square miles. Some ash fell as far as 900 miles away from Katmai Volcano; the presence of volcanic dust in the air was reported from North Africa.

When the island volcano of Krakatoa, in Indonesia, exploded in 1883, volcanic ash fell inches deep nearly a thousand miles from the scene of the explosion; some ash was deposited as far away as Holland. Quantities of volcanic dust were carried around the world, causing unusually brilliant sunsets.

It has been estimated that hundreds of cubic miles of volcanic ash have been blown into the air and spread out over the earth's surface within the comparatively brief span of time covered by historical records. This, however, is but a trifling amount compared to the enormously greater volume of dust that was transported during the preceding millions of years.

A great deal of volcanic ash falls on land, remains where it fell and becomes a part of the soil. It acts as a natural fertilizer, adding chemical elements needed by growing plants. A few years after the eruption of Katmai Volcano, the grass in the vicinity grew better than ever before; the berries were larger and more abundant. In some parts of the world the soil would be very poor indeed if it were not for the repeated additions of volcanic ash.

A certain amount of ash is drifted away by the wind, or washed away by running water. Sooner or later, much of it finds its way into streams and lakes, where it forms extensive deposits. Some of these are so thick and pure that they have been exploited as a source of scouring powder and for other purposes.

Some volcanic ash falls directly into the sea; to this is added the ash that has been washed down from the land by running water. In the course of long ages such deposits have formed distinct layers of sedimentary rock, known as tuff and bentonite. Rocks of this type have been found in many places; they were formed at various periods of the earth's history.

See also Vol. 10, p. 270: "Sculpture of Land."

SOME ANIMAL HELPERS OF MAN

Beasts of Burden in Many Lands

IN THIS modern age of science and industrial progress, man still remains dependent for much of his livelihood upon animals. The fact is that he still has his roots deep in nature, even though he is surrounded by synthetic materials — miracles of the chemist's art — and does so much of his work with power transmitted through machines. It is perfectly natural for him, therefore, to utilize the animals he finds about him for his daily needs. To be sure, there is less dependence on animals as beasts of burden and labor in some countries than in others. But domesticated animals are still used to supply food and clothing even in highly civilized countries.

In many lands man would perish if he were suddenly dispossessed of his animal allies. Not only would he lose important sources of food, clothing and goods for trade, but he would be deprived of the beasts that serve on the farm, draw vehicles and serve as mounts and pack animals.

The domestication of animals was one of the most important civilizing influences in man's early history. When he began to maintain large numbers of tame animals — cattle and sheep and poultry — he generally had to give up his nomadic life and settle down in an area that would support his newly acquired allies. He not only had to raise crops and build a home for himself, but he had to provide food and shelter for his animals. This stabilizing process strengthened community life.

Like the plants that were put under cultivation, animal servants also came from the wilds. The dog was derived from the wolf; the horse was a direct descendant of the wild horses of central Asia. Man's cattle, sheep and goats came from wild ancestral stock of Eurasia and Africa. He went to the jungles of southeastern Asia for his poultry (except for the guinea fowl, which came from Africa); he obtained his geese from northern Europe.

Man's mastery upon the earth is based in large part on the animals he has pressed into his service. He taught the dog, the fleet-footed cheetah, the ferret and birds of prey to hunt for him. He raised oxen, sheep, pigs, goats, rabbits, poultry, turkeys, geese and ducks for food. He acquired milk from his cattle and goats. For clothing he utilized the products of his cattle, sheep, pigs, goats, rabbits and alpaca. When the wild, fur-bearing animals became scarce, he raised fox, mink, chinchilla, raccoon and skunk for their pelts. To serve him as mounts and beasts of burden, he trained the horse, ass, ox, water buffalo, elephant, camel, reindeer, dog, yak and llama. He raised silkworms so that they might spin silk for his garments; he raised bees for their honey and wax. The guinea pig, the rabbit, the rat and the monkey have served him in recent times as experimental animals.

Through efficient management and control, man has adapted animals to his own needs and whims. First he selected animals with traits that would serve a certain purpose; then he strengthened these traits through the mating of favored specimens and the elimination of the unfit. Many breeds of animals have resulted from this purposeful tampering with nature. The many kinds of dogs (hunters, work dogs or fancy breeds), the various types of cattle (bred for meat or for milk production) and almost all the species of sheep that are valued for their wool and their flesh have been deliberately selected and developed by man.

The vicuña, found in the Andes range.

The elephant, largest of living land animals, has been trained for heavy work since ancient times. It inhabits India, the Malay Peninsula, Indo-China, Ceylon, Sumatra, Burma and Africa (south of the Sahara). The Indian, or Asian, elephant, which is somewhat smaller than the African variety, has been particularly useful to mankind. It has served as a means of transportation and also as an animated derrick, piling logs and immense blocks of stone with unfailing patience and almost unbelievable skill. The logging industry in the teak forests of India and Upper Burma depends largely upon the strength and skill of the elephant. The animal has also been used in warfare, carrying warriors into battle and hauling quantities of arms and stores for considerable distances.

· The trunk enables the elephant to perform various tasks. With this organ the animal can pick things up from the ground, reach fruits or leaves high above its head and convey water to its mouth. The trunk also serves as a weapon. The incisors of the upper jaw form the tusks of the elephant; they often weigh from 150 to 200 pounds each. The molars are extremely large and effectively grind the vegetation upon which the huge animal feeds. The skin of the elephant is exceedingly thick and is sparsely covered with hair.

The elephant is not bred in captivity, but is captured and trained. It is relatively easy to catch the animal, because its hearing and eyesight are poor. Prodigious quantities of food and careful attention are necessary to keep a captive elephant in a healthy condition. The life span is usually forty-five to sixty-five years; however, some captive elephants have been known to live a hundred years.

The gestation period lasts from eighteen to twenty-two months and the young are suckled for two years. The female elephant generally has one calf at a single birth; occasionally, however, two are born.

The ox was one of the first draft animals

The ox — the castrated adult male of domestic cattle — was among the first animals to be domesticated by man for draft purposes. This bulky creature is almost ideal as a beast of burden. It is docile and easily handled; it is hardy; it is immensely strong. Its chief defect as a draft animal is its slowness.

A special breed that is called the Africander ox has been developed by the farmers of South Africa. The hard, flinty feet of this sturdy animal permit it to trek for long distances over the veld — grassland containing scattered trees and shrubs. It can subsist entirely on the vegetation of the veld and can survive droughts.

The water buffalo is useful in rice cultivation

The lumbering domestic water buffalo is particularly useful to man in regions where rice is cultivated. It is very powerful and is capable of wading through mud almost up to its thighs, dragging a plow behind it. It can work in ground too wet and soft for other cattle, and it can feed upon coarse and marshy grasses. It is bred in Japan,

China, southern Asia, North Africa, southeast Europe, southern Italy and Sicily. There are about 20,000,000 buffaloes in India at the present time and about 1,000,000 in the Philippine Islands.

The thick hide of the buffalo is made into an excellent and long-wearing leather. The milk of the female is used in India to make the semifluid butter called ghee. Ordinary buffalo butter is melted and cooled; then the more liquid portion, which is the ghee, is poured off.

The camel is the most valuable beast of burden in Asia and Africa. The species with one hump is the Arabian camel; the Bactrian camel has two humps. A dromedary is any camel of the Arabian species bred for its speed and ease of gait. A camel can be bred for the saddle, or as a baggage carrier or as a draft animal. The Arabian camel is more common than the Bactrian species; it is found domesticated in India, western Asia, Arabia, northern and eastern Africa and, to a small extent, in Spain.

The animal is particularly well adapted for desert life. It has the ability to close its nostrils at will, protecting itself from the sharp and stinging sand. It is also pro-

tected from sand abrasion by callous pads on its feet and chest; it rests on these pads. The camel's long neck allows it to reach for food on the ground or in the trees. The animal is capable of seeing great distances, and its sense of smell is so acute that it can scent water a long way off. These traits have proved very valuable to travelers journeying through unfamiliar desert. The Arabian camel will carry a load of 500 pounds at 25 miles per day for several days without drinking water. This is about twice the load a mule is capable of transporting. The camel has been known to carry a traveler 100 miles a day — that is, with a light pack. Camel corps have played a considerable part in desert warfare.

Camels are usually thought of as docile, but sometimes they become very surly and show violent bursts of temper. The male camel is extremely savage during the rutting season. The female produces one baby camel at birth after a gestation period of about eleven months. The young camel does not reach maturity until its sixteenth or seventeenth year. The coat of the young animal gradually changes from a light color to the darker color of the mother's coat. Camels generally live to the age of forty or fifty years.

The meat of the camel is edible. The female camel yields rich milk from which butter is made. The milk cannot be used in tea or coffee because it tends to curdle.

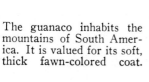

The guanaco inhabits the mountains of South America. It is valued for its soft, thick fawn-colored coat.

N. Y. Zool. Soc.

The massively built yak has long served man as a beast of burden.

The buttermaking is an extremely simple process. Separated cream is put into a skin bag and hung upon the saddle at the beginning of a day's journey. By the time that the resting place has been reached, the cream has become butter, churned by the action of the camel in walking.

The food of the Arabian camel is of the most meager variety. The animal can eat whatever vegetation the inhospitable desert affords; its wedge-shaped cutting teeth are well fitted for cropping the shrubs and other stunted plants of the desert. Not long ago the writer of this article watched a camel set to graze in a field of grass. The animal made straight for the edge of the field and began to browse contentedly upon the thorny vegetation that grew there; it ignored the grass.

The hump of the Arabian camel consists almost entirely of fat and it provides the animal with a reserve supply of food. If the camel draws extensively upon this reserve supply, the hump almost disappears; a rest of three or four months with plenty of food is necessary to restore it. Special cells lining the stomach of the camel absorb and hold considerable reserve supplies of water. By drawing on these supplies, the animal can travel for several days without drinking.

The camel has other remarkable adaptations that fit it for life in the desert. We have already mentioned the pads on its feet. It also has pads on its knee joints;

Llamas are still widely used as pack animals in Bolivia and Peru.

Camels on their way to graze in a desolate region of Saudi Arabia.

it rests upon these pads when rising, kneeling or lying down and in this way protects its joints from abrasion by the sharp particles of sand. The animal's long eyelashes protect it from the glaring sun and from the drifting sand of the desert.

The hair of the camel is woven by the Arabs into cloth; its bones are sometimes used as a substitute for ivory; its hide is made into leather. Its dried dung serves as fuel when no wood is available. It is rather interesting to note that the peculiar odor of a certain high-priced tobacco is derived from the fumes of burning camel dung, over which the tobacco is cured. The tracks of the camel give the nomads of the desert important information concerning the movements of neighbors or the raids of enemies.

The Bactrian camel, a native of central Asia, is readily distinguished from the Arabian camel because it has two humps instead of one. It has shorter and stouter limbs, smaller and tougher feet and longer and finer hair. Like its Arabian cousin, the Bactrian camel has wonderful powers of endurance. It can survive the terrific summer heat of Iran and the bitter cold of the Hindu Kush passes. It can feed upon the sparse herbage of the steppes; it delights in brackish water that very few animals can drink. It can make its bed in the snow as comfortably as the Arabian camel can in the sand. It can carry a load weighing half a ton upon its back and can travel at the rate of twenty-five miles a day. For many years, through all kinds of weather, caravans of sturdy Bactrian camels, sometimes numbering many thousands of animals, were almost the only means by which merchandise was transported between Russia and China. Camel caravans are still widely used in these regions for transportation purposes.

The Bactrian species occurs in the wild state in certain remote areas of Turkistan. It is thought that these wild animals may be descended from the same stock that man first tamed, perhaps thousands of years before the dawn of history.

Asia and Africa give us the largest members of the camel family. The South American members of the family — the vicuña, the guanaco, the alpaca and the llama — are considerably smaller. These animals have shorter ears, smaller feet and a shorter tail; they have no humps. Like the camels, though, they are generally ill-tempered beasts. The llama in particular has one exceedingly distressing habit. When it meets someone whom it dislikes — generally some entirely unsuspecting person — it is likely to discharge its ill-smelling saliva in his face.

The llama and the alpaca are the domesticated members of the group. It seems almost incredible, but the llama was the only beast of burden in America when the Spaniards first landed. The sight of a horseman filled the Peruvians with terror.

The llama has served man in the highlands of South America to much the same extent as the camel in North Africa and Asia. This South American camel was probably domesticated by the Peruvians soon after they began to colonize the Andean uplands — more than two thousand years before the Spanish conquest of the sixteenth century. In truth, the rich culture of the Peruvian Empire owed much of its existence to the domestication of the llama. The strong and sure-footed creature was used as a pack animal; it carried all kinds of articles of commerce, especially ores from the gold and silver mines. An excellent system of roads was purposely built for the heavy llama traffic. The hair of the llama was woven into textiles or twisted into rope. The animal was sometimes killed for meat, and its hide was used for shoes; candles were made from its fat. Even the droppings of the llama were dried and utilized as fuel.

A highly interesting use of the llama was its employment as a message carrier. The Peruvians conveyed information by a device called the quipu (see Index), which consisted of colored strings knotted in various ways to give meaning. The long hair that hung from the underside of the llama could be tied in a series of meaningful knots and could thus be utilized as a quipu. Sometimes messages were sent hundreds of miles through the Peruvian Empire by such an ingenious system.

Today the llama still serves as a beast of burden, particularly in rough mining country where it may travel twelve to fifteen miles a day, carrying ores to and supplies from the sea coast and waterways. Depending upon the size of the animal, the load varies in weight from sixty-five to two hundred pounds. Saddlebags of woven llama hair, in which the load is placed, are slung on each side of the animal and are fastened with rawhide llama ropes. If the load seems too heavy, the animal sits down and refuses to move, even if beaten, until the burden is lightened. As in the past, only the male llamas are employed; the females are maintained solely for the purpose of breeding.

The llama's ancestor, the guanaco, still exists as a wild animal from the mountains of Peru, Bolivia and Chile to the prairies of Patagonia. South American Indians hunted it for its meat and pelt. Today guanacos are raised on farms and are sheared of their soft, woolly hair or killed for pelts, which are made into short fur coats or utilized as trimmings. The guanaco is also the probable ancestor of the alpaca, a domesticated animal smaller than the llama and somewhat like a long-necked sheep in appearance. Its thick, long hair is woven into durable cloth.

The vicuña, a small llamalike animal living in the highlands of the central Andes, has never been successfully domesticated to any degree. Because of its hair, which is not particularly thick but is of a fine, silky quality, it was hunted almost to the point of extinction. Fortunately the animal is now rigidly protected by the Peruvian Government.

The reindeer has had a long association with man

Between twenty-five and thirty thousand years ago, stone-age man drew outstandingly good likenesses of reindeer on his cave walls. At that time reindeer were extremely numerous and were probably man's chief support. It is difficult to determine exactly when the animal was first domesticated, but we know that it has well served the people of the Eurasian Arctic region for many hundreds of years.

Though the reindeer is only a small deer, standing three and one half feet at the shoulder, it has great stamina and strength. As a steed, it can carry a man of average weight for days on end. It can also draw a loaded sled, twice its weight, over the snow, averaging forty miles a day. Reindeer meat is of excellent quality, and the skin can be converted into a high-grade leather, which, when the hair is left on, is used for sled blankets and such articles of clothing as shoes, trousers, gloves and parkas. The sinews serve as thread and the stomach membranes as food containers. The hair is woven into cloth and used to stuff mattresses. Reindeer milk is exceed-

The zebra is not particularly reliable as a domesticated animal.

ingly rich, requiring dilution with water before it can be drunk; cheese and an alcoholic beverage are made from it. The Lapps, perhaps more than any others, are truly reindeer people. They travel, as nomads, with the deer, from the summer to the winter feeding grounds and back again, and gain almost their whole living from these animals.

Domestic reindeer, imported from Siberia, were introduced into Alaska in the last decade of the nineteenth century; since then thousands of their descendants have been killed for food and clothing. The North American caribou, a larger relative of the reindeer, has never been domesticated by the Eskimos and Indians but has served these people admirably. All parts of the animal are used, from the antlers, which are fashioned into fishhooks and other utensils, to the blood, which is put in soups. Even the chewed and partially digested lichens and moss, found in the caribou's stomach, are valuable as a nutritious human food.

In Tibet, the all-purpose animal is the yak, a bisonlike beast with long, smooth hair growing down over its body. Wild yaks roam in small numbers on the more remote Tibetan plateaus; the smaller domestic variety, herded just like other cattle,

has been in association with man for centuries. In the mountain passes, the beast is employed as a sure-footed and hardy steed and a pack animal that can carry extremely heavy loads. It is also harnessed to the plow. Yaks are milked, and the rich cream is added to tea; rancid yak butter is widely used. The animal is slaughtered for its high-quality meat, and its hair is put to a variety of uses. The Tibetans formerly sent communications by tying knots in the hair hanging from the yak's belly, just as the ancient Peruvians did in the case of their llamas.

Zebras, which are members of the horse family, cannot truly be regarded as animals amenable to domestication. They have been occasionally trained to drive in harness but are not to be trusted nor depended upon as steady working animals. They tire quickly and usually refuse to work for man. Sometimes a zebra and a horse are crossed, but the hybrid does not serve as well as either the horse, ass or mule. The quagga, a now-extinct small zebra, seems to have been easily tamed, and reports tell of the Dutch colonists who kept quaggas among their herds of cattle. The quaggas defended the herds by kicking and trampling hyenas and various other carnivorous animals that attacked the cattle.

VOLCANIC CONE —
*BUILT UP OF LAYERS
OF ASH AND
LAVA*

VOLCANIC ROCK—
HARDENED LAVA

— MAGMA

SILL

MAGMA RESERVOIR

Cross section of a volcano. The magma that makes its way to the surface in the form of lava is molten rock material; it contains various volatile elements that are driven off as the magma solidifies. The cone of a volcano is formed as solid materials accumulate around the opening from which hot lava, rock fragments and gases are ejected.

VOLCANIC CATASTROPHES

Historic Eruptions in Many Lands

IN A later volume we discuss the origin, nature and evolution of volcanoes, as well as the effects of their eruptions upon their surroundings. In this chapter we shall give an account of some of the great volcanic catastrophes of the past.

One of the most notable eruptions of historic times was that of Vesuvius in 79 A.D., when the cities of Pompeii and Herculaneum were buried under masses of cinders, small stones and ashes belched forth by the great volcano. It was not until the nineteenth century that these cities were dug out from the overlying deposits, revealing priceless relics.

The volcano had slumbered for centuries before the great eruption. About 155 years before it took place, the gladiator Spartacus and seventy followers took refuge in its crater; when they were besieged by the praetor Claudius Pulcher, they escaped by climbing down with the aid of ropes made of vine branches. No one thought of the mountain as a volcano, with the exception of a few observant men, such as the geographer Strabo and the historian Diodorus Siculus, who noticed the cindery appearance of the surrounding country. Fields, vineyards, orange groves and villages extended almost up to the crater of the volcano; at its base were the cities of Pompeii and Herculaneum, with their villas, theaters and temples.

The inhabitants had had a number of warnings before the disaster took place; the land in the vicinity of the volcano had been shaken by a series of earthquakes, beginning with the year 63 A.D. Finally a violent shock, on August 24, 79 A.D., was followed by the historic eruption.

This disaster was graphically described in a letter of Pliny the Younger, an eyewitness. He told first how Vesuvius began to emit showers of cinders and ashes.

"Being got at a convenient distance from the houses," he added, "we stood still in the midst of a most dangerous . . . scene. The chariots which we had ordered to be drawn out were so agitated backwards and forwards, though upon the most level ground, that we could not keep them steady, even by supporting them with large stones. The sea seemed to roll back upon itself, and to be driven from its banks by the convulsive motion of the earth. It is certain [that] the shore was considerably enlarged, and many sea animals were left upon it. On the other side a black and dreadful cloud bursting with an igneous serpentine vapor darted out a long train of fire, resembling flashes of lightning but much larger. . . . The ashes now began to fall upon us, though in no great quantity. I turned my head, and observed behind us a thick smoke, which came rolling after us like a torrent.

"I proposed, while we yet had any light, to turn out of the highroad, lest we should be pressed to death in the dark by the crowd that followed us. We had scarce stepped out of the path when darkness overspread us, not like that of a cloudy night, or when there is no moon, but of a room when it is shut up and all the lights extinct. Nothing then was to be heard but the shrieks of women, the screams of children, and the cries of men; some calling for their children, others for their parents, others for their husbands, and only distinguishing each other by their voices; one lamenting his own fate, another that of his family; some wishing to die from the very fear of dying; some lifting their hands to the gods, but the greater part

imagining that the last and eternal night was come which was to destroy the gods and the world together. Among them were some who augmented the real terrors by imaginary ones and who made the frightened multitude falsely believe that Misenum was actually in flames.

"At length a glimmering light appeared, which we imagined to be rather the forerunner of an approaching burst of flame, as in truth it was, than the return of day. However, the fire fell at a distance from us. Then again we were immersed in thick darkness, and a heavy shower of ashes

rained upon us, which we were obliged every now and then to shake off; otherwise we should have been crushed and buried . . .

"I might boast that during this scene of horror not a sigh or expression of fear escaped from me, had not my constancy been founded on that miserable though strong consolation that all men were involved in the same calamity, and that I imagined I was perishing with the world itself.

"At last this dreadful darkness was dissipated by degrees, like a cloud of smoke; the real day returned, and soon the sun appeared, though very faintly and as when

Ewing Galloway

Amid the ruins of Pompeii: the Temple of Apollo with Vesuvius in the background. Pompeii was overwhelmed by an eruption of Vesuvius in 79 A.D.

with water and hardened into volcanic rock. As far as is known, no lava was emitted during this eruption.

The survivors of Pompeii returned to the town and by digging and tunneling were able to recover various valuables. Later, however, the very site of the two stricken cities was forgotten. They were rediscovered in the modern era and yielded many priceless relics of antiquity.

The records of later eruptions of Vesuvius are quite confused. Among the most violent were those of 203 and 472, during which fine ash was carried as far as Constantinople, and the one of 512, which devastated the Campania. Noteworthy eruptions also took place in 685, 993, 1036, 1139 and 1500. After this last eruption there was a long period of quiescence; the mountain was overgrown with vegetation, and cattle grazed along its slopes. Then it became active again. Six months of earthquakes in 1631 were followed by a terrific eruption, which blew off the whole top of the mountain and scattered ashes for hundreds of miles. Seven streams of lava poured from the crater, destroying several villages. Among these was Resina, which had been built on the site of Herculaneum.

This ushered in the modern period of fairly numerous eruptions, highlighted by particularly violent ones in 1794, 1872, 1906 and 1929. The eruption of 1906 was described in great detail by the distinguished American volcanologist Frank Alvord Perret. On April 4 of that year a massive cloud of steam, mixed with dark ash, shot skyward. By midnight lava was issuing rapidly from a fissure on the southern slope of the cone at a low level. At 8 A.M. on April 6, a new vent opened on the southeastern side of the cone only 1,800 feet above sea level; from this came a flood of very fluid lava. There was much earthquake activity and many explosions in the crater. Clouds of hot gases carrying relatively small quantities of ash were emitted, forming a huge "cauliflower" formation seven miles high. On April 8, the clouds became so charged with volcanic ash that they became black, casting a pall about Naples and the surrounding country.

an eclipse is coming on. Every object that presented itself to our eyes . . . seemed changed, being covered over with white ashes, as with a deep snow."

The great eruption of 79 A.D. completely destroyed the flourishing towns of Pompeii and Herculaneum. Pompeii was buried beneath layers of ashes, cinders and stones, in some places sixty or seventy feet deep. At Herculaneum these materials were drenched

An eruption in June 1929 caused the splitting and collapse of the central cone of Vesuvius. Lava overflowed from the crater and ran down the outer slopes toward the town of Terzigno; it halted at last only 400 yards from the houses of the town. There was a minor eruption on March 13, 1944; as a result, the ruins of Pompeii were buried under nearly a foot of ash.

The eruption of the Japanese volcano of Bandai-san in 1888 resulted in a frightful catastrophe. Bandai-san, in Fukushima prefecture, north central Honshu, Japan, is a group of four volcanic cones, which represent the remains of a vast ancient volcano; the highest cone rises to a height of 5,968 feet. According to tradition, the original mountain was split by a great explosion that buried fifty villages. The volcano of Bandai-san had been dormant for ten centuries; there were probably some small upheavals during this time, but nothing like a great eruption. The inhabitants of the area had been lulled into a sense of complete security. Peasants worked daily on the mountain slopes without the least foreboding of a fiery doom.

The volcano of Stromboli, a small island of the Lipari group. The volcano rises to a height of 3,038 feet.

A terrific eruption took place on July 15, 1888. First there was a severe earthquake; then a dense cloud of steam and dust shot into the air. Explosion followed explosion; darkness covered the land; lightning flashed. An avalanche of mud, earth and rocks roared down the mountain, burying villages and devastating an area of twenty-seven square miles.

A Japanese priest, Isurumaki, who had a miraculous escape from death, gave the following account of his experiences: "The morning of the 15th, which was the fatal day, dawned with a bright and pleasant sky . . . At about eight o'clock, however, there was a fierce convulsion of the ground, and we all rushed out of the house. In about ten minutes, while we were fearfully wondering what was the matter, a terrible explosion suddenly burst from the slope of [Bandai-san], about a quarter of a mile above a place from which steam had been issuing from time immemorial. This was followed by a dense mass of black smoke, which rose into the air and immediately covered the sky. At this time showers of large and small stones were falling all about us. To these horrors were added thunderous sounds; the rending of mountains and forests presented a most un-

Ewing Galloway

earthly sight, which I shall never forget as long as I live. We fled in all directions, but before we had gone many yards we were all thrown prostrate to the ground.

"It was pitch dark; the earth was still heaving beneath us. Our mouths, noses, eyes and ears were all stuffed with mud and ashes. We could neither cry out nor move. I hardly knew whether I was dead or in a dream. Presently a stone fell on my hand, and I knew that I was wounded. Imagining that death was at hand, I prayed to Buddha. Later I received wounds on my loin, right foot and back. After the lapse of an hour, the stones ceased to rain, and the atmosphere had cleared from darkness to a light like moonlight. Thinking this a fine opportunity to escape, I got up and cried: 'Friends, follow me!' but nobody was there. When I had descended about half a mile, there was a second explosion and a quarter of a mile further on a third, and ashes were ejected but no stones."

The most striking feature of the eruption was the immense mass of fragmental

A volcanic eruption on San Benedicto Island, one of the Revillagigedos, in the Pacific off the coast of Mexico.

material that was discharged. Part of it was hurled into the air; part of it slid down in an avalanche of mud, earth and rocks, with huge boulders here and there. As is usual in volcanic explosions, a great quantity of dust was shot forth, covering the country for miles around.

The solid matter that had been cast up from the volcano filled up all the ravines and gorges, engulfed all familiar landmarks, dammed up several rivers, converted twenty-seven square miles into a desert, buried four villages and partly covered several others. The death toll was 461, including 92 persons who fled from a safe hamlet and were all struck down within a few yards. The blasts of air produced by the explosions leveled houses and tore up trees by the roots or stripped them of their branches. Another interesting feature was the multitude of basinlike holes made in the ground by falling stones.

The island volcano of Krakatoa, in the Sunda Strait. An eruption in 1883 blew away a part of the island.

The 1883 eruption of the volcano on the islet of Krakatoa was probably the greatest in historical times. Krakatoa is in the Sunda Strait, between Sumatra and Java. This island and several others in the vicinity are the remnant of an ancient volcano which was probably about 10,000 feet high and had a crater about 25 miles in circumference. For more than 200 years there had been no volcanic activity on the island; it was covered with rich vegetation.

In 1880, Krakatoa was rocked by a series of earthquakes; these were no slight tremors for they were felt as far off as northern Australia. The eruption proper began on May 20, 1883, with a series of explosions loud enough to be heard a hundred miles away. The next day ashes were scattered on both sides of the strait. An investigating party visited Krakatoa and the neighboring islands on May 26 and found them covered with snowlike white dust; a column of vapor rose 10,000 feet in the air, scattering showers of pumice stone and dust. This state of affairs continued through June, July and August.

The culmination of the eruptions was reached on August 26 and 27, when a series of mighty blasts rocked the island. The whole northern portion of Krakatoa was blown away; the land was replaced by a huge submarine cavity. The forests of the surrounding islands were buried under thick layers of fragmentary materials.

Officers and passengers on ships that happened to be in the Sunda Strait at the time of the eruption provided details. Captain Wooldridge of the *Sir Robert Sale* wrote that on the 26th the sky presented "a most terrible appearance" and that the cloud above the mountain was like an immense pine, the stem and branches formed by volcanic flashes. After sunset the cloud looked like a "blood-red curtain with edges of all shades of yellow," and lightning zigzagged through it. Captain Watson of the *Charles Bal* told of chains of fire that rose in the sky and of "balls of white fire" that continually rolled down the mountain. Startling electrical phenomena occurred. A peculiar pinkish flame streaked from the clouds, and balls of fire studded the mastheads and yardarms of the ships.

The mainmast conductor of the Batavian ship *Gouverneur-Generaal Loudon,* fifty miles away, was struck by lightning six times. There was a continuous downpour of phosphorescent mud, which soon covered the masts, rigging and decks of the ship.

All day on August 27 the *Northam Castle* and the three vessels mentioned were in pitch darkness and under a continual rain of pumice stone and dust. So violent were the explosions that they were heard three thousand miles away. They caused some of the most tremendous air waves that have ever been known. Windows were broken, walls cracked and lamps overthrown a hundred miles away and some air vibrations traveled several times round the world. By the sudden dislodgment of vast volumes of rock below water, huge waves were sent sweeping over the adjacent shores of Java and Sumatra at the rate of nearly four hundred miles an hour. Towns and villages were destroyed, two lighthouses swept away and nearly 40,000 people drowned.

The incredible tale of a fortunate survivor

The town of Anjer, in Java, was overwhelmed by the sea. The Rev. Phillip Neale, one of the few survivors, gives the following account of his escape. He says: "About six A.M. I was walking along the beach. There was no sign of the sun, as usual, and the sky had a dull, depressing look. Some of the darkness of the previous day had cleared off, but it was not very light even then. Looking out to sea, I noticed a dark, black object through the gloom traveling towards the shore. At first it seemed like a low range of hills rising out of the water, but I knew that there was nothing of the kind in that part of Sunda Strait. A second glance — and a very hurried one it was — convinced me that it was a lofty ridge of water many feet high, and, worse still, that it would break upon the coast near the town. There was no time to give any warning, and so I turned and ran for my life. My running days have long gone by, but you may be sure I did my best. In a few minutes I heard the water, with a loud roar, break upon the shore. Everything was engulfed. Another glance around showed that the houses were being swept away and the trees thrown down on every side. Breathless and exhausted, I still pressed on. As I heard the rushing waters behind me, I knew that it was a race for life. Struggling on, a few yards more brought me to some rising ground, and here the torrent of water overtook me. I gave up all for lost, as I saw with dismay how high the wave still was. I was soon taken off my feet, and borne inland by the force of the resistless mass. I remember nothing more till a violent blow aroused me. Some hard, firm substance seemed within my reach, and, clutching it, I found that I had gained a place of safety. The waters swept past, and I found myself clinging to a cocoanut palm tree. Most of the trees near the town were uprooted and thrown down for miles, but this one, fortunately, had escaped, and myself with it.

"The huge wave rolled on, gradually decreasing in height and strength until the mountain slopes at the back of Anjer were reached, and then, its fury spent, the water gradually receded and flowed back into the sea. The sight of those receding waters haunts me still. As I clung to the palm tree, wet and exhausted, there floated past the dead bodies of many a friend and neighbor. Only a mere handful of the population escaped. Houses and streets were completely destroyed . . .

"Scarcely a trace remains of where the once busy, thriving town originally stood.

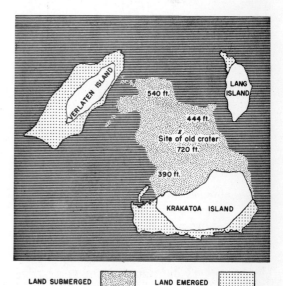

Changes produced by the eruption of the volcano on the islet of Krakatoa.

Unless you go yourself to see the ruin, you will never believe how completely the place has been swept away. Dead bodies, fallen trees, wrecked houses, an immense muddy morass and great pools of water are all that is left of the town."

The loudest sounds
ever heard on the earth

The sounds of the major explosions at Krakatoa were heard thousands of miles away — at Rodriguez, in the west Indian Ocean; at Bangkok, in Thailand; in Ceylon; in the Philippine Islands; in Western and South Australia. It is said that no other sounds originating on the surface of the earth have ever been heard so far away.

The dust of the explosion is said to have reached a height of seventeen miles. It was distributed far and wide by the upper air currents and gave rise to marvelous sunrises and sunsets all over the world for the following three years. Many witnesses gave glowing reports of these phenomena.

An extraordinary
sunset at Cannes

An English observer described the sky at Cannes, France, on January 10, 1884, as follows: "Orange . . . glow in southwest near horizon; above this a greenish-blue white arc, then a beautiful yellow band; then, up to the zenith, a very beautiful lilac tint. All of these colors were of extreme softness and, though not as striking as some of the sunsets in December [of the preceding year], in point of beauty they were quite unsurpassable and of superb magnificence in their . . . progress.

"Thirty-five minutes after sunset an arc was formed [with its] middle part yellow and the outer purple. Through the fringe of this, Venus shone beautifully. The horizon — about a quarter of the circle — was deep yellow. . . . In fifty-four minutes the primary glow was gone, having sunk in a deep red band. The eastern sky during the first part of the display was a glorious deep blue, then very dark purple-blue and lastly, only illumined by the silver moon."

A Frenchman wrote from San Salvador, in Central America: "Remarkable sunsets have been seen here since the last days of November, 1883. About half an hour after sunset and an hour before sunrise, the horizon is gradually illuminated with a magnificent coppery-red tint, very constant in color, very intense and lasting on the average twenty to twenty-five minutes.

"The moon, when circumstances allowed of it — that is, when its altitude did not exceed 15 degrees — was colored a magnificent emerald-green. It was extremely beautiful at the time of gray light, when its disc was of a pale green, with its crescent horn deep green in the midst of an immense crimson curtain. Only Venus was able to penetrate this curtain, and it too was green in color."

Green moons,
green and blue suns

Not only were there green moons, but also green and blue suns, because of the dust concentration in the atmosphere. On September 12, 1883, an English official wrote from Ceylon: "The sun for the last three days rises in a splendid green, when visible, i.e., about 10 degrees above the horizon. As he advances, he assumes a beautiful blue, and as he comes on further looks a brilliant blue resembling sulphur. When about 45 degrees, it is not possible to look at him with the naked eye, but even when at the zenith the light is blue, varying from a pale blue to a light blue later on, somewhat similar to moonlight, even at midday. . . . As he declines, the sun assumes the same changes, but vice versa. The moon, now visible in the afternoons, looks also tinged with blue after sunset, and as she declines, assumes a very fiery color 30 degrees from the zenith."

The eruption of
Mount Pelée in 1902

The eruption of Krakatoa was certainly unique in some respects. It caused much destruction and great loss of life, particularly because of the extensive sea waves that the disturbance created and that swept the shores of Java and Sumatra. But if we consider only the loss of life brought about in its immediate vicinity,

it is not to be compared with the May 8, 1902, eruption of Mount Pelée, which completely wiped out the city of St. Pierre. Mount Pelée is on the island of Martinique, in the West Indies. At the time of the disaster, St. Pierre, which nestled at the southwestern foot of the mountain, was a flourishing city with over 28,000 inhabitants; it was the leading commercial center of the island.

For some time before the disastrous eruption of May 8, Mount Pelée had been threatening. Throughout the month of April steam was seen issuing from the mountainside. On the 25th there was a discharge of ashes; there were minor earthquakes and explosions on the 30th. The inhabitants at first disregarded these warnings of impending danger. The chief Martinique newspaper, *Les Colonies* (The Colonies), advertised that an excursion to the summit of Pelée would take place on May 4, if the weather were favorable.

On May 2, there was a violent eruption with loud explosions and heavy clouds of condensed steam. The people now began to take alarm, and schools and stores were closed. Streets and gardens were soon covered with a layer of ashes; birds dropped dead, apparently smitten by poisonous gases. On the fifth of the month a torrent of lava and mud burst through the crater into the valley of the River Blanche,

An artists's conception of the fiery cloud that rolled down Mount Pelée and destroyed the city of St. Pierre.

completely destroying a sugar works and killing twenty-four persons. The governor of Martinique now appointed a commission of experts to report on the situation. After a remarkably rapid survey, the commission announced that St. Pierre was in no danger. The governor himself decided to remain in the city, in order to reassure the frightened inhabitants.

Finally the blow fell. At about a quarter of eight on the morning of May 8, there was a deafening series of detonations. A dense black cloud of superheated gases, charged with incandescent dust, came rolling down the mountainside and enveloped the doomed city of St. Pierre. In a moment or two the whole town was ablaze. All the ships in the harbor except two — the *Roddam* and the *Roraima* — were destroyed. Captain Freeman of the *Roddam*, who watched the tragedy from the sea, saw a few persons running about the beach in a panic; moments later they collapsed on the sands.

Only four persons were taken alive out of the stricken city. Of these, two died almost at once, and a third succumbed not long afterward in the hospital. Ironically enough, the sole survivor who ultimately recovered was a criminal who had been condemned to death for murder! This man, a Negro called Joseph Surtout, had been shut up in a cell so far underground that the flames and gases had not reached him. For four days he remained in his cell without food or drink; at last rescuers heard his cries for aid and he was carried out.

Torrential rains followed the eruption. As one writer put it: "At the end of this time, the city was laid in smoldering ruins, coated with ash paste and looking as if built of adobe plaster. What had before been the vivid coloring of houses of the tropics was now an ashen gray."

Only one of the villages in the vicinity of St. Pierre was spared; the others shared the fate of the city. It is estimated that 40,000 persons lost their lives as a result of the eruption. Such was the consternation of the inhabitants of Martinique that it was proposed to abandon the island; but the French government did not consent. Later, a new city was built.

The overwhelming flood of hot gases and dust, sweeping down from the volcano,

View of Halemaumau, the fire pit of Kilauea, while it is filled with lava. The lava disappears at intervals.

greatly impressed volcanologists. The name "Peléan" was given thereafter to eruptions in which hot gases and dust are blown out horizontally and rush down the side of the erupting volcano.

The Hawaiian Islands have been the scene of many spectacular eruptions that have sometimes caused great property damage. It is a curious fact, however, that there has been practically no loss of life in these eruptions, as far as we know, because persons in the vicinity have had time to reach places of safety.

The islands are chiefly of volcanic origin, though they include some atolls, or coral islands (see Index, under Atolls). There are fifteen large volcanoes in all; the only two that are active at the present time — Mauna Loa and Kilauea — are on the island of Hawaii, the largest of the group. Hawaii has five volcanoes about twenty miles apart and connected by saddle-like ridges formed by lava flows.

The highest mountain in the island of Hawaii is Mauna Kea (White Mountain), which is 13,825 feet above sea level. In one sense, it is the largest mountain in the world, since it rises from a vast plain

about 18,000 feet below the level of the sea. Mauna Kea has been extinct for a great many years; one of its old craters is now occupied by a lake.

Mauna Loa, the
largest active volcano

Mauna Loa (Long Mountain) lies to the south of Mauna Kea and, like it, rises from a plain 18,000 feet below sea level. Mauna Loa is 13,675 feet high; it is the largest active volcano in the world. Its dimensions are quite staggering; it is about sixty miles across and two hundred miles in circumference at sea level. At the summit there is a huge crater, which is called Mokuaweoweo. This crater is five miles in circumference; the cliffs around its pit are from 500 to 600 feet high. There is considerable volcanic activity in Mokuaweoweo, particularly before an eruption; white-hot fountains of lava rise to a height of several feet. However, there have been no flows of lava from this crater, at least in historic times. All the lava discharges from Mauna Loa come from the flanks of the mountain, at elevations of from 7,000 to 13,000 feet.

Some historic
Mauna Loa lava flows

Mauna Loa emits more lava than any other volcano at the present time. There have been numerous eruptions in the past hundred years or so, at intervals of a few years. The eruption of 1873–74 lasted eighteen months; that of 1880–81, nine months. In 1926, a lava flow from a crevice in the southeastern side of the mountain destroyed the village of Hoopuloa, on the coast. This flow was thirty feet high, and advanced along a front of about a hundred feet at the rate of three feet a minute. Thomas A. Jaggar, director of the Hawaiian Volcano Observatory, compared the flow to the lumbering advance of a caterpillar tractor. Said he: "An upper layer of boulders and paste is rolled forward on a viscous red-hot paste inside, tumbles down at the front in a debris slope and this is eternally overridden by the advancing mass for which it lays the track."

In 1935, the city of Hilo was threatened by a lava flow proceeding from the northwest flank of Mauna Loa. The United States Army Air Corps (it is now a separate military arm) came to the rescue. A fleet of bombers dropped 6,000 pounds of bombs from a height of about 5,000 feet above the lava. The bursting bombs opened up new channels for the lava flow, diverted its course, and saved the city.

The low-lying
volcano of Kilauea

The volcano of Kilauea, twenty-two miles from the crater of Mauna Loa, does not look like a mountain at all, but rather like a cuplike depression in an immense plain. It appears almost like a big hole in the side of Mauna Loa at a height of about 4,000 feet. But Kilauea is really a volcano in its own right; its periods of activity are quite independent of those of Mauna Loa.

Within the crater of Kilauea there is a vast pit, called Halemaumau, or House of Everlasting Fire; it is about 400 feet across and over 700 feet deep. In this pit, lava rises and falls in fairly regular cycles. It keeps rising until it breaks through a subterranean passage and drains out, only to begin the cycle anew. At times, therefore, Halemaumau is a heaving lake of lava; at other times it is simply a huge hole in the crater.

Generally the activity of Halemaumau is confined to the area within the crater of Kilauea. In 1924, however, there was an explosive outburst. A few months before the lava lake disappeared, boulders and ash were hurled thousands of feet in the air. No particular damage was caused.

Pele's hair —
a product of Halemaumau

The air above the seething lava lake of Halemaumau is sometimes filled with gossamer threads of lava; they are known as Pele's hair. (Pele is the Polynesian goddess of volcanoes.) These threads of lava are often carried away by the winds; birds use them as material for their nests.

See also Vol. 10, p. 271: "Volcanoes and Geysers."

THE DEFENSES OF THE BODY

Structures and Processes

That Guard Us

BY W. W. BAUER

THE human body, moving over a long span of life, through a world filled with hazards, possesses a remarkable system of defenses against attacks from without — against accidents (or other violence) and disease. It has an outer parapet, many internal defenses and a highly ingenious regulatory system. A distinguished American physiologist, Dr. Walter B. Cannon, has described this elaborate series of defenses as "the wisdom of the body."

The senses and nervous system provide warning

The organs of perception provide warning of violence, due either to accident or to intentional attack by man or animals. The eye can perceive, or the ear detect, the approach of danger. Man's sense of smell is less acute than that of many animals, but it is keen enough to help ward off danger of accident from fire or fumes or even to warn against attack. Skin sensation is useful in detecting contacts with crawling insects and other animals that may have a dangerous bite and in warning of excessive heat or cold. It also helps us to determine the texture of solid objects or the character of fluids with which we make contact.

Closely allied to the senses of perception is the action of the nervous system. A warning signal received through the eye,

the ear, the nose or the skin, is transmitted to the brain, which then directs appropriate action. If there is plenty of time, this may be a conscious action. You may be driving a car, and see a traffic jam in the next block. You slow up and look the situation over as you approach, then decide what to do next. But if the driver ahead of you suddenly puts on the brakes, you do the same before you have time to size up the situation. This is called reflex action. A reflex action takes place when you duck without thinking as a snowball is thrown at your head; in fact, the chances are that you will duck even if you know there is a heavy sheet of unbreakable glass between you and the oncoming object. Reflex action is at work, too, when you jerk your hand away, quicker than a wink, upon touching a hot stove.

This is how reflexes work. In the skin there are tiny organs called receptors, or receivers; there is a different one for each sensation: pain, heat, cold, pressure, texture and so on. From these, the nerve fibers go through various nerve trunks to a nerve center or ganglion, one of a chain which is just outside the backbone in the abdomen, chest and neck, and in various places in the head. Here the impulse is delivered to other nerve fibers, in much the same manner as a telephone relay carries the voice from station to station, and the message is conveyed to the brain. Action impulses then originate in the brain. They are transmitted to a nerve center in the spinal cord and from there, by other nerve fibers, to muscles that bring appropriate sense receptors, such as the eyes, nose or fingers, into play. Thus we look at, or smell or feel the object that had stimulated our skin receptors and learn what it is like.

But if there is an emergency, the whole action is short-circuited. We touch a hot stove, and the pain sensation flashes to the spinal-cord relay station. Here it is intercepted and transmitted to the appropriate nerve fibers bringing about action; these at once impel the appropriate muscles to get the hot contact broken without delay. The interception saves the longer time it would take for the danger signal to reach the brain, and there to be transferred through a num-

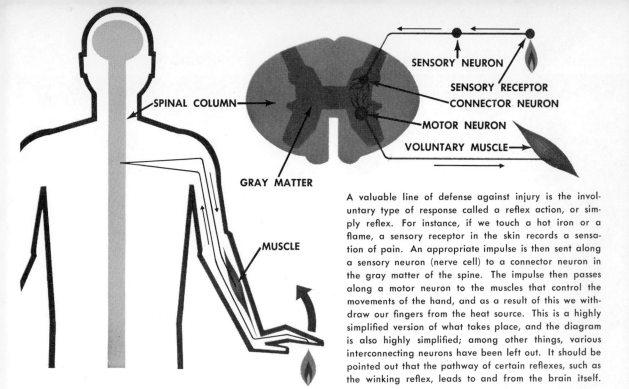

SPINAL COLUMN

GRAY MATTER

MUSCLE

SENSORY NEURON

SENSORY RECEPTOR

CONNECTOR NEURON

MOTOR NEURON

VOLUNTARY MUSCLE

A valuable line of defense against injury is the involuntary type of response called a reflex action, or simply reflex. For instance, if we touch a hot iron or a flame, a sensory receptor in the skin records a sensation of pain. An appropriate impulse is then sent along a sensory neuron (nerve cell) to a connector neuron in the gray matter of the spine. The impulse then passes along a motor neuron to the muscles that control the movements of the hand, and as a result of this we withdraw our fingers from the heat source. This is a highly simplified version of what takes place, and the diagram is also highly simplified; among other things, various interconnecting neurons have been left out. It should be pointed out that the pathway of certain reflexes, such as the winking reflex, leads to and from the brain itself.

ber of nerve paths and connections to the cells on the upper surface of the brain that ordinarily control voluntary actions.

Pain sensations occur in many parts of the body other than the skin and are manifested in different ways. The brain substance feels no pain, but the covering membranes do. The intestines are insensitive to cutting, but react painfully to twisting, pulling or distention with gas. Pains may be steady or intermittent, sharp or dull, shooting or burning. Pain is usually regarded as something to be avoided; actually it is a safeguard. It is a signal that something is wrong; properly interpreted and acted upon, it is a safety device, a protective function.

Closely related to pain are other uncomfortable sensations such as itching, nausea, dizziness and faintness, which indicate that some function is disturbed and needs attention. These sensations should be heeded, particularly if they are severe, if they tend to return at frequent intervals or if they are present for more than a few hours.

It is illogical to suppress these safety signals through the use of drugs. Merely dulling the pain does not remove its causes; it may even permit an illness which was mild in the beginning to become serious and perhaps even incurable.

The first line of defense —
the skin and mucous membranes

Diseases that attack the body from without are met with what military men would call defense in depth: that is, when one defense is breached, another stands ready to take over. The first line consists of the skin and the mucous membranes of the body. Attacking germs do not really get inside the body until they have penetrated either the skin that covers it or the mucous membrane that lines it. The digestive and breathing systems, commonly considered to be "inside" us, are really outside the body proper; they are covered with mucous membrane, which resists penetration.

The defenses supplied by the skin are of several different kinds. A natural barrier is provided by the outer layers of dead, flattened cells; they may be scuffed off without damage to the body and without pain. The oiliness of the skin, derived from the sebaceous glands, makes it waterproof and helps to prevent germ invasion. The blood vessels in the skin rush blood to the menaced area and thus bring to bear forces that fight germ progress.

Mucous membranes afford protection through the chemical composition of their secretions. The slightly acid saliva in the

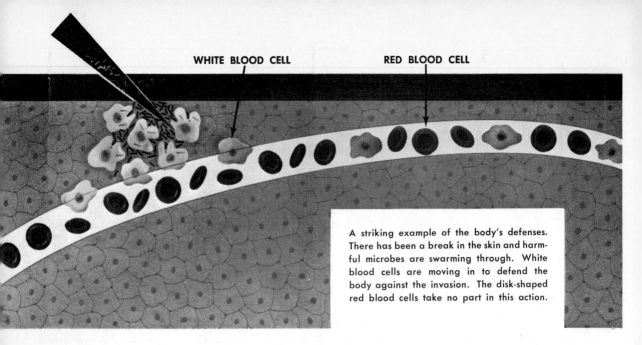

WHITE BLOOD CELL   RED BLOOD CELL

A striking example of the body's defenses. There has been a break in the skin and harmful microbes are swarming through. White blood cells are moving in to defend the body against the invasion. The disk-shaped red blood cells take no part in this action.

mouth and the more strongly acid stomach secretions both help to keep bacteria at a minimum. In the nose, the membranes have tiny cilia, or hairlike structures, which have a rhythmic waving motion toward the nostrils, thus moving secretion, dust and germs outward.

Invasion of the body by infectious agents

If the skin is broken, the way is prepared for the entry of infection. Small breaks heal quickly, first being sealed against invasion by the clotted blood, which not only prevents hemorrhage but also constitutes a defense against infection. If, however, infection has been introduced at the time of the injury, there may be a small center of inflammation around the affected area. It will become reddened; there will be pain or throbbing and heat; perhaps abscesses will form. A typical abscess on the surface of the body is a boil.

This is what happens when the infection sets in. More blood is rushed to the point of entry, accounting for the increase in local temperature, and the blood brings with it more white cells. These do not remain inside the blood vessels, but ooze through the thin and easily permeated capillary walls into the tissues, where they meet the invading organisms and attempt to engulf and destroy them. A cross section through an abscess in the early stages shows tissues swollen with blood and mak-

ing a protective wall around the site of entry of the infection. If the infection is held at bay, the center of the abscess liquefies; in ordinary language we say it "comes to a head," or "points." It is then ready to open. The pus that is discharged contains fluid serum, white blood cells and bacteria, both dead and alive. If we open an abscess too soon, we may cut through the wall of protective inflammation and cause the infection to spread into deeper tissues or into the blood. Squeezing a pimple may produce the same unfortunate effects.

The invasion of the body by infectious agents—bacteria, viruses, molds—has been likened to war. There are local skirmishes at the points of attack, followed later by generalized battles. In this "war," the cells and the chemical constituents of the blood, tissue cells and body fluids represent the defending forces, or "armies," against the invaders. The white cells may be compared to the infantry, which does the hand-to-hand fighting and which mops up after the action. The outcome recalls the classic description of the successful campaigner — the one who "gets there fustest with the mostest." Either the enemy is repelled or the defenders are overwhelmed.

The invading bacteria direct a chemical attack against the body tissues; they produce chemical substances, called toxins, or poisons, which may be damaging or even fatal. These substances may be picked up from the site of the local invasion and may be carried through the whole body; they may do harm to vital organs, particularly the heart, the nervous system and the kidneys. A local infection, such as diphtheria or septic sore throat, may affect the entire body, producing fever and a feeling of illness.

Antibodies
to the rescue

Against the "poison attack," there is chemical defense. The invaded tissues soon begin producing substances called antibodies, which counteract the effects of the invading germs. There are several kinds of antibodies. The antitoxins directly neutralize the toxins; the agglutinins cause the invading bacteria to clump together and become inactive; the opsonins "soften up" the bacteria for the attacks of the white cells; the bacteriolysins dissolve the germs.

Lymph nodes as
a defense line

Suppose, however, that the body loses the local skirmish. The defending forces may be overwhelmed; or a person may squeeze a pimple or try amateur surgery on a boil or carbuncle. The local defenses are bypassed; the invading forces make a "breakthrough." Fortunately, the body has a second line of defense. Part of the tissue fluids drain through a series of vessels that parallel the veins and eventually convey these fluids (lymph) into two large veins. These lymph vessels lead through collections of spongy tissue known as lymph nodes, which constitute traps for infection. Lymph nodes are strategically placed at circulation "bottlenecks," which correspond to the easily defended posts, such as mountain passes, where a military commander would station his reserves. The tissue fluids from the hand, for example, pass through lymph nodes near the elbow, and again in the armpit; those from the foot, through nodes near the knee and in the groin. These nodes are filled with a special type of white blood cells, the lymphocytes. They may break down in the process of handling heavy infections and form abscesses; but they protect the main body structure and the vital organs. An infection which has overcome local defenses may often be traced by angry red lines running from the point of injury up the arm or leg; these are the inflamed lymph vessels. There are systems of lymph nodes and drainage vessels that serve the internal organs in the same manner as the superficial nodes serve the extremities.

In certain instances of tuberculosis, the attack upon the lymph nodes is less violent but more persistent. If all goes well, nodules, called tubercles, will develop. These nodules will trap and hold the invading bacilli, sometimes permanently; ultimately they will become encrusted with calcium deposits. If, later, the body is weakened by fatigue, malnutrition or further infection, the walls of the nodules may break down, releasing the imprisoned bacilli. This may take place many years after the first attack.

Defensive substances that
keep circulating in the body

When the body forces have been mobilized for defense and have won the action, the local site of the skirmish clears up; the healing process takes place, and a scar often remains. Once they have gone into action,

Lymph nodes aid the body in its fight against infection by trapping microbes and removing them from the blood stream. Within the lymph nodes, the microbes are attacked by lymphocytes—cells that engulf invading germs.

LYMPH VESSEL WITH VALVE

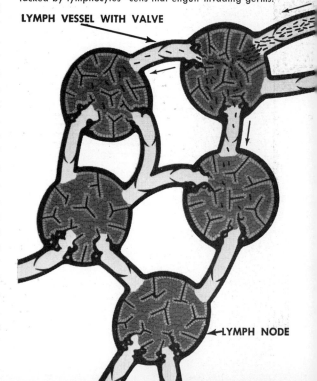

◄LYMPH NODE

the body defenses remain ready for further trouble. The tissues that have reacted against an invasion by a communicable disease do not cease reacting when the emergency has passed; they go on manufacturing the defensive substances, which circulate in the blood. For example, after a case of diphtheria, the body continues to manufacture antitoxin, which circulates in the blood for many years. The blood of a person who has recovered from typhoid fever contains agglutinins, which cause living typhoid bacteria to clump together. The presence of such antibodies gives a person protection which wears out only after a period of years, varying with the disease and the individual. That is why second cases of communicable disease are unusual; persons who do suffer such diseases a second time have defective powers of defense and may become more seriously ill than the normal person, or may even succumb.

Immunization
against disease

Man has taken advantage of the slowly unfolding knowledge of natural immunity provided after recovery from disease. An English country doctor, Edward Jenner, led the way in 1798 with his famous INQUIRY INTO THE CAUSE AND EFFECTS OF THE VARIOLAE VACCINAE (cowpox). In this monograph he showed how he could prevent smallpox by vaccination — that is, by inoculation with cowpox. Nowadays the practice of vaccination is universal among enlightened people, and smallpox exists only where vaccination is neglected. The same basic principle, with variations in detail, is used in immunizing children against diphtheria, whooping cough, tetanus (lockjaw) and poliomyelitis (polio). In areas where water supplies may be contaminated or the milk supply is of dubious quality, typhoid vaccination may be advised. (Nowadays the word "vaccination" refers to any type of preventive inoculation.) Persons traveling into certain infected areas, particularly in the tropics, the Orient and the Near East, may be advised or required to be vaccinated against cholera, typhus fever, yellow fever or other prevalent diseases for which there

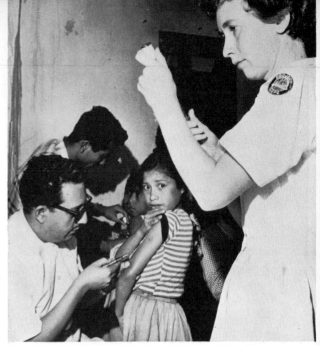

Project Hope

is available protection. The search for a polio vaccine has been one of the fascinating dramas of modern medicine: it is described in some detail in another chapter (see Index, under Poliomyelitis).

Certain diseases can be warded off by individual immunization, but not on a mass basis. For example, until recently measles was preventable only by immunizing with gamma globulin, or GG, a protein contained in the blood. Measles was so prevalent, however, and the supply of GG so limited that mass advance immunization was impracticable, even if the GG measles immunity were lasting, instead of being limited to a few weeks. Now, a practical measles vaccine for mass immunization is available.

There is no advance immunization against rabies. When a person has been bitten or licked by a dog which is known to be, or possibly might be, rabid, the Pasteur vaccination procedure begins. Usually it is in time to save the individual from the disease, which is invariably fatal.

Efforts to prepare advance immunization against the common cold have met with scant success. There is a cold vaccine, and its use is sometimes recommended, but both doctors and patients are divided in their opinions as to whether it is useful. It is harmless and may be tried if desired; but one should not expect too much from it.

There are various kinds of immunity. If it occurs as a result of having a disease

The ravages of poliomyelitis (more familiary known as polio) have been largely stemmed by immunizing children and adults with antipolio vaccine. At the left, we see children in the Peruvian town of Trujillo receiving shots of Salk antipolio vaccine, containing killed viruses. This type of vaccine must be administered by injection. The Sabin antipolio vaccine contains live viruses; it is not injected but is given by mouth. The above photograph shows students at East High School, Rochester, New York, taking doses of the Sabin vaccine.

and overcoming it, it is called natural; if it is acquired by a man-made procedure, it is called artificial. Both types of immunity depend upon stimulation of the body's own defenses; they are both described as active and they may last for many years.

Suppose that a child exposed to serious disease has not had previous immunization. Then, as in diphtheria, we may give the child a preventive injection of antitoxin. This does not stimulate the body mechanism, so this kind of immunity is called passive; its effect is only temporary.

The substances used to provide immunization are called antigens. Some, used in protecting against smallpox and rabies, are virus vaccines, containing living but greatly weakened viruses. Some are bacterial vaccines — killed bacteria in suspension. Some, called toxoids, are greatly weakened toxins. All have the property of stimulating the tissues to form antibodies, but they do not produce the disease; all provide active, artificial immunity. The substances known as antitoxins produce temporary protective immunity; they are not antigens.

Some persons find it hard to distinguish between vaccines, toxoids, serums and antitoxins. A vaccine is prepared from a virus or a culture of bacteria. A toxoid is prepared from the liquid culture medium in which bacteria or viruses have grown and produced their toxins. A serum is the liquid portion of human or animal blood that contains immunizing substances. An antitoxin is the purified immunizing agent derived from a serum. Use of serums is decreasing because of the danger that they may trigger allergic attacks. (See Index, under Allergies.)

General infections are particularly dangerous

Some infections are so massive or so violent that they immediately overwhelm all local and secondary resistance and enter the blood stream; in some cases infection is introduced directly into the blood. Unless there has been previous immunization, such general infections, as they are called, are often serious and sometimes fatal. They include rheumatic fever, typhoid fever and chronic infection of the heart valves by a streptococcus. They may be overcome by the body's own immunity mechanism, through the gradual development of antibodies. The course of the disease indicates the struggle between infection and resistance; the result is victory (recovery), defeat (death) or stalemate (chronic disease).

Recovery from disease, or injury or surgery

Even when the patient has recovered, certain organs or tissues may have been more or less seriously damaged. In such cases, the body may continue to function quite well, as long as there are no unusual

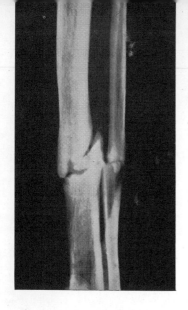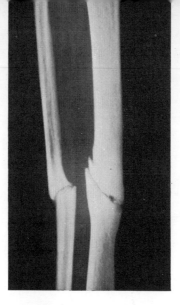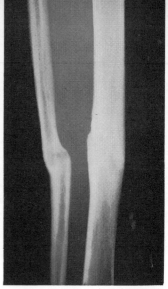

General Electric

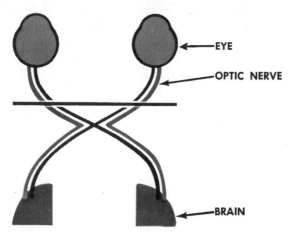

EYE

OPTIC NERVE

BRAIN

In some cases, there is complete recovery from disease or injury. Thus bone heals well and full function is generally restored. Above, we see three stages in the healing of a fracture of the tibia and fibula (bones of the leg), with the photo at the extreme right showing the completely healed bones. There are cases, however, in which specialized tissue, once destroyed, is not replaceable by healing. Thus vision is permanently lost when the optic nerve, which leads to the eye (see diagram, left), is destroyed.

stresses. Scarlet fever, for example, before the days of modern drug treatment, frequently did serious injury to the kidneys, destroying some cells and damaging others. Under proper treatment, recovery would take place in some instances. If the injury to cells had not been too great, they would recover and resume their function. The cells which had been killed would be partially replaced by other cells of similar character or by fibrous scar tissue.

If the end result of the healing process, following disease or injury or surgical removal, leaves too much scar tissue and too little cell replacement, there is limitation of function. There is a final line of defense, even against such limitation. There is so much tissue reserve that in many cases there will be enough functioning tissue to meet the ordinary needs of the body. As a result, many persons with damaged organs or tissues may live long and well by exercising

due care and remaining under competent medical supervision.

Many tissues are replaceable after injury. Muscles, skin and even bone heal well, and full function is restored except in very severe injuries. Vital organs such as the heart have a high rate of recovery after serious injury as from diphtheria toxin, rheumatic fever or even a heart attack. There is much more kidney and liver substance than is required for ordinary living. The mucous membranes of the lungs, stomach, intestines and other internal organs are still able to carry on after they have been extensively damaged.

To some extent, certain organs may even take over the activity of others that have been impaired. The stomach can be almost completely removed, and digestion still goes on in the intestines with hardly diminished efficiency, if a few eating habits are modified. Removal or failure of glands can now be remedied by supplying the hormones they would normally secrete; thyroid feeding by mouth and insulin replacement by injection are prominent examples.

The more specialized tissues are not replaceable by healing, once they have been destroyed. When the nerve of hearing is gone, that function is lost forever; vision is lost when the optic nerve is destroyed. When cells of special sense nerves and of the brain and spinal cord are killed, they are not replaced by other nerve cells but by scar tissue. Thus when a brain area concerned with a certain function, such as muscle control in the hand or the foot, is destroyed, the hand or foot will be paralyzed, unless other muscles can take the place of the lost muscle. It is true that sometimes recovery or at least partial recovery may occur after a brain injury; but this is because the nerve cells have not been destroyed, but only injured. As they recover, function returns.

Many nerve cells have long filaments carrying impulses to a distant point or bringing them from that point. If such a filament is cut but the cell remains intact, function will return when a new filament has grown along the path of the injured one.

In the healing process involving the less specialized tissues, new cells are formed only to the extent that the tissue is made whole again and resumes its function. In the normal individual, regrowth and scarring after an injury or surgery seldom get out of control. In a few instances a scar may overgrow and form a structure called a keloid, which can usually be corrected by minor surgery.

Other defenses
of the human body

In addition to all the defenses that we have considered hitherto, there are psychological and mental defenses, arising out of instinctive fears and acquired experience. The baby is born with fear of falling and of loud noises; as we grow older, we develop various patterns of cautious procedure that help to keep us out of trouble and reasonably intact most of the time.

See also Vol. 10, p. 276: "General Works."

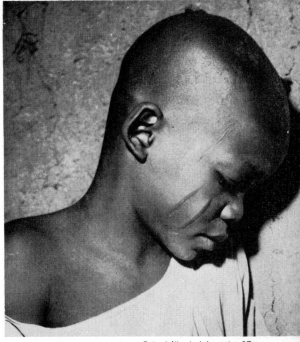

Federal Nigeria Information Office

The scar that is left after injury to the skin consists of specialized fibrous tissue and does not represent replacement of the original cells. This scar tissue lacks elasticity, and the epidermis is not firmly joined to it. In some cases, scars represent deliberate patterns which are brought about by means of incisions in the skin in accordance with tribal rites. The incisions may be superficial, resulting in comparatively slight scars, like those on the face of the boy, above, belonging to the Yoruba tribe in Nigeria. If the incisions are deep, the corresponding scars are prominent, as in the case of the Western Australian aborigine shown in the lower photo.

Australian News and Information Bureau

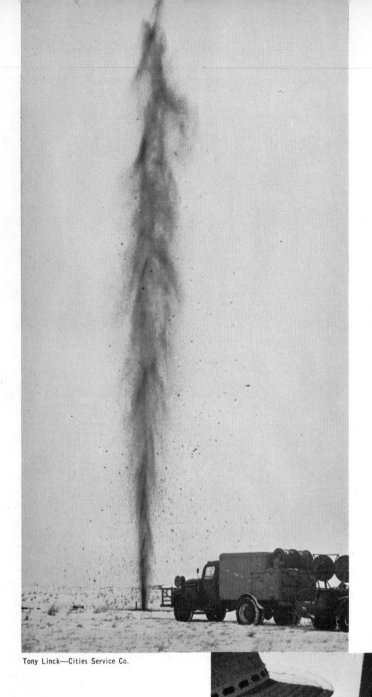

Above: a seismic exploration party, prospecting for oil, sets off a dynamite charge near Lewiston, Montana. Shock waves from the blast are reflected from underground rock layers and are recorded by a seismographic recording truck. A good idea of the contour of the rock layers is obtained and the search for oil can be narrowed down. In the right-hand photograph, we see a reflection seismographic record being inspected by an operator after a shot.

MODERN METHODS
OF PROSPECTING

How Hidden Ore and Oil
Deposits Are Located

FOR the past few generations, industry has been draining our mineral resources at an ever increasing rate. It is a matter of great importance, therefore, to find new sources of coal, iron, petroleum and the other minerals that are used in such vast quantities in peace and war.

Until comparatively recently, prospecting for new deposits was usually limited to a study of rock outcrops, shallow test pits and drill holes. Rock outcrops furnish the best key to what lies beneath the surface; unfortunately, in most mining and oil districts they cover less than 2 per cent of the total surface. Furthermore, in the temperate zone large areas are more or less masked by a thin veneer of soil; in the tropics the rock has been rotted so deeply and the vegetation is so dense that a geological report is sometimes impossible. In the northern United States and in Canada the continental glacier left behind an immense amount of debris, which in many cases directly conceals the rock crust. This crust is hidden, in other cases, under thousands of lakes and swamps occupying hollows in the glacial material.

In those regions where dependable rock outcrops are plentiful, a geologist can rather accurately foretell the underground structure; in most cases, however, such territories have already been most carefully studied. The really important areas for present and future prospecting are the hidden and as yet unidentified portions of the earth's crust. Prospectors seeking mineral treasure welcome any economical method that will pierce the covering mantle and reveal the presence or absence of ore deposits, as well as provide information about the structure of the underlying rock.

Perhaps the first method used to locate hidden deposits was a forked stick, like the hazel twig that is supposed to locate subsurface water. When a gold nugget was tied to the forked stick and the prospector walked over the territory that was to be examined, the stick was supposed to dip suddenly if it neared a lode of gold. Petroleum deposits were sought by means of a vial of petroleum attached to a forked stick. To the scientifically minded person, of course, all this seems absurd.

Yet in the present century some of the shrewdest oil operators in the United States actually took seriously a modern counterpart of the forked stick. This device, which was called a wigglestick, consisted of two coiled aluminum springs joined together so as to form a fork. A platinum vial was inserted in the fork; it contained a secret chemical that was supposed to have an affinity for oil. Needless to say, this curious device was a fraud.*

Today, oil prospectors seldom rely on direct methods for detecting petroleum; they seek structures that may hold oil. On the other hand, solid ore deposits are located directly by truly scientific devices. In general, the scientific detection of hidden ore bodies is based on the differences between the properties of the ore mass and of the enclosing medium. Where such differences do not exist, prospecting instruments can furnish no clue.

Where, however, the ore body differs from the surrounding rock in density, in electrical conduction, in magnetism or in the way it transmits earthquake shocks,

---

* However, chemical methods are employed in geological prospecting; for example, samples of mineral are tested, or assayed, for their content. Oil deposits may be detected by analysis of surrounding rock and soil for any hydrocarbons that have seeped into them from the oil.

modern scientific methods enable the geologist to locate hidden ore deposits. By way of illustration, we might compare the locating of ore deposits underground with the hunting of submarines underwater. During World War I, the detection of a motionless, submerged submarine was a difficult matter, since the propeller was not turning and therefore making no noise. However, experiments soon revealed that a submerged steel vessel has electrical and magnetic properties different from those of sea water and that it reflects sound waves. Ingenious instruments, based on all these discoveries, were then devised. It became a fairly easy matter, by the close of the war, to detect the presence of submerged underwater craft.

Modern methods of locating valuable hidden deposits work on much the same principles. They have nothing at all in common with the "wigglestick" and "forked stick" for which no scientific basis has yet been discovered. They make use of differences in certain physical factors — differences that are caused by the presence of ore deposits. Important factors in geophysical prospecting include magnetism, density, elasticity, electric conductivity, thermal (heat) conductivity and radioactivity. Since nature provides an infinite number of combinations of these factors in each situation, it would be unfair to expect all the methods to work equally well. Consequently, it is customary to make a preliminary geological survey to determine which method is best suited to the existing local conditions.

Magnetic methods

A compass needle free to swing around a vertical axis will point toward the earth's north magnetic pole in the Northern Hemisphere and toward the south magnetic pole in the Southern Hemisphere. These poles do not coincide with the geographic ones. Our globe is like a giant magnet, with many local variations in the intensity of its magnetism, due to differences in the composition of the earth's crust. The compass needle is used to indicate the horizontal variations in the earth's magnetism. Vertical variations, known as "dip," are indicated by means of a needle mounted on a horizontal axis. Determinations of such variations have been made for many years at various stations. Specially prepared maps show points of equal magnetic intensity over all of the earth's crust.

Magnetic-intensity changes are detected more accurately by the deflection of a magnetometer, a device suspended on a knife edge. There are magnetometers for measuring the horizontal or the vertical component of the earth's field.

As we have said, the earth behaves like a giant magnet producing lines of magnetic force that radiate from the poles. These lines of force are affected by the composition of the earth's crust; they can penetrate some rocks better than others. Rocks that are very easily magnetized tend to warp or distort the magnetic lines of force, or magnetic gradient, as they are sometimes called. Each type of rock produces its own distortion pattern. The magnetometer measures the magnetic gradient in a given region; the distortions of the gradient make it possible to detect valuable deposits.

Although most of the earth's minerals are not very magnetic, a few are. One example is the mineral known by the ancients as "loadstone" and now called magnetite (Fe_3O_4), one of the oxygen compounds of iron. Magnetic substances distort the magnetic gradient by pulling the lines toward them. In Figure 1, we can see how a deposit of iron is acting like a small magnet. It has formed two distinct magnetic poles around which the lines of force have formed. A map such as this would probably indicate a workable ore.

Sometimes a valuable ore is commonly associated with a less valuable magnetic mineral. Certain ores are located by this means. In this case, the desired ore causes no magnetic distortion, but the surrounding material distorts the lines of force around the deposit. Magnetic maps show this deviation from the normal magnetic gradient. The valuable deposit is silhouetted by the magnetic distortion of the surrounding material. In this way, other types of deposits are found.

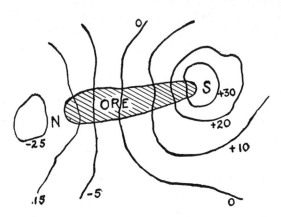

Fig. 1. This highly simplified plan shows several lines of equal magnetic attraction.

The magnetic exploration method is sometimes used in prospecting for oil deposits. These deposits are often found in places where there are salt domes within the crust of the earth. The domes, consisting of rock salt, have been forced up through various rock strata from salt beds lying deep in the earth. If certain strata have not been penetrated by the salt but have been pushed up as shown in Figure 2 (B and C), they provide fine reservoirs for petroleum. The magnetic method is used to detect salt domes because they are diamagnetic (not easily penetrated by magnetic lines of force) and therefore bring about a definite weakening of the magnetic field of the earth at its surface.

Magnetic prospecting instruments installed in planes and helicopters have been used in recent years to carry out exploratory surveys in various parts of the world. The instruments used record variations in magnetic intensity on a moving strip of paper as the plane or helicopter flies back and forth over the area that is being surveyed. This makes it possible to complete a survey in much less time than would otherwise be required.

Seismic methods

When an earthquake occurs or when a charge of dynamite is exploded, the sound travels through the air and the shock moves through the ground. Both travel in the form of vibration impulses, or waves. Shock waves traveling through the various

strata of the earth form the basis of the seismic, or earthquake, method of exploration. This method depends upon the measurements of the time intervals that elapse between the firing of charges of dynamite and the arrivals of vibration impulses at a number of sound detectors set up on the surface of the ground.

The speed with which a rock formation transmits the vibrations will depend upon the elasticity and density of the rock. The vibrations travel more rapidly through hard and solidly compacted rocks than they do through formations that are loose or entirely uncemented. They may go through loose sediment at the rate of only 1,000 feet a second, while they may attain a velocity of 17,000 feet per second when they make their way through salt domes.

It may take the shock waves less time to go from one point on the earth's surface to another by way of an underlying rock formation than by way of the earth's surface. In Figure 2, A represents the place where a charge of dynamite has been set off; D the place where the wave arrives. If D is far enough away from A, the wave going along the path ABCD will arrive at D before the wave that travels along the short cut AD. The reason is that the first wave will pass through much denser formations, including a salt dome.

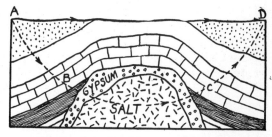

Fig. 2. How the different strata are arranged in an area containing a salt dome.

In seismic exploration the dynamite charge, varying in weight from a few pounds to about a ton, is put in a hole drilled in the ground; this hole must be deep enough so that it will be below the soil or below the area containing fragments of weathered rock. The time when the charge is set off is recorded when the explosion

Fig. 3. The X-ray diffractometer identifies rock samples and it also aids in mapping various structures deep within the earth.

breaks an electric circuit carried through a wire imbedded in the dynamite charge.

The detectors, called geophones, are in small metal cylinders; they are set upon the ground at intervals. As the vibrations hit a geophone, they are converted into electrical impulses; these are transmitted through wires connected with a recording station installed on a truck. In the recording station the impulses are amplified and transmitted to a mirror galvanometer. This records the vibrations on moving photographic paper by reflecting a spot of light upon the paper. Impulses from a number of geophones can be recorded in a single station at the same time.

In the refraction method of seismic exploration, the shock waves are refracted, or bent, as they reach formations of greater or lesser density. Experts analyze the differences in speed of waves traveling through the different formations. In the reflecting method, waves are reflected back to the surface from underlying rock formations; the detectors are set comparatively near the place where the charge is fired.

Gravitational methods

According to the law of gravitation, the attraction between two bodies is proportional to the product of their masses and inversely proportional to the square of the distance between them. By measuring the force of gravitational attraction at a given area, the geological prospector can derive a good deal of information about the nature of the underlying rock formations, since these formations differ in density. We know, for example, that sedimentary rocks, such as mudstone, sandstone and shale, are not so dense as igneous rocks, such as granite and basalt. Obviously the gravitational attraction will be greater where the underlying rock formations are igneous than where they are sedimentary.

Any instrument that measures very small changes in gravity is extremely valuable in locating hidden deposits and in deciphering the underground structure of the earth's crust. Such an instrument is the torsion balance; with it, changes in gravity of one part in a million may be detected.

The torsion balance usually consists of a more or less horizontal bar that is suspended by its center from a vertical wire; at each end of the bar is a weight. Both weights are of equal mass and, as a rule, are set so that each one is at a different height above the (horizontal) surface of the ground. (In one form of balance, the bar itself hangs slanted at an angle to the horizontal; the directly attached weights are thus at different altitudes.) The torsion balance is free to move horizontally around the vertical wire axis that passes through its center.

When the balance is brought above an area of ground where the underlying rocks show differences in density, it rotates. This happens because the density variations have warped the gravitational field above the surface of the ground. In effect, the torsion instrument has become unbalanced with respect to the gravity prevailing in that region. Since the balance can move only horizontally to compensate, it rotates and twists the wire to which it is attached. The amount of torsion developed in this wire is recorded photographically and indicates the quantity of gravitational change in the area. Although a highly sensitive instrument, the torsion balance requires much time to operate and has been replaced in the United States by the gravity meter, or gravimeter.

Unlike the torsion balance, the gravimeter measures the vertical force of gravity directly. There a number of types, but most instruments employ a suspended metal spring with an attached weight or arm; the spring responds to rock-density and gravity changes by lengthening or shortening or sometimes by rotating (Figure 4). Since this motion is extremely small, it is magnified by means of electrical, optical or mechanical devices, which also record the changes involved.

At first, the gravimeter was not so efficient as the torsion balance, but it has since been perfected. It is now extremely sensitive as well as easy to use. The gravimeter is placed on the ground or is held above it on supports; it is also used on airplanes and ships. Light portable models have been developed to allow numerous readings over a wide range of country in a single day. Gravity meters are used extensively in prospecting and research, because of their sensitivity to density variations in rock — a widespread phenomenon.

At one time the pendulum was commonly used in the United States in gravitational prospecting. The period of motion of a freely swinging pendulum is affected by changes in the gravitational force to which the device is subjected. These time changes are recorded and measured. Today, surveys based on the use of the pendulum are restricted mainly to purely scientific work.

Electrical methods

There are several methods of prospecting by means of natural and induced electrical currents and waves. They have been successful to a limited extent, particularly where metallic ore bodies are sought. Electrical procedures have also been applied to petroleum exploration, as we shall note later on in this article.

The so-called self-potential technique relies on natural electrical currents generated by ore bodies. Electrochemical activity often results when a metal-sulfide mineral in the ground comes in contact

Fig. 4. These simplified diagrams show two types of gravimeters. The force of gravity is measured in the first of these devices (A) as a spring is lengthened or shortened; in the second one (B) as the spring rotates.

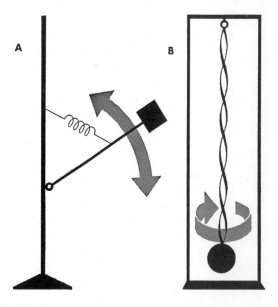

with mineral-water solutions. The associated differences in electrical potential (voltage) produced in different parts of such a mineral mass can be detected by sensitive instruments.

Wherever the rocks are not electrically active, current must be artificially generated in them, in the applied-potential methods. The simplest way is to place a pair of electrodes in the ground and pass current between them. More elaborate methods employ copper conductors up to thousands of feet long placed parallel to each other

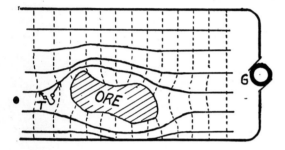

Fig. 5. Diagram showing how the applied-potential method of electrical prospecting is employed. A portable alternating-current generator sends currents flowing through two bare copper conductors about 3,000 feet long and about 1,000 feet apart. In an area of equal conductivity, current flows through the ground between the wires in straight lines (dotted) at right angles to the wires. Lines of equal potential (solid) are also straight but parallel to the conductors. The area between the two conductors is searched with two movable electrode pegs, connected to an amplifier and earphones. If no sound is heard in the earphones, both pegs are on one equipotential line of force. The equipotential lines in the area can be mapped out in this way. An ore body, with different conductivity from that of the rest of the area, will affect the lines as shown.

along the surface of the ground and electrified by means of an alternating-current generator of low voltage, as shown in Figure 5. Ground currents then flow between the conductors at right angles to them. Naturally, along each electrical current there is a progressive change in electrical potential, measured in volts. Imaginary lines connecting regions of the same potential are known as equipotentials, generally running at right angles to the directions of the earth currents and parallel to the copper conductors (Figure 5).

The lines of equipotential are plotted by means of movable electrodes with earphones stuck here and there into the ground. As long as the electrodes are on the same equipotential line, no current flows through them, and the earphones are silent. But as soon as the electrodes are placed on different equipotentials, current flows into the phones, and the prospector hears a noise. In this way, the potential and current patterns are plotted; wherever distortions in these lines occur, a mass with marked electrical properties is probably responsible. It may be a body of ore, as shown in Figure 5.

The self-potential and applied-potential methods discussed above permit only a relatively shallow penetration of the earth. Other electrical operations probe deeper into the ground and provide a somewhat clearer picture of conditions there. Deep penetration of the earth can be obtained by the investigation of what are known as natural earth, or telluric, currents — broad sheets of electricity that sweep along the crust of the globe daily. Special instruments determine the patterns of earth currents; they have been used in detecting deep rock structures that may contain petroleum. Various layers in the crust react differently to the passage of the current; from this, it is possible to build up a picture of the situation at relatively great depths, up to several thousand feet.

Other deep-penetration methods use applied electrical currents. One of them measures the resistivity of various rock layers to the passage of electricity. Current is sent through the ground by means of two widely separated electrodes stuck into it. A second pair of electrodes is moved around between the current electrodes and traces the distribution of electric currents and potentials. Each rock layer affects, or "refracts," the paths of the applied currents differently. From all these measurements, the investigator calculates the resistivities of beds thousands of feet down to the transmission of the currents. The farther apart the current electrodes are located (sometimes over a hundred miles), the deeper the electricity penetrates the earth; however, there is a practical limit to the separation of the electrodes, for the currents would them become too weak. The entire procedure bears some resemblance to seismic refraction surveys, in that the paths of artificially induced disturbances (here, electricity instead of dynamite-explosion waves) are characteristically distorted by particular kinds of rock.

One electrical prospecting method uses electrical pulses sent underground; these pulses are then reflected back up from different rock layers somewhat as earthquake waves are in seismic reflection surveying. Another technique involves sending radio waves into the ground; in response, the latter may emit secondary radio waves if it contains metallic ores. In still another method, a magnetic field is induced in the ground by means of an electrified surface cable; the nature of the rocks affects this field, which is measured by means of detecting coils at the surface.

Electrical measurements in rocks are also made in wells being driven for petroleum — a process called electric well logging. The resistivities of various rock beds that have been drilled are taken by means of one or more electrodes lowered into the borehole. Current is passed between the electrodes and a source of electricity on the ground surface near the well; the rock surrounding the well (plus any metal casing in the hole) conducts the current. The electrical characteristics of the beds at different levels are taken in successive readings; they indicate the nature of the rocks. In some cases, current is not applied; the self-potential of the rocks is sufficient for measurement and study.

Radioactivity methods

Prospecting for radioactive minerals, containing uranium, thorium and radium, involves certain special techniques. Since radioactive materials give off radiation, they are most readily found by devices that detect this emission. Among them is the Geiger counter. This instrument consists essentially of a gas-filled tube containing a pair of electrodes. Normally the gas inside does not conduct electricity; but when radioactivity from an outside source passes through the tube, it electrically charges the gas. The latter then allows current to discharge between the electrodes, producing a loud click when it is amplified. The rate of clicking indicates the concentration of the radioactive mineral that is emitting the radiation.

A far more sensitive (and costly) radiation detector is the scintillation counter, or scintillometer. In this instrument, a crystal responds to atomic emission striking it by producing minute light flashes, or scintillations. These flashes are magnified and recorded by electronic tubes and meters. There are various types of Geiger counters and scintillometers on the market, suitable for the different conditions that are encountered in prospecting.

Radioactivity measurements are also taken in wells that are being bored for oil. A detector is lowered down the hole; as it passes through various formations of rock, it records the intensity of radioactivity found in each. This gives a clue to the nature of the earth below the surface.

Summary

We have sketched some of the more important scientific techniques involved in prospecting for mineral wealth. There are numerous variations of these methods, which are constantly being improved, and new procedures are being developed.

The selection of appropriate operations in a given area often poses a problem. Much depends on the extent to which the region has been previously explored (if at all) by conventional surface-geology methods, as well as on the expense involved and on just what is being sought. Seismic surveying is costly; magnetic and gravimetric methods, less so. Seismic, gravimetric and electrical techniques are often employed in petroleum work. The various methods may corroborate each other; sometimes they conflict or do not agree with the conventional geological picture. At least, they may indicate something unusual or unsuspected below the surface of the earth. Much depends on the correct geological interpretation of raw instrumental data.

STREAMLINING RECORD COLLECTIONS

A graphic illustration of the space savings made possible by LP (long-playing) records. This man is holding in vinylite LP records the equivalent of the pile of "78" shellac records shown at the left.

MACHINES THAT TALK

The Story of the Phonograph
and of Magnetic Recording

SOUNDS — the noise of waves breaking on a beach, the melodious tones of a trumpet, the crash of thunder — are really vibrations in the air or in some other medium. When they strike the eardrums of a listener, they cause nerve impulses to arise and these impulses are transmitted by way of the nerves to the hearing centers of the brain. The brain then interprets the nerve impulses, and as a result we *hear* the pounding of the waves, the blaring of the trumpet and the peal of thunder.

Until the last part of the nineteenth century there was no way of recording and re-creating the vibrations of sounds. Nobody will ever be able to hear again the beautiful voice of Jenny Lind, the exciting piano playing of Franz Liszt or the glorious tones that Nicolò Paganini drew from his Guarnerius violin. But man has now devised a way — or rather several ways — of recording the vibrations that make up sound and of re-creating these vibrations at will. Through the medium of the phonograph and of the tape recorder, generations yet unborn will be able to listen to the ringing tenor of Enrico Caruso, the polished artistry of the violinist Jascha Heifetz, the amazing technique of the pianist Vladimir Horowitz and the thrilling tone patterns spun by our best symphony orchestras.

The phonograph was the invention of perhaps the greatest inventor who ever lived — the American Thomas A. Edison, to whom the world owes the incandescent electric light, the motion-picture camera, the microphone and hundreds of other useful devices. Curiously enough, the phonograph was an unexpected end product of a youthful invention that was not intended, originally, to reproduce sounds. Edison had become an operator in an Indianapolis telegraph office; to help him with his work, he had developed a device that would speed up the recording of telegraph messages.

The device consisted of two old Morse receiving instruments, which he converted into a kind of tape machine. A strip of paper was run through the first of these instruments; as the dots and dashes of a message came through the receiver, they made indentations on the paper. The paper was then run through the second instrument at much slower speed; the indentations were converted again into the clicks — the dots and dashes — of the original message. In this way the machine succeeded in slowing down the transmission of telegraph messages from forty words a minute to something like twenty-five words a minute. Edison and another lad who was working with him were able, by means of this instrument, to keep pace with the swiftest dispatchers in the country. The manager of the Indianapolis telegraph office and the other operators could not imagine how the boys could do such amazing work; for they had kept their recorder hidden when it was not in use.

But one night their ingenious system failed. It was the eve of a presidential election; messages kept pouring into the telegraph office at the top rate of speed. The young operators were swamped; they fell two hours behind with their work of transcription. Newspapers relying on the service provided by the office sent in frantic complaints. The manager made a hurried investigation and Edison's secret was discovered. He was not allowed to use his automatic recorder any more.

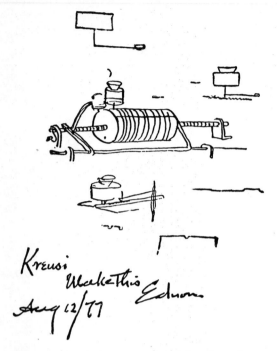

Kreusi
Make This   Edison
Aug 12/77

Edison's sketch for the model of the first phonograph, shown on another page of this article.

However, he kept his machine for converting telegraph signals into marks on paper and then changing the marks back into sounds again. He continued to improve the instrument until in 1877 it was quite dependable. He had connected an electromagnet to an embossing point. When the telegraph circuit was closed, the magnet caused the point to make an indentation on a revolving paper disc. The disc could then be played back at any desired speed.

One day, as Edison was experimenting with the device, he began to rotate the disc rapidly. He found that when the marks on the rotating disc came in contact with the embossing point, a musical tone was produced. At that time, Edison had been experimenting with Alexander Graham Bell's telephone. His mind was filled with theories of sound vibrations and their transmission. On hearing the musical note coming from the revolving disc, Edison was struck with the idea of recording and reproducing sound.

In order to try out his idea, Edison hastily rigged up an instrument composed of a membrane connected to a steel point.

He pulled a paraffin-coated paper strip through the device and shouted against the membrane which vibrated the point in unison with the sound waves. He then pulled the paper through the device again and listened breathlessly. A faint, but distinct, sound was heard. On July 18, 1877, Edison made the following entry in his laboratory notebook:

"Just tried experiment with diaphragm having an embossing point and held against paraffin paper moving rapidly. The speaking vibrations are indented nicely and there is no doubt that I shall be able to store up and reproduce automatically at any future time the human voice perfectly."

On August 12, 1877, Edison made a rough sketch of the machine he had devised and wrote on it: "Kreusi, make this." (The sketch is reproduced on this page.) He gave it to John Kreusi, one of his workmen, with the comment, "Here's an eighteen-dollar job for you." Looking at the drawing, Kreusi asked, "What are you going to do with this?" Edison replied that it was intended to be a machine that would record and reproduce sound. "You're crazy this time," laughed Kreusi, but he made a model from the drawing.

In a few days, Kreusi brought the model to Edison and then stood by with a grin on his face. It was a crank-driven cylinder with an embossing needle, attached to a diaphragm, at the top of the cylinder. With much deliberation, Edison fixed a sheet of tin foil around the cylinder, adjusted the metal point and then shouted into the funnel:

"Mary had a little lamb,
Its fleece was white as snow."

Then he adjusted the reproducing diaphragm and on turning the cylinder again heard the words come back — a little squeaky but quite distinct. Edison was astounded, for he had not expected such perfect results upon the very first trial of the first model. The astonished Kreusi turned pale and exclaimed in German, "Mein

Gott in Himmel (my God in Heaven) !" Thus the phonograph came into existence.

Edison was not the first to attempt to record sound. Leon Scott in 1855 had in-invented the "phonautograph," a device that "wrote" sound. The instrument had a speaking horn which was shaped like a small barrel and made of plaster. Con-nected to the lower end of the horn was a short brass tube about four inches in diam-eter. A flexible membrane was stretched over the end of the tube. This membrane would vibrate in unison with the sound waves that struck it. A hog's bristle was fastened to the membrane and acted as a stylus, or pen. The end of the bristle pressed against a cylinder covered with smoked paper. When the cylinder was ro-tated by a hand crank and sound waves struck the membrane, the bristle traced a wavy line on the sooty paper. This line was an exact representation of the forms of the sound waves that had struck the membrane. However, there was no way to produce sound from the wavy lines that appeared on the smoked paper.

Edison's phonograph, on the other hand, was not merely a sound-writer; it was also a sound-producer. When the little steel pen traveled over the indentations it had made in the tin foil wrapped around the cylinder, it caused the membrane to vi-brate in response to its movements, just as the membrane of a telephone receiver vibrates in answer to fluctuations of the electric current — fluctuations set up by human speech at the other end of the wire. As the membrane vibrated, it produced sound waves in the air. In other words, the membrane of Edison's phonograph did the work first of the human ear and then of the human voice. First it vibrated to the air-waves of sound and then it trans-mitted them, by means of its stylus, to the tin foil. Finally it vibrated in obedience to the up-and-down motion of the stylus as it traveled along the indented lines of the tin foil.

Marvelous as they were, the first talking machines that Edison devised were little more than scientific curiosities. The re-produced sounds were distressingly "tinny" and unmusical. It was difficult to remove the tin foil from the cylinder or to replace it without distorting the material and in-juring the indentations. Consequently, it was necessary to have a separate cylinder for every new record. Moreover, the mo-tion given to the cylinder by turning a handle was not regular. The pace at which the cylinder is rotated has an important effect on the quality of the sound produced. Too high a speed makes the note sharp; too slow a speed makes it flat. Both make it very unmusical, false and distorted.

Alexander Graham Bell improves the phonograph

After a few months of work and experi-ment on the phonograph, Edison laid it aside as he busied himself with the incan-descent lamp and with electrical devices for producing heat and power. He saw, however, the possibilities of the phono-graph and contributed an article in the NORTH AMERICAN REVIEW of June 1878, prophesying the future of the device with startling accuracy. Alexander Graham Bell, too, saw possibilities in this talking toy. Assisted by his brother Chichester and by Charles S. Tainter, he produced a machine called the graphophone, in which clockwork was used instead of hand mo-tion. The recording and reproduction of sounds were much improved. Instead of the tin foil, Bell and his colleagues employed a thin mixture of wax on light paper cylin-ders. The new machine was more con-venient and effective than the rather crude original phonograph.

As soon as Edison had some time to spare from his pioneer work on the incan-descent lamp, he also began to improve on his original sound-reproduction idea. He produced a new machine in which a special wax cylinder was employed. The record was cut by means of a tiny agate or sapphire point, which made minute de-pressions in the wax. The reproducing point was also of sapphire. It passed over the indentations and communicated its movements to the diaphragm of the re-cording machine by means of a delicate combination of weights and levers.

In Edison's phonograph and all its varieties, the record in the wax was formed of a series of little hills and dales, running round and round the cylinder or disc. Thus the sound-waves were represented by actual waves in the wax, each wave sloping upward and downward like the rippled surface of the sea. When in 1886 Bell and his partners patented their graphophone, they kept to what is now known as the "hill and dale" groove. But another inventor, Emile Berliner, who had been working on a new device in connection of thin, flat discs. A single movement of a turntable spun the disc around and worked the instrument. In the cylinder type of the phonograph, on the other hand, there were two movements. First the cylinder had to be turned round; then, while it was turning, a secondary motion had to be imparted to the pen and the diaphragm that travel with a sideways movement over the revolving cylinder. This secondary motion, by means of a separate feed-screw, was not needed in the gramophone.

Brown Brothers

Thomas A. Edison and his first talking machine. It utilized a rotating cylinder instead of a disc.

with the telephone, became interested also in experimenting on talking-machines and departed from the "hill and dale" method by causing the engraving tool to move from side to side in the groove. This was particularly adaptable to a disc type of record and is known as the "lateral cut." The type of machine invented by Berliner in 1887 was called the gramophone, which later was merged into the Victor.

Externally the chief difference between the gramophone and the phonograph was that the gramophone records consisted

Another difference between the gramophone and the older instruments lay in the manner in which waves of sound were translated into waves of wax. Berliner thought that both Edison and Bell were at fault in adopting the "hill and dale" method of recording sound. He claimed that the needle might jump from the top of one hill to another, and miss the dales. Berliner claimed that he avoided this by inventing something quite different. In the Berliner cut, the depth of the grooves in the record was always the same.

HOW A PHONOGRAPH WORKS

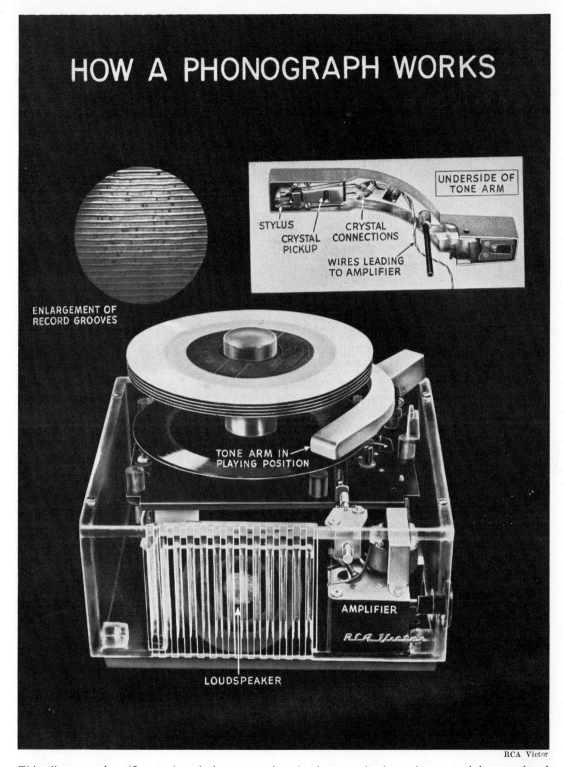

ENLARGEMENT OF
RECORD GROOVES

UNDERSIDE OF
TONE ARM

STYLUS
CRYSTAL
PICKUP
CRYSTAL
CONNECTIONS
WIRES LEADING
TO AMPLIFIER

TONE ARM IN
PLAYING POSITION

AMPLIFIER

LOUDSPEAKER

RCA Victor

This diagram of a 45-rpm (revolutions per minute) phonograph shows how sound is reproduced from a phonograph record. The undulating grooves cut into the record are an accurate representation of the original sound. These undulations are produced by the vibration of the recording stylus or "needle" in response to the undulations of the original sound waves picked up by the recording microphone. The reproducing stylus in the tone arm of the phonograph follows the undulations in the walls of a given groove. The resulting vibration activates the crystal pickup immediately adjoining the stylus; the pickup generates electrical impulses that correspond to the movements of the stylus. These electrical impulses are carried by wire to the amplifier, which increases their strength and feeds them to the loudspeaker. The loudspeaker then reproduces the original sound, music or spoken words.

In the course of the years that followed, various improvements were made in both recording and reproducing. By the early years of the twentieth century a fairly standard sort of sound box had been developed for reproduction purposes. The discs were made to revolve by a clockwork machine. A mica diaphragm transmitted the sound waves to the ears of the listener through a big horn, which was fixed by a bracket to the sound box. Later the horn was inverted and placed inside the cabinet.

Phonographs still had very definite limitations. In the earliest recording methods, many tone values which the musicians produced were not captured on the record. An impressive forward step was taken when the electronic method of recording was adopted. This method, which has been in use many years now, is based on much the same sort of system as radio. First of all, sound waves are converted into electric current by means of a microphone, or several microphones. The varying intensities of the waves are matched by changes in the strength of the electric currents. These changing currents are enormously magnified by means of an electronic amplifying system. They are then used to operate the cutting tool, or stylus, which cuts the record.

When a record is to be played, the vibrations picked up by the phonograph needle are converted into electric currents. These are passed through the amplifying system of the electrical phonograph or radio-phonograph combination and are then transformed into sound waves again in the loudspeaker.

An entirely different principle is used in magnetic recording, discovered in 1900 by Valdemar Poulsen in Denmark. In this method, sound vibrations are changed into varying electric impulses, much as they are in radio. These impulses affect the strength of an electromagnet (see Index). Wire or tape is drawn lengthwise past this magnet and becomes magnetized. Since the strength of the big magnet changes constantly, the amount of magnetism in the wire or tape that travels past it will vary accordingly. Hence the wire will carry, strung along its length, a complete history of the many changes that have taken place in the strength of the electromagnet. The varying electric currents in this device represent converted sound waves; consequently the magnetized wire or tape constitutes an accurate record of the waves. Anything producing sound — an orchestra, a large choir, a violin, a single human voice or an animated group at a family gathering — can be recorded on this magnetic device.

To play back the recording, the magnetized wire or tape is moved past a coil, consisting of many turns of fine wire. As the wire or tape passes by the coil, a small electric current is formed. This current flows into a loudspeaker, where it is converted into sound. The volume will depend on the strength of the magnetized sound-wave record passing by the coil. A strong record will produce a loud sound; a weak record will produce a soft one. The pitch of the sound will depend on how quickly the poles of the magnet pass by the coil. If the poles change rapidly from north to south, the sound will be high in pitch; if the poles change slowly, the sound will be low.

The wire or tape, wound on a spool, is easily stored away, and it never wears out. Moreover, it is possible to erase the recording from the wire or tape so that a new one may be made. In this case, the previous magnetization of the tape is overcome by a direct current passed through the coil or by a high-frequency circuit. The same set of poles and coil can be used for recording, reproducing and erasing.

Magnetic tape is used in making recordings, which are then transferred to discs. It also serves for transcriptions of radio programs and for business dictation. Home magnetic-recording sets are becoming increasingly popular. Using a machine of this kind, a musician, professional or amateur, can make fine and inexpensive recordings of his performances, making it possible for him to analyze his progress. With magnetic tape, a music lover can make recordings of good music from the radio. In this way he can acquire a most extensive record library at moderate cost.

Before the 1950's, the sounds produced by the most advanced recording methods differed considerably from those heard in a concert or recital. This was because radios and phonographs were able to reproduce sound frequencies of from only 200 to 5,000 cycles per second. This was approximately one-third of the range of 20 to 15,000 cycles that can be heard by most persons. People were annoyed, too, because the ordinary record played for only a few minutes on each side; the constant changing of records plagued music lovers. These problems were to be solved to everybody's satisfaction in the course of time.

First, in the late 1940's, the automatic record-changer was devised; it could play a number of records in succession. Then in 1948 and 1949, two new types of long-playing records for the home were introduced. One of these revolves on the turntable of the phonograph 33⅓ times a minute, as against the 78 revolutions per minute of the older discs; each side of a 12-inch record can play up to 20 minutes. The other type of long-playing record revolves 45 times a minute. It is much smaller than the old "78" disc, but it plays just as long. The "16½" disc, a later development, turns only 16½ times a minute and plays up to 40 minutes per side. It has not enjoyed as great popularity as the "33⅓" and the "45" types. All the long-playing records save storage space. They are made of vinylite, a lightweight plastic material which is not fragile and which produces a minimum of surface noise.

With the introduction of the long-playing record, interest in developing better phonographs spread. Phonographic records have a potential sound-reproduction range equal to the total range of human hearing. This advantage is lost, however, if they are played on a phonograph with a limited sound-reproduction range. Engineers spent years developing more sensitive instruments. Finally, in 1954, the high-fidelity phonograph components were introduced to the public. The present high-quality set can produce the full audible range of sound frequencies. A high-fidelity system is characterized by three separate components —

the turntable and pickup, an amplifier and a speaker, or speakers. A diamond or sapphire needle is used, since it is less wearing on the record groove than the old metal ones. The amplifier can deliver to the speaker an amplified audio voltage large enough to reproduce sound with a minimum of distortion. Highly sensitive speakers are able to reproduce all audible frequencies. There may be one speaker unit or several speakers, each designed to handle a specific portion of the total audible frequency range. For example, in a two-speaker system, the bass speaker, or woofer, handles the low frequencies and the treble speaker, or tweeter, the high frequencies.

In order to get the best sound reproduction from a high-fidelity system, the position of the set in the room must be carefully considered, since sound reflections from the walls may either hinder or improve the quality of sound.

The latest development in high-fidelity sound reproduction is the stereophonic sound system, which was introduced in 1958. This method provides an astonishingly high degree of realism in recorded music. The sounds from the various instrumental groups in a symphony orchestra — the string section, the wood winds, the brasses and so on — seem to come from different parts of the room.

In making a stereophonic recording of a symphony, two separate microphones are set up at either side of the conductor's podium to capture the musical performance much as your ears do in the concert hall. Two recordings are cut into the same record in two separate channels in the groove. One channel is cut at the bottom of the groove; the other, at the side. The stereophonic needle picks up both recordings and reproduces them through two separate amplifying and speaker systems, which are placed some distance apart. The sound then seems to fill the room and to "surround" the listener, as it were. Stereophonic sound is especially satisfying to the opera lover, since he can readily visualize the positions taken by the different singers on the operatic stage.

See also Vol. 10, p. 287: "Sound Recording."

The French physician Laënnec, inventor of the stethoscope, using the instrument in auscultation. This drawing has been adapted from a painting by Chartran.

SCIENCE AND PROGRESS
(1815-95) VII

BY JUSTUS SCHIFFERES

NEW DIRECTIONS IN MEDICAL RESEARCH

AT the beginning of the nineteenth century, medicine had come a long way from the unscientific mumbo jumbo that had disgusted the sixteenth-century anatomist Vesalius (see Index). Yet much remained to be done. The basic causes of disease were still unknown. Medical men were ignorant of just what happened in vital processes like respiration and digestion; they knew little about the true functioning of the liver; the pancreas was a fascinating mystery to them. When in doubt, they had recourse to bleeding (the drawing of blood by surgical instruments) and leeching (the application of the blood-sucking worms known as leeches). Safe anesthesia and antiseptics were unknown. Surgery was still a pretty crude affair. Surgeons were often considered to be no more proficient than the barber-surgeons of former times; their workshops — the hospitals — were feared as death houses.

In the nineteenth century remarkable progress was made in almost every field of medicine. Improved methods of diagnosis were developed. New light was thrown on the functioning of the body organs. The germ theory of disease revolutionized the treatment of many human ailments and gave promise of controlling epidemics of communicable diseases. The introduction of anesthesia and the use of antiseptics made surgery safer, and as a result its scope was greatly widened.

The groundwork for these spectacular advances was laid in the first five or six decades of the century. For one thing, this period saw the development of an instrument — the stethoscope — that has since become indispensable in diagnosis.

Its inventor, René-Théophile-Hyacinthe Laënnec (1781–1826), was a native of the French province of Brittany, a regimental surgeon in the French Revolution, a physician in two famous hospitals in Paris and a professor of medicine at the College of France.

Laënnec has told us the story of his invention of the stethoscope. "In 1816," he wrote, "I was consulted by a young woman presenting general symptoms of disease of the heart. Owing to her stoutness, little information could be gathered by application of the hand and percussion [tapping the chest with flexed fingers]. The patient's age and sex did not permit me to resort to direct application of the ear to the chest.

"I recalled a well-known acoustic phenomenon; namely, if you place your ear against one end of a wooden beam, the scratch of a pin at the other end can be distinctly heard. It occurred to me that this physical property might serve a useful purpose in the case with which I was then dealing.

"Taking a sheet of paper, I rolled it into a very tight roll, one end of which I placed over the chest in front of the heart. To the other end I put my ear. I was both surprised and gratified at being able to hear the beating of the heart with much greater clearness and distinctness than I had ever done before by direct application of my ear."

On the basis of this principle, Laënnec proceeded to build his stethoscope (from two Greek words meaning "chest-observer"). In its original form, it was a short wooden tube, up to twelve inches in length, and widening somewhat toward each end. (The modern version has two ear tubes with flexible attachments.) In 1819 Laënnec published a treatise in which he described the use of the stethoscope.

Through the use of Laënnec's invention, physicians could now make more accurate diagnoses of diseases of the heart and lungs; they could follow fairly well the spread of these diseases. They could also detect the presence of the fetus in a pregnant woman, since the heart of the fetus beats almost twice as rapidly as that of an adult woman.

There were a number of outstanding developments in physiology in this period. Our scene now shifts to an American frontier army post, where a United States Army surgeon, William Beaumont (1785–1853), has just taken charge of a serious gun-shot case. To quote the good doctor: "Alexis St. Martin [the victim] was a Canadian of French descent, about eighteen years of age, of good constitution, robust and healthy. He had been engaged in the service of the American Fur Company . . . and was accidentally wounded by the discharge of a musket on the sixth of June 1822 . . .

"I saw him in twenty-five or thirty minutes after the accident occurred; and on examination found . . . a protrusion . . . which proved to be a portion of the stomach, lacerated through all its coats and pouring out the food he had taken for his breakfast through an opening large enough to admit the forefinger."

By skillful surgery and constant attention, Beaumont effected a cure. When the wound healed, the stomach remained exposed to the outer air through an aperture about two and a half inches in circumference. It dawned upon Beaumont that he now had a unique opportunity to study the functioning of the stomach, since food could be readily introduced into St. Martin's stomach through the aperture and could be just as readily withdrawn. He therefore proposed to the young French Canadian that he should serve as a living laboratory; in return he agreed to provide the young man with food (and drink) and

lodging. St. Martin agreed, and in 1825 Beaumont began the experiments that were to win him lasting fame in the annals of physiology.

He has thus described one of these experiments: "August 1, 1825. At twelve noon I introduced through the opening into the stomach articles of diet, suspended by a silk string and fastened at proper distances, so as to pass in without pain. [These "articles of diet" included highly seasoned beef à la mode, fat pork, lean beef, salted beef, a piece of stale bread and a bunch of raw, sliced cabbage.] . . . At 1:00 P.M. withdrew and examined them. Found the cabbage and bread about half digested; the pieces of meat unchanged. Returned them into the stomach. At 2:00 P.M. withdrew them again. Found the cabbage, bread, pork and boiled beef all clearly digested and gone from the string; the other pieces of meat but very little affected. Returned them into the stomach again . . . At 3:00 P.M. examined again. Found the à la mode beef partly digested; the raw beef was slightly macerated on the surface. Returned them again." There was little change the next time Beaumont withdrew the string. But St. Martin decided that his patron had done enough experimenting for that day; and so the doctor set to work analyzing his notes.

A classic of medical literature

St. Martin proved to be an exasperating sort of laboratory. He was constantly running away and had to be brought back at great trouble and expense. Furthermore, Beaumont found it hard to carry on his experiments, since War Department orders kept him moving frequently from one post to another. However, he persevered and at last, in 1833, presented his findings in a little book — OBSERVATIONS ON THE GASTRIC JUICE AND THE PHYSIOLOGY OF DIGESTION — published at his own expense. It was perhaps the most important single work ever written on the digestive process. It established once and for all the essential chemical nature of this process; it provided basic information

William Beaumont withdrawing gastric juice from the stomach of Alexis St. Martin. From a painting by Cornwell.

about the secretions of the stomach and its functioning. All modern studies of nutrition, all modern diet tables are based upon this remarkable little book. The author's experiments, performed under backwoods conditions, represented something more than a landmark in the history of physiology. They proved that the progress of science does not necessarily depend on the amassing of great funds for research and on the construction of big laboratory buildings.

Another great name in the history of physiology is that of Claude Bernard. This distinguished French doctor was born at Saint-Julien in 1813. He studied medicine at Paris, becoming a pupil of the eminent physiologist François Magendie. Bernard completed his medical schooling in 1843; ten years later he became a doctor of science. In the year 1855 he succeeded Magendie as professor of physiology at the College of France. Bernard was an accomplished teacher and lecturer, but his achievements in physiological research were even more outstanding.

He was the first to show how the pancreas functions; in the course of time this discovery pointed the way to the control of diabetes, a disease in which the pancreas fails to function properly. Bernard also made some remarkable discoveries concerning the functioning of the liver. He demonstrated that the liver modifies the glucose (a form of sugar) that reaches it, turning it into a substance called glycogen, which simply means "sugar-maker." This substance is then stored in the liver. When and if the concentration of glucose in the blood falls below a certain level, the liver transforms part of its store of glycogen into glucose and releases it in the blood stream.

Bernard called the conversion of glycogen into sugar an "internal secretion." Today physiologists would no longer so label it, since sugar is not a specific product of the liver. But Bernard's researches pointed the way to the discovery of bona fide internal secretions — those of the pituitary gland, the thyroid gland, the parathyroid glands, the pancreatic islets of Langerhans and the adrenal glands.

Bernard also did important research on the vasomotor nerves, which control the size of the blood vessels. He demonstrated that vasodilator nerves cause the arteries to dilate, and that vasoconstrictor nerves cause these blood vessels to contract.

This great physiologist won many honors in his lifetime. He became an officer and later a commander of the Legion of Honor; he was appointed to the renowned French Academy and he became the founder and first president of the French Biological Society. Upon his death in Paris, in the year 1878, he was given a truly magnificent public funeral.

The first half of the nineteenth century saw a frontal assault on the disease known as puerperal, or childbed, fever, which attacks women in childbirth. This ailment had long taken a fearful toll. The death rate among women admitted to lying-in hospitals was particularly high. One of the first physicians to call attention to the contagious character of this disease and to the manner in which it was carried from one victim to another was the Yankee medical man, poet and essayist Oliver Wendell Holmes (1809–94).

At a meeting of the Boston Society for Medical Improvement, in 1843, Holmes read a paper, The Contagiousness of Puerperal Fever. He pointed out that the disease "is so far contagious as to be frequently carried from patient to patient by physicians and nurses." And why? Because these doctors and nurses did not clean themselves properly before attending their patients. To Holmes this seemed a particularly blameworthy offense. "The woman about to become a mother, or with her newborn infant upon her bosom, should be the object of trembling care and sympathy . . . God forbid that any member of the profession to which she trusts her life . . . should hazard it negligently, unadvisedly or selfishly!"

Holmes' paper aroused the indignation of Dr. Charles Meigs, professor of obstetrics at the University of Pennsylvania. Meigs took Holmes' remarks as a direct slur upon the members of the medical profession as a whole. Holmes answered Meigs with another paper called Puerperal Fever as a Private Pestilence; in this, he pointed out that a certain obstetrician called Senderein had scrubbed his hands with chloride of lime before attending his patients, and that as a result the mortality rate among them had dropped amazingly. Not long afterward Holmes became professor of anatomy at the Harvard Medical School. He seems to have become so absorbed in his new duties that he lost interest in the controversy over puerperal fever.

The war against this disease was to be carried on by the man whom Holmes had called Senderein and whose name was really Semmelweis. Ignaz Philipp Semmelweis (1818–65) was a native of Budapest, Hungary, who received his medical training in Vienna. After completing his schooling, he became an assistant in the lying-in hospital in Vienna. It consisted of two different departments, or divisions.

OLIVER WENDELL HOLMES

PUERPERAL FEVER

IGNAZ PHILIPP SEMMELWEIS

Medical students were taught in the first division; in the second, the women who were to become midwives received their training. In the first division, over a period of six years, there had been 99 deaths per 1,000 births; in the second division only 33. Semmelweis came to the conclusion that puerperal fever was a wound infection transmitted by the hands of the physicians and medical students who examined women during and after childbirth. Sometimes they would turn from one patient to another without troubling to wash their hands. If they did take the trouble, they used only soap and water, and they never quite succeeded in removing all the germ-bearing particles that adhered to their hands.

Semmelweis, who was in charge of the first division, now insisted that all the students should wash their hands in a solution of chloride of lime before examining a patient. The results were truly startling. At the time when Semmelweis issued his order, the death rate from puerperal fever was 120 per 1,000 births; after seven months, the death rate had dropped to 12 per 1,000 births.

Semmelweis' hasty temper had aroused the dislike of his superiors, and he was continually harassed by petty persecution.

At last he could endure no more, and he returned to Budapest. In time he became the director of the lying-in hospital in his native city. Here, too, he was the object of much criticism; as a result, his mind was affected. In 1865 he left for Vienna to consult a specialist in mental diseases. In the course of a routine examination it was discovered that Semmelweis had contracted the same type of infection that he had spent his life in combating. Soon afterward he died of the infection.

The validity of the position taken by Holmes and Semmelweis was not fully appreciated until the germ theory of disease was later demonstrated. Today puerperal fever is no longer a menace in maternity wards. No longer is infection carried, as Holmes had put it, "from bed to bed as rat-killers carry their poison from one household to another."

As a young man Holmes had studied in Paris under Pierre-Charles-Alexandre Louis (1787–1872), the founder of medical statistics. This is an exceedingly important branch of medicine. It is difficult to prove that one method of treatment or control of disease is better than another; often, only accurate statistics can provide this information. Louis showed the possibilities of the statistical method.

A MEMORABLE VICTORY OVER PAIN

On the 16th of October, 1846 — celebrated in Massachusetts as Ether Day — a great victory was won over one of man's most terrible scourges — pain. That day saw a public demonstration of the first safe anesthetic, ether — or sulfuric ether, to give it its full name — in the Massachusetts General Hospital, in Boston. The discovery of the anesthetic effects of ether was due to three Americans — Morton, Jackson and Long — and it represents one of America's most notable contributions to the progress of science and the welfare of mankind.

Before Ether Day, it is true, anesthetics of sorts had been administered to patients about to undergo an operation. The soporific (sleep-producing) and anesthetic

(pain-killing) effects of intoxicating liquids and of the poppy seed — source of morphine, cocaine and laudanum — were well known. The luckless victim of accident, war or disease was sometimes drugged with alcohol or morphine before he took his place on the dirty operating table. But this did not happen as often as one might imagine, for surgeons had discovered that patients drugged with alcohol or morphine were particularly likely to die during or after operations. Often, therefore, the surgeon refused to administer any opiate, and the patient had to bear his pain as best he could. Necessarily cruel, surgery had to be mercifully swift. Naturally, operations lasting hours at a time — like some that take place quite frequently today —

were impossible under such circumstances.

The discovery of a safe anesthetic changed all this almost overnight. Not only did ether make it possible to abolish pain during a surgical operation, but the surgeon no longer had to hurry in order to reduce to the minimum the sufferings of his patient. Many operations now became feasible that would have been quite unthinkable before this time.

Ether was known to chemists years before it was put to practical use to deaden pain. In the 1840's students at the Harvard Medical School occasionally indulged in "ether frolics" — that is, they would inhale the vapor of ether for the mild form of intoxication that it produced. One of the teachers at the Medical School, Dr. Charles Thomas Jackson (1805–80), observed that while under the influence of ether the students seemed to be oblivious to pain. Jackson now began to experiment with sulfuric ether as an anesthetizing agent. One of his students, Dr. William T. G. Morton (1819–68), a dentist who was studying medicine at the Harvard Medical School, became greatly interested in these experiments.

Morton had long sought an effective painkiller that he could use in his dental work. His friend and former partner, Horace Wells, had used nitrous oxide ("laughing gas") to deaden pain in dental treatments, and Morton had been greatly impressed. But unfortunately one of Wells' patients had died from the effects of this anesthetic; shocked, Wells withdrew from practice. Morton realized that he would have to seek a safer anesthetic than nitrous oxide. It now occurred to him that ether might prove to be ideal for this purpose. He experimented with the substance in his own home, first putting his dog to sleep and then testing the effects of the anesthetic on himself. Finally he employed ether on a patient; he removed a tooth without causing any pain.

Morton thought now of the possibility of using ether as an anesthetic in surgical operations. He persuaded Dr. John Collins Warren (1778–1856), senior surgeon of the Massachusetts General Hospital, to let him give ether to a patient who was about to undergo an operation. The date of this demonstration was set for October 16, 1846. A young medical student, Washington Ayer, has left the following eyewitness account of what happened:

"The day arrived; the time appointed was noted on the dial when the patient was led into the operating room. Dr. Warren and a board of the most eminent surgeons in the state were gathered around the sufferer ... It had been announced that 'a test of some preparation was to be made for which the astonishing claim had been made that it would render the person operated upon free from pain.'

"Those present were incredulous, and, as Dr. Morton had not arrived at the time appointed and fifteen minutes had passed, Dr. Warren said with significant meaning: 'I presume he is otherwise engaged.' This was followed by a derisive laugh and Dr. Warren grasped his knife and was about to proceed with the operation.

"At that moment Dr. Morton entered a side door. Dr. Warren turned to him and in a strong voice said: 'Well, sir, your patient is ready.' In a few minutes he was ready for the surgeon's knife, when Dr. Morton said: '*Your* patient is ready, sir.'

"Here the most sublime scene ever witnessed in the operating room was presented, when the patient placed himself voluntarily upon the table, which was to become the altar of future fame ... That was the supreme moment for a most wonderful discovery. Had the patient died under operation, science would have waited long to discover the hypnotic effects of some other remedy [than ether] of equal potency and safety. It may be properly questioned whether chloroform would have come into use as it has at the present time.

"The operation was for a congenital tumor [a growth existing at birth] on the left side of the neck ... The operation was successful; and when the patient recovered, he declared he had suffered no pain. Dr. Warren turned to those present and said: 'Gentlemen, this is no humbug.'"

Morton and Jackson, joining forces, tried to patent ether under the name of

Letheon. They proposed to issue permits for a considerable fee to physicians who wished to use the new anesthetic. This scheme was doomed to failure as soon as doctors recognized that Letheon was simply sulfuric ether. Morton and Jackson soon parted company. Each of the former partners insisted that he was the sole discoverer of anesthesia. That honor was also claimed by Horace Wells, who, as we saw, had used nitrous oxide as an anesthetic.

In 1849, Morton petitioned Congress to give him a substantial reward for his work in developing a safe anesthetic. Jackson also pressed his claims before Congress; the friends of Wells, who had died, likewise entered the lists. For several years the controversy raged. At last a measure appropriating $100,000 for Morton was being actively pushed in the Senate, when one of the senators rose to protest. He pointed out that Dr. Crawford W. Long (1815–78), of Georgia, had used ether in a surgical operation in 1842 — four years before the historic demonstration at the Massachusetts General Hospital. Long's patient, James Venable, had declared under oath in 1849:

Dr. William T. Morton administering ether to a patient before an operation.

"In the early part of the year [1842] the young men of Jefferson and the country adjoining were in the habit of inhaling ether for its exhilarating powers. I inhaled it frequently for that purpose and was very fond of its use.

"While attending the academy, I was frequently in the office of Dr. C. W. Long, and having two tumors on the back of my neck, I several times spoke to him about the propriety of cutting them out, but postponed the operation from time to time. On one occasion . . . I agreed to have one tumor cut out and had the operation performed that evening after school was dismissed. This was in the early part of the spring of 1842.

"I commenced inhaling the ether before the operation was commenced and continued it until the operation was over. I did not feel the slightest pain from the operation and could not believe the tumor was removed until it was shown to me."

Congress took no further action on the $100,000 appropriation after Venable's affidavit had been revealed. To this day the question of priority in the matter of the discovery of anesthesia is shrouded in doubt. It is important to observe that Venable's affidavit was not given until 1849, three years after Morton's public demonstration. The only compensation that Morton ever received for his work on anesthesia was an honorary degree from an American university and a gold medal from the French Academy of Sciences. He died poor and embittered in 1868.

The news of the successful use of ether as an anesthetic flashed quickly around the world. British surgeons, notably James Syme, the tutor of Lister, and James Y. Simpson (1811–70), professor of obstetrics at the University of Glasgow, hastened to adopt this new medical procedure; so did eminent surgeons in other lands.

On November 4, 1847, Simpson announced that chloroform, which had been discovered by the French chemist Jean-Baptiste-André Dumas, could also be employed as a safe anesthetic; he had used chloroform to ease the pains of a woman in childbirth. Simpson was greeted by a perfect hail of criticism. He was denounced by clergymen and others who claimed that painless childbirth was a violation of God's will. They quoted Scripture (Genesis, III:16): "Unto the woman [God] said, I will greatly multiply thy pain and thy travail; in pain thou shalt bring forth children." Simpson answered these objectors in 1847 in a famous paper called Answers to the Religious Objection against the Employment of Anesthetic Agents in Midwifery and Surgery. He failed to silence his critics. But then, in April 1853, Queen Victoria consented to take chloroform during the birth of her seventh child, Prince Leopold. The opposition to pain-killing in childbirth collapsed as if by magic. Chloroform was often administered to women in childbirth thereafter. Nowadays, however, most obstetricians favor so-called "natural childbirth."

THE GERM THEORY OF DISEASE

For many centuries mankind had lived in constant fear of epidemics — a fear that was aggravated by a sense of utter helplessness. The germ theory of disease did much to free man from this terrible nightmare of dread. The theory was one of the great scientific triumphs of the nineteenth century — indeed, of all time.

It had a strange and humble origin — the study of fermentation. Men had been making and drinking alcoholic beverages since time immemorial. Yet not until after the middle of the nineteenth century did anyone know the why and wherefore of fermentation, which turns grapes into wine, and hops, malt, rice, barley and other cereal grains into beer, ale and mead. The chemists of the early nineteenth century, led by Liebig, had said that fermentation was a chemical reaction.

The French chemist Louis Pasteur showed that it was due to the activity of certain living organisms, called bacteria. Continuing his study of bacteria, he launched his truly epoch-making theory that some of these tiny creatures — also called microbes, or germs — could overwhelmingly invade the human or animal body and thus cause disease.

Today this theory is almost universally accepted. But until Louis Pasteur entered the scene, certain learned professors were still maintaining that earthquakes brought on the plague; and others attributed fevers to such factors as marshes, night air and bad odors. A few men, indeed, had at least some idea of the true facts in the case. The sixteenth-century Italian physician Fracostoro (see the Index) had expressed his belief in "seeds of contagion." John Snow (1813–58), a Yorkshire-born London physician, almost hit upon the truth when he traced a cholera epidemic in London in 1854 to the polluted water of a cer-

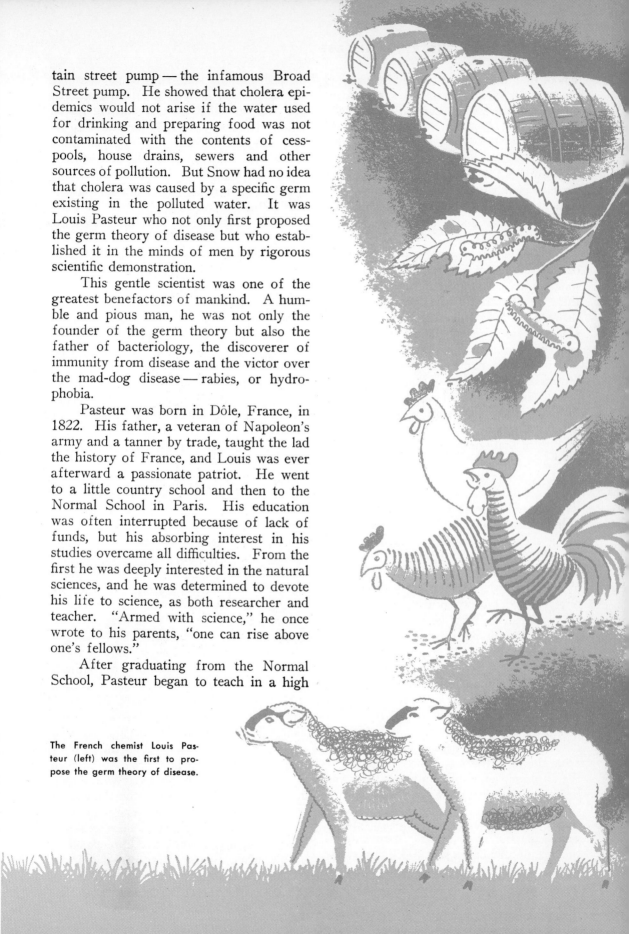

tain street pump — the infamous Broad Street pump. He showed that cholera epidemics would not arise if the water used for drinking and preparing food was not contaminated with the contents of cesspools, house drains, sewers and other sources of pollution. But Snow had no idea that cholera was caused by a specific germ existing in the polluted water. It was Louis Pasteur who not only first proposed the germ theory of disease but who established it in the minds of men by rigorous scientific demonstration.

This gentle scientist was one of the greatest benefactors of mankind. A humble and pious man, he was not only the founder of the germ theory but also the father of bacteriology, the discoverer of immunity from disease and the victor over the mad-dog disease — rabies, or hydrophobia.

Pasteur was born in Dôle, France, in 1822. His father, a veteran of Napoleon's army and a tanner by trade, taught the lad the history of France, and Louis was ever afterward a passionate patriot. He went to a little country school and then to the Normal School in Paris. His education was often interrupted because of lack of funds, but his absorbing interest in his studies overcame all difficulties. From the first he was deeply interested in the natural sciences, and he was determined to devote his life to science, as both researcher and teacher. "Armed with science," he once wrote to his parents, "one can rise above one's fellows."

After graduating from the Normal School, Pasteur began to teach in a high

The French chemist Louis Pasteur (left) was the first to propose the germ theory of disease.

school in Paris; in due course he married the principal's daughter. His researches on the structure of crystals attracted wide attention and won him a professorship of chemistry at the new University of Lille. As a university professor, Pasteur was a civil servant. The Government did not hesitate to call upon him to help businessmen and others who had problems that could be solved more or less readily by chemical methods.

Soon Pasteur was requested to find out why the wine in Orléans was turning sour. After careful research, he discovered that fermentation was brought about through the activities of certain minute organisms. He showed the vintners how to keep their wine from spoiling by gently heating it to a temperature of 55 degrees centigrade — the process that we now call pasteurization. The heating destroys the tiny organisms that sour wine, beer and milk. "I found," wrote Pasteur, "that all real fermentations depended on the presence and multiplications of organic beings [that is, bacteria]."

The chemists of the old school were unconvinced; they maintained that fermentation was a chemical phenomenon and that living things like microbes had nothing to do with it. "Where did these microbes come from?" they asked. "They are found in the air," was Pasteur's response. He now devised an experiment that silenced his opponents. First he caused air to be sucked through a tube filled with cotton wool. The dust particles that collected in the tube were transferred to a sterile sugar solution in a gooseneck flask. The neck was then sealed. In a few days great colonies of bacteria were found inside the sealed flask. Obviously these colonies had descended from the original bacteria drawn from the air. Pasteur showed also that if the dust and sugar solution in the flask was boiled, no bacteria grew. He won a prize — and many adherents to his new germ theories — as a result of the crucial experiments.

Another practical problem was now thrown into his lap. The silkworms in Alais (Alès) were dying of a mysterious disease and the silk industry of southern France was in danger of extinction. His old teacher, the chemist Jean-Baptiste-André Dumas, begged him to investigate the problem. "I have never had anything to do with silkworms," protested Pasteur; but in the year 1865, he packed his trusty weapon, the microscope, and went south to attack this new problem.

By the year 1868, Pasteur had discovered that the silkworms of southern France were dying not from one disease but from two. What was even more important, he had demonstrated conclusively that each of these diseases was caused by a specific microorganism. Pasteur devised a method for preventing contagion as well as for detecting diseased silkworms — and the silkworm industry was saved.

Pasteur now turned his attention to the study of the germs causing chicken cholera and anthrax. He cultivated the germs of chicken cholera in a kind of chicken soup — a culture medium in which the microbes could find suitable food. One day he injected an old, stale culture into a few hens. The fowl became slightly sick but soon recovered. Then it occurred to him to inject fresh, germ-full cultures into the hens. Nothing happened. In some unique way, the old, thinned out cultures had strengthened the hens' defenses against infection with fresh germs. This is an example of the phenomenon of immunity, or the mobilization of the body's defenses against bacterial invaders.

In May 1881, Pasteur and his associates proved that immunity to anthrax could also be acquired. In a demonstration at a farm near Melun, twenty-five sheep and six cows were injected with vaccines made from old cultures of the anthrax bacillus (germ). An equal number of cows and sheep were not vaccinated. Then all the animals were infected with living anthrax germs. To the astonishment of the veterinarians and farmers who were following the experiment, the vaccinated sheep and cows remained in excellent health, while all the unvaccinated animals quickly sickened and died.

Pasteur's researches in the dread dis-

ease called hydrophobia, or rabies, made a particularly dramatic story. Hydrophobia attacks many animals, including dogs, cats, wolves, foxes and jackals; it is transmitted to man by the bite of infected animals. The most striking feature, perhaps, of the disease is shown at an advanced stage. The patient is tormented by thirst and yet is seized with a violent choking fit when he attempts to swallow water. In fact, the very thought of water may bring on such an attack; hence the name "hydrophobia" ("fear of water," in Greek). Until Pasteur, hydrophobia victims always died after suffering the most frightful agonies.

After extended experiments with saliva from infected animals, Pasteur came to the conclusion that the germ — or as it later turned out, the virus — of the disease was concentrated in the nerve centers of the victims. He found, for example, that when he introduced matter from the spinal cord of an infected dog into a healthy animal, he could produce the symptoms of hydrophobia. He also discovered that he could prevent dogs from dying of the disease by a series of injections of the dried spinal cords of infected rabbits.

On the morning of July 6, 1885, a nine-year-old Alsatian lad, Joseph Meister, who had been cruelly bitten by a mad dog, was brought to Pasteur's laboratory in Paris for treatment. Pasteur had never tried his vaccine on a human being, and he hesitated to do so now. "The death of this child," he wrote afterwards, "appearing to be inevitable, I decided, not without lively apprehension, as may well be believed, to try upon Joseph Meister the method that I had found constantly successful with dogs." The treatment proved to be entirely successful; the boy speedily recovered.

Soon Pasteur was besieged by patients from all over the world, seeking treatment for the bites of mad dogs and wolves. A grateful world honored the great scientist's name by contributing funds to build a series of Pasteur Institutes, where vaccine could be manufactured, treatments given and research on immunization undertaken.

Pasteur died quietly, full of years and honors, in 1895. His career emphasizes an important point: that efforts to solve immediate and practical problems of industry may lead to the discovery of immensely valuable theories in pure science. Thus Pasteur's work on the problems of vintners and silkworm-growers in France led to the establishment of the germ theory of disease, the science of bacteriology and the doctrine of immunization.

The world of science and the general public eagerly accepted Pasteur's demonstration of the thesis that tiny organic beings cause disease — that is, the germ theory of disease. In the first flush of enthusiasm, it was thought that all diseases were probably germ diseases; but twentieth-century physicians have come to recognize that this is not the whole story. There are many important classifications of disease for which no "germ" can be found; for example, heart disease, cancer, and a host of psychosomatic (mind-body) ailments, in which the effect of emotions on bodily states is the predominating factor. We know now that the causes of disease are multiple. To understand and control them, we must study the host of the disease (that is, the sick person), the agent of disease (which may be, but is not always, a germ) and the environment of both.

After Pasteur, the science of bacteriology, dealing with the study of bacteria, moved ahead at an exciting pace and for perhaps half a century dominated all thinking about the nature of disease and the means of preventing it. There were two trails leading to the conquest of communicable diseases — one the bacteriological trail, uncovering the specific agents of specific diseases; the other, the epidemiological trail, noting and finding means of preventing the transmission of disease. Often these trails overlapped as the microbe

ROBERT KOCH

hunters discovered both new agents of disease and unexpected methods of transmission.

Perhaps the greatest microbe hunter after Pasteur was Robert Koch (1843–1910), a German doctor. He was an exceedingly painstaking man of science, who established a series of postulates, or rules, for determining the identity and the specific character of disease germs. He discovered how to stain bacteria with dyes so that they could be seen and studied under the microscope. He isolated and described the anthrax germ in 1876; in 1882 he discovered the tubercle bacillus, the germ that causes tuberculosis; in the following year he identified the comma bacillus, the germ of Asiatic cholera.

Koch electrified the whole world in 1890 by announcing that he had discovered a substance — tuberculin — that could cure tuberculosis. Koch had prepared this substance by sterilizing the medium in which tubercle bacilli were grown in the laboratory, filtering it and concentrating it by evaporation. It soon became evident that tuberculin could not live up to the extravagant claims that Koch had made for it as a cure for tuberculosis. Eventually, however, it proved to be useful in diagnosing the disease, and it is employed for this purpose at the present time. Despite the failure of his tuberculosis "cure," Koch continued to enjoy the esteem of the scientific world. He investigated many other

diseases — notably African sleeping sickness. In 1905 he was awarded the Nobel Prize in medicine.

Other microbe hunters made notable finds in the exciting final quarter of the nineteenth century. Armauer Gerhard Henrik Hansen discovered the bacillus causing leprosy in 1874; this germ is often referred to now as Hansen's bacillus. In 1879, Albert Ludwig Siegmund Neisser isolated the gonococcus, the germ that causes gonorrhea. In the following year the bacillus of typhoid fever was identified by Karl Joseph Eberth. In 1881, George Miller Sternberg in the United States and Karl Fraenkel in Germany discovered the bacterium that causes lobar pneumonia; the great Pasteur isolated this germ at about the same time. The bacillus causing diphtheria was first described by Edwin Klebs in 1883 and isolated by Friedrich August Johannes Loeffler in 1884. Arthur Nicolaier discovered the bacillus of tetanus (lockjaw) in 1894.

Once the agents that caused communicable diseases were discovered, it became imperative to find out how they were transmitted. It was found that certain hosts will harbor the bacilli of specific diseases. It was also demonstrated that certain kinds of food and drink could carry and transmit disease germs. Thus, the burly British doctor David Bruce (1855–1931) found out in 1887 that the microbes of Malta fever, or undulant fever, can be carried in goat's milk. The disease is transmitted to humans when such infected milk is drunk. Undulant fever can be transmitted from cow's milk; milk derived from tubercular cows can also infect human beings with tuberculosis. That is one reason why we insist on pasteurized and certified cow's milk today.

It was discovered by nineteenth-century microbe hunters that insects also transmit disease germs. Men had long suspected that the fly was a vector (carrier) of disease; but the guilt of the insect was first definitely established in 1869, when it was demonstrated that the fly carries the germs of anthrax. Four years later an English nurse, who was in the Near East during a cholera epidemic, noted that the epidemic abated when flies vanished. In Washington, D. C., it was shown that these insects carried the germs of typhoid fever from insanitary privies to kitchens and parlors. A United States Army Medical Commission report revealed that flies carrying typhoid fever had killed more soldiers than bullets had killed during the Spanish-American War.

Fleas turned out to be another vector of disease. In 1894 the Japanese Shibasaburo Kitazato and the Swiss Alexandre-Emile-John Yersin discovered the bacillus of bubonic plague in the fleas that infest rats. International health activities have done much to protect the world from this ancient scourge; but it has not yet been exterminated. Plague bacilli have also been found in the fleas that infest the rats' fellow rodents, particularly the ground squirrel.

The story of how the tick was proved to be the vector of Texas fever is particularly engrossing. It carries us to the Old Chisholm Trail, over which Texas cattle were driven north for fattening and marketing. The hero of this tale was a city-born doctor, Theobald Smith (1859–1934), an employee of the United States Bureau of Animal Industry in Washington, D. C.

The bureau had been called upon to solve a serious problem confronting the cattlemen of the West. When southerners brought their cattle northward and set them to graze together with northern cattle, the southern cattle remained fat and healthy. But the northern cattle, after a month or so, took sick with the disease called Texas fever. They stopped eating, lost weight and died. Why? Smith went out west to solve the mystery. Cowpunchers, he found, claimed that the disease was due to ticks. Smith sought to find out whether or not this was the case.

It was a strange laboratory in which he worked; it consisted of six dusty fenced-

off fields, in which there were northern cows, southern cows and ticks. Some ticks were so small that they could be detected only with a magnifying glass; others — big female ones gorged with cow's blood — were half an inch long. Through microscopes Smith and his co-workers spotted little pear-shaped microbes in the blood cells of sick cows and in the bodies of the ticks. But ticks have no wings; they do not jump from cow to cow; they live their whole lives on a single host. How then can ticks carry Texas fever from a Texas cow to a Montana cow? Smith pondered the question.

One hot summer day in 1890 he found the answer. After infesting southern cattle, adult female ticks dropped to the ground where they laid microbe-infected eggs. In about three weeks, the eggs became baby ticks whose bodies and stingers harbored the microbes of Texas fever. When a baby tick crawled up a northern cow's leg and bored its way in, it carried the fever germs with it. The southern

Disc-shaped rat guards are placed on the mooring lines of ships coming from plague-infested ports. The rats on board may harbor the fleas transmitting bubonic plague.

cows, source of the infection, did not become sick because they had developed an immunity to the disease.

In 1893 Smith issued a report on his findings, called INVESTIGATIONS INTO THE NATURE, CAUSATION AND PREVENTION OF TEXAS OR CATTLE FEVER. His solution of the Texas fever problem was a simple one. "Dip your cattle clean; kill the ticks," he said, "and the disease will disappear." Later investigations have shown that ticks also carry germs of diseases — like Rocky Mountain spotted fever — that attack man.

Smith had opened the way to a whole new field of inquiry in the matter of disease control — the sciences of epidemiology and preventive medicine. He put David Bruce on the trail of the tsetse fly, which turned out to be the vector of African sleeping sickness in horses, wild animals and men. Most important, he influenced the men whose work put the mosquito at the top of the list of insect vectors of disease.

The mosquito is the vector of the most prevalent disease on the face of the globe — malaria. This fact was brilliantly demonstrated, just before 1900, by the British poet-physician Sir Ronald Ross (1857–1932), serving in the Indian Medical Service, and the Italian physician Giovanni Battista Grassi (1854–1925). Mosquito

control is the essence of malaria control. Only a few species of mosquitoes, of course, carry disease.

A dramatic demonstration of the role of the mosquito in carrying disease occurred in the case of yellow fever, sometimes called yellow jack. Havana, Cuba, was infested with the disease. A Cuban physician, Carlos Juan Finlay (1833–1915), suspected that the mosquito might be the carrier of the disease, but he could never definitely prove that his theory was correct. Following the Spanish-American War of 1898, the United States took over Cuba, formerly a Spanish possession, and faced the problem of combating yellow fever.

The United States Army sent a distinguished commission to Cuba to investigate the disease. It was headed by Major Walter Reed (1851–1902) and included Dr. Jesse William Lazear, who died a martyr to the investigation. At a crucial point in the proceedings, it became necessary to find human volunteers who would risk contracting yellow fever. Two soldiers and one civilian employee of the Army offered themselves without pay for this experiment, acting, as they said, "solely in the interests of science and for humanity." With the help of these volunteers and other devoted men, Reed and his associates were able to prove conclusively that the mosquito called *Aëdes aegypti* is the carrier of the virus that causes yellow fever.

A unique engineering triumph — the building of the Panama Canal — was due in large part to this discovery. The French had tried to build the canal in the 1880's under the guidance of Ferdinand-Marie de Lesseps, the promoter of the Suez Canal. The appalling death rate among workers because of yellow fever was one of the factors that brought operations to a standstill. When the American Government decided to go ahead with the canal in the

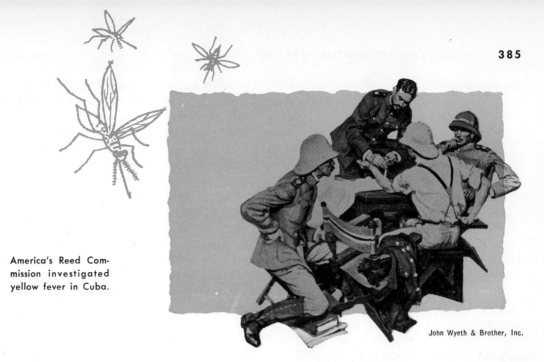

America's Reed Commission investigated yellow fever in Cuba.

John Wyeth & Brother, Inc.

early 1900's, it was recognized that the first job to be tackled was that of cleaning up the Canal Zone. The task was entrusted to Colonel William C. Gorgas (1854–1920). He knew how yellow jack was carried; he kept down the mosquito population and the canal was pushed through.

We can look upon the Panama Canal as a monument to the devoted men who demonstrated that *Aëdes aegypti* carries yellow fever; to Theobald Smith, who showed how insect vectors could be tracked down; to Louis Pasteur, who gave the world the germ theory of disease.

SURGICAL OPERATIONS BECOME SAFER

Modern surgery is based upon anesthesia and asepsis. The word asepsis means complete exclusion of disease-producing bacteria in order to prevent infection. This idea, based on the germ theory of disease, was new to the nineteenth century. Before asepsis came antisepsis — the process of actually killing microbes by application of chemicals.

The man who brought antiseptic methods into surgery was a gentle, clear-eyed and high-minded Quaker wound-dresser, Joseph Lister. He was born at Upton, in the English county of Essex, in 1827, the son of a wine merchant who devoted his leisure hours to making microscopes. Lister attended the Quaker schools at Hitchin and Tottenham; he then studied medicine at University College, in London. After completing his medical course, he went to Edinburgh armed with an introduction to the eminent Scottish surgeon James Syme (1799–1870), professor of clinical surgery at the University of Edin-

burgh. Lister became a house surgeon under Syme; three years later he married Syme's daughter.

Lister was appointed professor of surgery in Glasgow in 1860. He was particularly interested at that time in the infections that originated in wounds and that caused the formation of pus — "laudable pus," the surgeons of that time called it, because it was supposed to be a sign of healing. What was the source of the putrefaction — that is, the infection — that caused the formation of pus?

In 1865 a colleague of Lister called his attention to Pasteur's demonstration that putrefaction is caused by microbes present in the air. It occurred to Lister that, if such were the case, one might be able to prevent surgical infection and the consequent formation of pus by killing the microbes. For this purpose, Lister decided to use carbolic acid, which had served previously to disinfect sewage. He painted the wound of a patient suffering from a compound fracture with undiluted carbolic acid and then dressed it with cloths that had been dipped in diluted acid. The wound healed beautifully. In the course of the following two years Lister used this method on many patients with excellent results.

On August 9, 1867, he reported on the method at a meeting of the British Medical Association in Dublin. "I left behind me at Glasgow," he said, "a boy, thirteen years of age, who, between three and four weeks previously, met with a most severe injury to the left arm, which he got entangled in a machine at a fair . . . Without the assistance of the antiseptic treatment, I should certainly have thought of nothing else but amputation at the shoulder joint . . . Now I did not hesitate to try to save the limb by wrapping the arm from the shoulder to below the elbow in the antiseptic application . . . Before I left, the discharge was already somewhat less, while the bone was becoming firm . . . I feel sure that if I had resorted to ordinary dressing when the pus first appeared, the progress of this case would have been exceedingly different." In other words, the injured boy would have lost his arm.

Lister ended his address by pointing to the revolution that the introduction of antisepsis had brought in hospital methods of treatment. "Previously to [the introduction of antisepsis]," he said, "the two large wards in which most of the cases of accident and operation are treated, were among the unhealthiest in the whole surgical division of the Glasgow Royal Infirmary . . . I have felt ashamed when recording the results of my practice to have so often to allude to hospital gangrene and pyemia [pus in the blood stream] . . .

"But since the antiseptic treatment has been brought into full operation, and wounds and abscesses no longer poison the atmosphere with putrid exhalations, my wards, though in other respects under precisely the same conditions as before, have completely changed their character. During the last nine months not a single instance of pyemia, hospital gangrene or erysipelas [inflammation of the skin] has occurred in them."

Since undiluted carbolic acid irritates the tissues of the body, Lister mitigated its effects by blending it with shellac. In

Joseph Lister supervising the use of dilute carbolic acid in the form of a spray, in one of the earliest "antiseptic" surgical operations.

Bettmann Archive

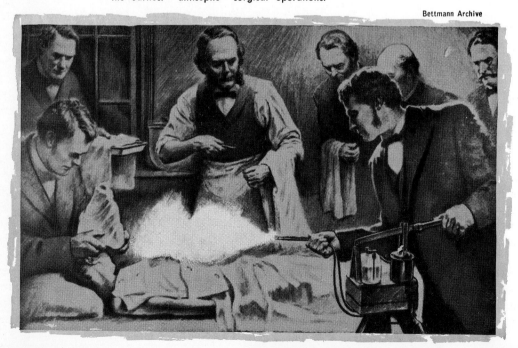

order to get rid of the germs that might be floating around in the operating room, he used a spray apparatus to disinfect the air with a dilute solution of carbolic acid. Lister and other surgeons soon saw that it was not essential to kill germs in the air — a possible but unlikely source of wound infection in the operating room. It was only necessary to take sensible precautions to keep germs away from the operative wounds and the surgical field. This could be done by sterilizing all the instruments used in operations, by using sterile clothing and by keeping the hands aseptically clean. Surgery moved from antisepsis to asepsis.

Lister won the respect and esteem of the men of his generation. He was President of the Royal Society from 1895 to 1900; in 1897 he was raised to the peerage as Baron Lister of Lyme Regis. When he died, in 1912, he was buried in Westminster Abbey, reserved for Britain's greatest men. His chief monument, perhaps, is the Lister Institute of Preventive Medicine in London, modeled after the Pasteur Institute of Paris.

With the advent of anesthesia, antiseptics and scrupulous cleanliness in the operating room, surgeons all over the world began cutting into the human body more deeply than ever before. They penetrated the abdomen; many surgeons devised elaborate operations which they named after themselves. Soon all the organs in the abdomen — the stomach, the gall bladder, the pancreas, the intestines and the appendix — could be operated on successfully. Countless human lives were saved as a result. For example, before Lister's time, appendicitis (inflammation of the vermiform appendix) had probably been about as common as it is today. But surgeons had never operated on the appendix, and the patient had always died. With the perfection of abdominal surgery, an appendectomy (removal of the appendix) became a relatively simple operation.

Later developments made it possible for the surgeon to invade the chest cavity, where the lungs and heart are located. In the twentieth century, surgeons found safe ways of getting deep into the skull cavity.

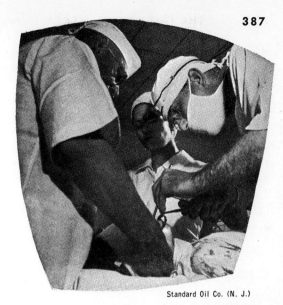

Standard Oil Co. (N. J.)

A surgeon removing a patient's tonsils in a modern hospital.

The great American surgeon Harvey Cushing (1869–1939) finally reached the pituitary body, a small gland that lies at the very base of the brain. The ancients had considered this gland or a nearby one — the pineal — as the "seat of the soul." Actually, the pituitary gland exerts a great effect upon the chemistry of the human body — and particularly on the reproductive system.

The advancing science of surgery gave new impetus to pathology — the study of diseased tissues. Toward the end of the eighteenth century, the problem of tissue changes in disease had been attacked by Marie-François-Xavier Bichat (1771–1802), who had served as a surgeon to the armies of the French Revolution. An appointment as physician to a famous hospital, the Hôtel-Dieu in Paris, gave him the opportunity to perform numerous autopsies. In a period of six months he opened some six hundred bodies in order to study the changes brought about in human tissues by disease.

The man who put pathology on a modern footing was a vigorous little German anatomy professor, Rudolf Virchow (1821–1902). He was perhaps the greatest medical figure in Berlin in the last half of the nineteenth century. He made many important contributions to histology (the science of tissues) and to the study of various diseases. In his book CELLULAR PATHOLOGY, published in 1858, Virchow gave a striking account of what happens in disease states.

Virchow was probably the first to point out that white blood cells — leucocytes — mobilize to defend the injured part of a body. The Russian-born bacteriologist and zoologist Elie Metchnikoff (1845–1916) later explained that the white cells gobble up germ invaders. He found a fine mouth-filling word for this process: "phagocytosis" ("cell-eating process," in Greek). Metchnikoff's work on immunity won him a Nobel Prize in medicine (with Paul Ehrlich) in 1908. This many-sided scientist made a study of syphilis and prescribed calomel ointment as a protection against it. He was also interested in gerontology — the science that studies the process of growing old. It has taken on added importance in our own day. A world made safer by surgery and sanitation contains far more old people than ever before.

"THE QUALITY OF MERCY IS NOT STRAIN'D"

"The quality of mercy is not strain'd,
It droppeth as the gentle rain from heaven
Upon the place beneath."

These magnificent lines from Shakespeare's MERCHANT OF VENICE apply admirably to the noble profession of nursing. It is hard to realize that until about the middle of the nineteenth century, nurses (except for women in religious orders) were held in almost universal scorn. As the London TIMES put it, in 1857: "Lectured by committees, preached at by chaplains, scowled on by treasurers and stewards, scolded by matrons, bullied by dressers, grumbled at and abused by patients, insulted if old and ill-favored, talked flippantly to if middle-aged and good-humored, seduced if young — they are what any woman would be under the same circumstances." They were, generally, disreputable, dowdy and, all too often, drunken.

This sad state of affairs was changed chiefly through the efforts of a high- (and strong-) minded Victorian lady, Florence Nightingale (1820–1910). Through her ideas and ideals, her widely heralded errands of mercy and her skill as an administrator she made nursing a highly honored profession.

A gentlewoman by birth, Florence Nightingale received an excellent classical and mathematical education at home with her father as teacher. Instead of settling down thereafter to the conventional life of an aristocratic young lady, she shocked her friends by showing a deep interest in nurs-

FLORENCE NIGHTINGALE

ing. She underwent a regular course of training as a nurse at the Institute of Protestant Deaconesses, located at Kaiserswerth, Germany, and operated by Pastor Theodor Fliedner. After she had completed the course, she underwent further training in hospitals in London and Edinburgh and studied nursing problems in London.

Miss Nightingale's opportunity to show the world the need of scientific methods in nursing came in 1854 during the Crimean War. The scene of the fighting was the Crimea, a peninsula on the Black Sea; the British and French were pitted against the Russians. After the manner of poets, Tennyson later glamorized a relatively minor episode of the war — the charge of the Light Cavalry Brigade at Balaklava. But as a matter of fact, there was precious little glamour in the Crimean War. The fighting was bitter; losses were heavy; the sufferings of the wounded and sick on both sides aroused the horror of the world. "The soldier knew some one had blundered," wrote Tennyson of the order that sent the gallant Light Brigade "onward into the valley of death." But the chief English blunderers were those who had failed to provide suitable medical care for the British Army.

Conditions at the British base hospital at Scutari (Üsküdar, Turkey) were abominable. The alleged hospital was a dirty, stinking yellow barn, infested with rats and vermin and without any comforts or conveniences. When British war correspondents revealed these conditions, there was a great public protest, and a royal commission of inquiry was set up. Florence Nightingale wrote to her friend Sidney Herbert, the War Secretary, offering her services and he gladly accepted them. Accompanied by thirty-eight trained "sisters of mercy," she set sail on October 21, 1854, and two weeks later arrived at Scutari. Miss Nightingale and her associates went to work immediately; they scrubbed, they washed, they re-arranged, they cooked for the sick and wounded. The sight of Florence Nightingale walking through the wards late at night with a lamp in her hand, inspecting and comforting, became an inspiration to the patients and to the nurses who worked with her.

The "Lady with the Lamp" had much to contend with. The military authorities resented what they considered her interference. They did what they could to thwart her, but she had the backing of the War Secretary and she overcame all obstacles. The results were truly astounding. In February 1855 the death rate in the hospital was 42 per cent; by June it had dropped to 2 per cent.

Upon her return to England in 1856, Florence Nightingale founded the Nightingale Home for Training Nurses at St. Thomas' Hospital, with a fund of £50,000 ($250,000), which had been raised by subscription in recognition of her services. Soon the nurses who had been trained at St. Thomas' were in great demand, and other nurses' training centers were set up

in the United Kingdom. In time the ignorant and dirty slatterns who had served in the hospitals were replaced by well-trained and self-respecting women who inspired in the patient the will to live.

War, serving as a whip to mercy, had provided the impetus for Miss Nightingale's reform of the nursing profession. It also led to the formation of the Red Cross societies — those wonderful national organizations that seek to alleviate and prevent human suffering. The founding of the International Red Cross goes back to the year 1859, when the French and Sardinians fought a bloody battle at Solferino, Italy, against the Austrians. A young Swiss banker, Jean-Henri Dunant (1828–1910), was an eyewitness of the battle; after the fighting was over, he organized a corps of volunteers to search out and nurse the wounded left unattended on the field of battle. The experience touched him so deeply that in 1862 he struck off a pamphlet, called MEMOIR OF SOLFERINO, in which he proposed the formation of an international society to provide aid to the wounded and sick in wartime.

In 1864, a conference for this purpose was held at Geneva, with twenty-six delegates from sixteen governments attending. This conference, known as the Geneva Convention, witnessed the official birth of the Red Cross organization. Its identifying symbol was to be a square red cross on a white field — the flag of Switzerland reversed. This emblem was to be a sign of impartial help to both sides in a war; it was never to be fired upon. Since that time, the Red Cross societies have widened their scope by peacetime activities. They have provided relief in disasters; they have worked for public education in health matters, for public-health nursing and for improved health conditions generally.

Many other voluntary health agencies have also entered the lists to battle against disease and needless suffering. They bear eloquent testimony to the medical progress of the nineteenth century — progress that has brought new hope for happier, healthier living to mankind.

Continued on page 251, Volume 8.

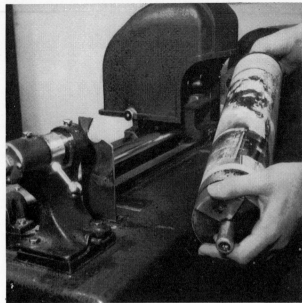

The Associated Press Wirephoto transmitter. The photo shows a cylinder being put in position on the transmitter. Under the metal hood is an electric eye that scans the picture as the cylinder revolves. Electric impulses from the scanning process flash over the Wirephoto network to receiving units in all parts of the United States

FACSIMILE
TRANSMISSION

Western Union's Desk-Fax. To send a telegram over the machine, the sender wraps his message around a metal cylinder and pushes a button. An electric eye scans the message and sends it automatically to Western Union. Telegrams are received with the same machine.

Operator demonstrating the mounting of print paper on the receiving cylinder of a Wirephoto recorder. In actual operation the paper must be mounted in a darkroom.

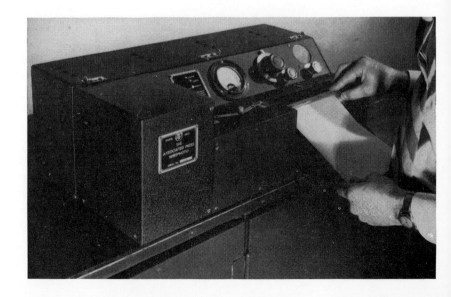

FACSIMILE MAGIC

How Pictures and Printed Matter Are
Transmitted by Wire and Radio

A PLANE crashes in a residential district in the outskirts of Chicago and causes terrible havoc. An hour or so later a photograph of the flaming wreckage is wrapped around a horizontal cylinder in a Chicago office. An attendant pushes a button, and a facsimile of the picture is sent at lightning speed through the wires of telephone circuits to hundreds of stations on a press-association network. In a matter of minutes newspaper offices in New York, Boston, Pittsburgh, New Orleans and San Francisco will have an almost perfect reproduction of the original picture in the form of a photographic print. An hour or so later hundreds of thousands of readers will be looking at the photograph in their favorite newspapers. Facsimile transmission like this would have seemed like black magic a hundred years ago; today we take it as much for granted as we do radio broadcasting, television, jet planes, antibiotics and other modern miracles.

Most facsimile transmission is carried over telephone circuits. Suppose we wish to transmit a facsimile of a photographic print over such a circuit. First the print is wrapped, picture side up, around the horizontal cylinder of the sending machine. The descriptive material accompanying the picture is typed at the sending station on a strip of gummed paper and pasted to one end of the photograph. This typed matter is transmitted together with the photograph itself.

When an attendant presses a button on the transmitting machine, the cylinder begins to rotate. A beam of light is thrown through a tiny aperture one-hundredth of an inch square and traces a continuous spiral over the picture as the cylinder turns.

This process is called scanning. It is somewhat like the scanning system used in television but much slower. It takes a television camera only one-thirtieth of a second to scan a complete picture; it takes a facsimile machine several minutes.

The beam of light is reflected from the photographic print to a photoelectric cell mounted on a carriage that moves alongside the cylinder. (In certain facsimile machines it is the cylinder that moves.) The photoelectric cell transforms the reflected light rays into electrical impulses; the lighter a given area of the picture is, the stronger the impulse will be. The impulses are greatly amplified before they are sent out over a telephone circuit.

The receiving machine converts the electrical signal into the blacks and whites and grays of the original picture. For high-quality facsimiles, a photographic process is used. Unexposed photographic film is wrapped around the receiving cylinder, which starts to turn at the same time as the sending cylinder.

The electrical impulses from the sending machine pass to a glow lamp, which brightens or darkens in proportion to the strength of the current. The light from this lamp is focused through a lens upon the film on the rotating receiving cylinder. The film is exposed as it is struck by the light, which is now bright, now relatively faint. When the transmission is completed, the film is taken to a darkroom a few steps away from the machine; here the negative is developed and a print is made. Fifteen minutes from the time the sending machine begins to transmit the picture, the facsimile is in the hands of an editor perhaps several thousand miles away.

In another receiving process, the glow lamp is replaced by a stylus pressing upon moist paper that has been treated with molybdenum salts. The amplified facsimile signal is passed through the stylus; the molybdenum salts are decomposed by the electric current and marks are made on the moist paper. These marks are in different shades of gray, depending upon the strength of the current. In the so-called dry process, the electrical signal coming over the wires produces marks on dry paper as a result of the heat generated by the passage of current through the stylus. In still another process, carbon paper is wrapped around plain white paper. A metal unit, moved by the electrical signal through the agency of electromagnets, causes the carbon paper to press upon the white paper and produces markings in this way. Facsimile reproductions made by these nonphotographic processes are not as faithful as those produced by the photographic method. They are cheaper, however, and they serve certain purposes.

Picture transmission over a network of telephone lines was begun in the United States on an experimental basis in 1925. In 1935 the Associated Press put into operation its Wirephoto picture-transmission system over a network covering 24 newspaper offices; this network has been greatly expanded. Other press associations transmit facsimiles over their own networks.

The apparatus used to transmit pictures and other matter over radio links is much like that which we have just described. In this case, however, a different kind of recording system is used. The picture at the receiving end of the apparatus is made up of dots that are spaced at regular intervals from each other; these dots are shorter or longer according to the light or dark areas of the picture that is being scanned. This kind of picture transmission is called radiophoto.

The facsimile transmission system is used nowadays to speed the delivery and pickup of telegrams. Western Union has developed a miniature facsimile machine called Desk-Fax, which has been installed in thousands of business offices. The sender places his message on the cylinder of the Desk-Fax machine and presses a button. A tiny electric eye then scans the telegram and flashes an exact picture of it to a Western Union office. Desk-Fax machines can also receive telegrams.

There are several adaptations of the Desk-Fax principle. A facsimile machine called a Telefax is installed in hotel lobbies. Guests send telegrams by depositing a telegraph in the slot of the machine and pressing a button; messages for guests are also received over Telefax. An automobile called a Telecar, equipped with radio facsimile transmission apparatus, receives messages from the main office of a city while it is cruising in suburban areas. It proceeds to the appropriate addresses and delivers the messages.

Western Union has also developed a high-speed facsimile machine, called High-Speed Fax. The cylinder of this apparatus rotates before a scanning photoelectric eye at the rate of 1,800 revolutions a minute. An entire 90-page issue of a magazine can be flashed by High-Speed Fax to a distant point in just one hour!

Western Union

The Western Union Telecar, shown above, receives messages over radio transmission apparatus and delivers them to the appropriate addresses.

FROM CLAY TO POTTERY

The Story of One of the World's Oldest Industries

A GREAT variety of useful and ornamental objects can be fashioned from plastic clay, which is first molded and then fired (baked) in an oven. The art of preparing these objects is known as ceramics (from the Greek word *keramos,* meaning "earthenware"); the objects themselves are called either pottery or ceramics. Ceramics are not easily affected by heat, and they resist water, acids, salts and other erosive agents. For these reasons they make ideal food containers — jars, plates, cups, saucers and the like. They are also widely used as building materials (such as tile), as furnace linings (fire clay), as insulators for electrical devices of many different kinds, as special refractories to be used as combustion chamber linings for jet engines and for various other purposes.

The manufacture of pottery goes back far into prehistory. Earthenware utensils occupy a prominent place among the earliest relics of prehistoric men that have come down to us. Excavations by archaeologists have shown that these utensils were in common use among the ancient cave dwellers of central France, the lake dwellers of Switzerland and the cliff dwellers of the western United States.

At a very early stage of his development, primitive man must have felt the need of some container in which he could store his beverages and other articles of food. Perhaps he noted that clayey soil was molded by his feet, and that his footsteps, imprinted in the clay, held water. He might have observed, too, that when the sun dried up the water, the clay became hardened. What could be simpler, then, than to scoop up the soft clay, knead it and mold it and set it to dry and harden in the sun? After the clay became hard, it could be used as a cistern for holding water (rain water, perhaps) or as a convenient storage place for food.

At first, probably, no tools were required to fashion earthenware from clay; human hands sufficed to make many useful utensils. Later, certain ingenious workers found that when they began to build up a clay pot on a flat stone, they could turn the stone so that each part of the pot would be brought under the hand in turn; in this way, perhaps, the potter's wheel came into being.

In time it no doubt became evident that fire would quicken the drying of pottery and that fire-baked pottery is harder and more durable than the sun-dried variety. Hand-shaped clay vessels, carefully baked, became objects of pride to an individual or to a family.

It is thought that the first artistic pottery was created to honor the dead or to propitiate whatever spirits were supposed to have custody of the dead. The sacred urns created for this purpose were often beautifully decorated. The sides of these urns were ringed round with ornamental bands marked in the clay; between the bands, designs were drawn in short, straight lines, inclining toward each other at varying angles.

So simple is the process of making pottery that it evolved independently almost everywhere, even among the least inventive races. In this widespread industry, the borrowing of ideas and methods seems to have played a relatively unimportant part, at least in the early stages of development. For this reason it is often hard to determine which people should be given the credit for having been the first to develop a given process.

Most authorities believe pottery-making began in the Neolithic period of man's cultural development — about 12,000 years ago. The earliest pottery probably copied the shape of leather bags or baskets, which were invented even earlier. Some prehistoric craftsman may have daubed clay on a basket and dried it in the sun to make it watertight. This theory is suggested by a type of primitive pottery (from the fifth millennium B.C.), which was made of clay mixed with chopped straw and reeds.

Archaeologists study the shape and decoration of ancient pottery, as well as the manner of construction (including firing methods). Even shards, or fragments, may serve as a guide to the characteristics and contacts of ancient cultures. Each prehistoric period and people had its own types of pottery. Therefore, these ceramics can serve as a means of identifying their makers. We can even trace the movements of early peoples by their pottery. For example, the first appearance of the Hellenes on the Greek mainland is marked by the presence of pottery similar to that found at Troy. If an archaeologist found a particular type of pottery scattered over the burnt site of an ancient city, he might conclude that the makers of the pottery were the successful invaders of that city. Pottery also gives us evidence of cultural exchanges which took place long ago. Thus, the ware of the Han dynasty in China (206 B.C.–220 A.D.) shows influences which indicate the existence of trade between the East and the West at that time.

As we shall see, the story that covers the period from the sun-baked pottery of primitive man to the delicate ceramic art treasures of the Renaissance and modern times is quite long. Almost every nation, both ancient and modern, is in it.

The earliest pottery is undoubtedly Egyptian. Clay fragments have been found in Egypt in excavations that are estimated to be over 13,000 years old. Syria, Anatolia, Cyprus and Crete all were producing distinct types of pottery by 5,000 B.C..

The earliest firing of pottery was done in an open fire. Primitive man found that various clays fire to different colors, and he made use of this knowledge. The most usual colors of ancient pottery were tan, red or brown. If the oxygen was shut off from the fire during the firing process and the smoke penetrated the clay, the ware was gray or black. This method is illustrated by the black-topped red pottery of ancient Egypt. The top of the utensil was placed in the coals and became black, while the rest of it remained bright red. Pottery was decorated by daubing another clay of a different color onto the base clay.

The potter's wheel was a comparatively late invention. Before its development, the potters built up their bowls and jars by coiling long clay strings one on top of another and pressing them together. The vessel was then smoothed, both inside and outside. Utensils made by this method were often unsymmetrical. The potter's wheel changed this, making it possible to produce perfectly symmetrical jars. In its simplest form, it was a heavy disk pivoted at a central point and set going by hand. This type of wheel is known to have existed as early as 4,000 B.C. in both Mesopotamia and Iran. It was one of the much-needed inventions arrived at independently by many cultures. About the time of Christ, a wheel was developed which the potter could rotate with his foot; this was a great advance, because the artisan then could use both hands on the clay.

By 3,000 B.C., pottery-making had developed into a fine art. The Egyptians had invented the glaze, a liquid coating that was poured over a piece of ware and that gave it a lustrous, nonporous surface after it was baked. The earliest glazes were of colored glass which contained copper or iron. These were the green, turquoise and yellow glazes developed by both the Egyptians and the Assyrians. Glass glazes helped make the earthenware more watertight. The Egyptians were also the first to mix glass with clay in order to form a harder, less porous material. Somewhat later, they developed a clear soda-lime glaze. The Syrians were using closed kilns by the year 3,000 B.C. These kilns permitted indirect firing to be carried out at controlled temperatures.

In a kaolin pit of Cornwall, England. A blast has just broken up the kaolin into chunks.

The Chinese originated the art of painting in colored fusible glasses on the already fired glaze, after which the vessel was fired again at a lower temperature. This development gave a much greater variety of color. Most colors used under the glaze were altered by the high temperatures required for the first firing, so that the range of possible colors in the finished work had been very limited prior to this.

The ancient Assyrians and Babylonians also helped lay the groundwork for modern ceramic art. They excelled in bricks and tiles with lustrous glazes in several colors. Today archaeologists seek out writings inscribed on baked clay tablets among the ruins of Assyrian and Babylonian cities.

The Chinese seem to have had a particular genius for ceramics. Independent of foreign influences, they developed delicate, highly refined wares to a point unequaled in modern times. They possessed clays of extraordinary purity and excelled in glazes reproducing colors found in nature — the colors of fruit and flowers, of a clear blue sky reflected in an icy stream, of new grass, of ripening grain. Even the bottoms of their dishes were decorated by the conscientious Chinese artists. After 600 A.D., the fame of their fine porcelain spread to Europe, and the demand for it there greatly encouraged trade and cultural exchange between the East and the West. The great skill of the Chinese was due partly to their complex division of labor; many artisans worked on a single ceramic piece. Unfortunately, this system also led to stagnation as far as new ideas were concerned, for no one workman could carry through a new work as a whole. As a consequence, China, which had begun by influencing pottery-making all over the then-known world, ended up by being influenced by neighbors who were more progressive, though their art had developed later.

Fine pottery was also made in Korea from early times. Japan had a shortage of materials — clays, kaolin and feldspar — and so her ceramic art developed much later than that of China and Korea.

Through these centuries of pottery-making, several types of ware evolved.

The whole history of pottery-making is the story of man's struggle to find vessels that will be more serviceable and at the same time more aesthetically pleasing.

The earliest pottery was earthenware. This was brittle and still porous after firing. Lead glazes were usually used on it. Representatives of earthenware today range from the common flower pot or terra-cotta vase to white china.

Stoneware developed next. It was made of coarse, sandy, heat-resisting clays which became vitreous on firing. Hence stoneware was completely impenetrable to liquids. It was fired at a higher temperature than earthenware and was usually given a salt glaze. It is seen today in the form of crocks and jugs, especially those used in the older methods of food preservation, such as salting and pickling.

Porcelain represented the highest achievement of the ancient artisan's striving for aesthetic expression. It was light, delicate and translucent, yet extremely hard and smooth. Porcelain was made then, as now, of an infusible white clay — kaolin — and a fusible white granite — petuntse. (These materials are really the same rock in different stages of disintegration.) A feldspar glaze was usually used. There are two types of porcelain today: one is made of "soft" paste; the other of "hard" paste. The soft-paste porcelain is distinguished by being fairly easily scratched by a sharp tool; when it is broken, it shows a grainy body. Hard-paste, or true, porcelain cannot be scratched and when broken, gives a smooth, even surface along the broken edge. Unglazed porcelain is called bisque or biscuit. Tableware and decorative statuary made of porcelain has graced the homes of the world's wealthy and aristocratic for centuries. It still takes its place on well-set tables along with crystal and silver.

The world's early teachers of ceramic arts

The potter's craft was passed on to the Greeks by their Aegean and Cretan teachers. The Greeks developed pottery-making to its highest point between the sixth and the fourth centuries B.C. Highly stylized dec-

Photos, Pix, Inc.

Two steps in the processing of kaolin. Above: After being purified, the kaolin is pushed toward a press by a worker armed with a wooden "broom."

Right: After the kaolin has been pressed, it is spread out on the floors of immense drying sheds. The sheds are heated from below by big furnaces.

orations portraying their gods and heroes appeared on many of their wares. They made use of both slips and glazes. (Slips are thin solutions of different-colored clays used for underglaze decorations.) Even today their pottery is known for its delicacy, beauty of form and tasteful refinement in ornamentation.

Once the Greeks had learned the potter's craft, their pottery trade expanded. Increasing exports created a demand for containers for such articles as oil and honey in addition to those needed for household use. The Greeks used a great deal of wine, olive oil and honey, which took the place of sugar. These products were stored and transported in oenochoae (wine jugs), hydriae (water jugs), craters (large bowls in which wine and water were mixed), lecythi (oil flasks) and amphorae (double-handled jars for honey, grain and olive oil). The Greeks were ingenious in discovering new forms and shapes for their vases and drinking cups.

In the beginning this pottery industry had centered in Mycenae. After the collapse of the civilization of Mycenae, Attica gradually became the pottery center for the entire eastern half of the Mediterranean. Finally, during the middle of the sixth century B.C., the Attic city of Athens monopolized the industry.

The pottery of the Romans was characterized by the skillful use of molds; their earthenware was decorated with finely wrought mold-impressed ornamentation. In conquered areas of inferior culture, the advanced techniques of the Romans were adapted to existing primitive methods. The designs that resulted had little skill or grace, and the lead glazes used had a coarse quality. After the fall of the Western Roman Empire in the year 476 A.D., pottery-making skills were lost.

Egypt and the Near East carried onward their traditional ways of making glazed and painted pottery. In the Far East, the Chinese refined upon their techniques for glazed, hand-fired pottery. They developed white, translucent porcelain of the most exquisite beauty and elegance.

In the twelfth century, Italy, France and Burgundy became familiar with the technique of Saracen potters. Italian craftsmen were admitted into potters' shops in eastern Spain and an exchange of ideas and techniques developed.

During the fifteenth century, Chinese porcelain was imported into Europe. Its translucence was soon imitated by various potters working under the patronage of Francesco de' Medici (1541–1587). Artists of the Italian Renaissance brought enameled wares to a high state of perfection.

Bernard Palissy, born about 1510, either at Saintes or Agen in France, was so inspired by the loveliness and harmony of a white enameled cup that he determined to discover the secrets of its manufacture. He pounded different substances and broke up

Photos, Metropolitan Museum of Art

Egyptian baby's feeding cup of blue glazed faïence dating from the Twelfth Dynasty.

Blue faïence flower vase of the Twelfth Dynasty decorated with lotus blossoms.

Dresden porcelain, of which this is an eighteenth-century example, was first produced at Meissen, Germany, by Böttger.

earthen pots by the hundred and baked them afresh with his pounded substances and chemicals in a furnace of his own making. Unfortunately his many experiments yielded unsatisfactory results, and he had exhausted his money on earthenware and fuel.

In tile furnaces and glass furnaces he continued to experiment with his broken pots and chemicals whenever he had earned a little money by land surveying. At last, after years of effort and failure, he secured a single specimen of white enamel. He then built a furnace, baked some vessels he had fashioned and covered them with his enamel compound for a second baking. But the enamel would not melt. For six days and nights he fed the fire. Not only did the enamel fail to melt, but his pots were now spoiled. With borrowed money he bought more fuel and pots, covered them with his favorite compound and relighted the furnace; but the fuel disappeared and the enamel had not melted. The garden palings, household furniture and shelving from the walls were thrust into the furnace to raise the heat; and then while his wife rushed through the town crying that her husband was mad, the enamel melted, coating the rought pots with a beautiful white glaze. Palissy's long search had at last been rewarded.

What Palissy suffered during these years from his uncontrollable materials was scarcely less disheartening than the reproaches of his wife. Indeed, after the discovery of the enamel, some eight more years were spent in experiments before the potter's perfected discoveries could take the form of salable wares. Then the ornamental crockery for which he was both artist and craftsman began to be popular, and Palissy became "inventor of rustic figulines [figurines]" to the King. About 1562 he moved to Paris and set up his pottery works near the Tuileries. Though a Huguenot, he was protected by Catherine de Médicis. He worked in Paris for more than twenty years, becoming a philosopher, a writer and a scientific reformer. His best-known productions are his *pièces rustiques* — dishes decorated with crabs, frogs, snakes, shells, lizards and plants in their natural colors. His plates with mythological figures for their subjects are even more highly valued, however. The J. P. Morgan collection includes fine specimens of his work, which is also well represented in various European museums, especially the Louvre.

John Böttger was cast in a ruder mold and had a strain of the charlatan in him, but his story is as romantic as Palissy's, notwithstanding its discordant tone. He was born at Schleiz, in Thuringia, in 1685, and apprenticed to a Berlin apothecary, in whose laboratory he claimed to have made gold. When Frederick I wished to replenish his purse from the young alchemist's "gold," the lad fled, with a regiment at his heels, and took refuge in Saxony, where the Elector protected him. Again he was in danger because he failed to make gold for the Elector, but eventually he was put to work in the laboratory of Pschirnhaus, the chief chemist at the Dresden court. Here he succeeded in making a kind of red stoneware, capable of a high polish, and resembling porcelain in texture, but not in transparency.

THREE METHODS OF SHAPING POTTERY

Using the pottery wheel in Norway today. As the whirling disc rapidly spins the clay, the potter's able hands, working with inside and outside pressure, create a piece of ceramic art.

A skilled French artisan, using a method handed down for generations, shapes a vase.

A Zuñi Indian fashions pottery by an ancient method. Strips of clay are rolled out, and one coil is placed upon the other. The hollow pot takes shape by a careful working of the coils with the fingers.

At last he discovered the secret of a white porcelain, it is said, by using the powder on his wig as an ingredient in his experiments. Böttger's success was his undoing. He was guarded day and night to prevent his escape with the secrets of his search, and the Elector established a royal manufactory for exploiting his skill. Here Böttger was kept a prisoner and treated as a slave; under this cruel treatment he became a drunkard and died in his thirty-fifth year. The workmen in the factory, running away to avoid his fate, carried the secrets of the business with them, and so the making of porcelain spread through the royal factories of Europe.

Josiah Wedgwood, a master of the potter's art

Josiah Wedgwood began work as a potter at the age of ten. He was born at Burslem, in Staffordshire, in 1730, the son and grandson of a potter. Beginning with the study of chemistry, he observed closely the properties of different kinds of clays and earths. He made experiments with fluxes and glazes, till after numerous failures and heavy losses he produced earthenware and porcelain that spread the fame of English pottery throughout the world. Wedgwood knew the charm of classical art, valued form as well as texture and ornament and called to his assistance the best designers. The work done by this broad-minded and capable man is well summarized in his epitaph. He "converted a rude, inconsiderable manufacture into an elegant art and an important branch of national commerce." He became the most renowned potter in Europe and was so influential that, impatient with the slow transportation of his increasing volume of orders, he instigated the construction of better roads and the development of canals throughout England.

Types of pottery, and clays used in their manufacture

The methods of making pottery are in general much the same the world over but differ somewhat with different kinds of ware. Common red earthenware calls for a single clay, and so may gray stoneware, but both white earthenware and porcelain are made of mixtures of kaolin, ball clay, ground flint and ground feldspar. The exact proportions used are always a carefully guarded secret, which is often handed down from father to son.

For common earthenware and stoneware the clay is simply kneaded thoroughly before molding, but for white wares the ingredients (some of them previously refined by washing) are mixed with water to a creamy liquid, or slip. The slip is then forced through a filter press, and the doughlike mass of clay is removed for shaping by the potter.

Ever since the days of the ancient Egyptians, the potter has been known to form his wares on a wheel — a process requiring such skill and dexterity as to command the greatest admiration.

The potter's wheel aids in shaping pottery

The potter's wheel, or throwing machine, is a vertical shaft supporting a revolving horizontal disc upon which the clay is molded, either by hand or by tools, as it is whirled around. There is no more striking process in all the operations of manufacture than this ancient one of raising graceful shapes by the action of the hand, the eye and the flying wheel. The shaped vessel is then put aside to dry. When it is tough enough to be handled, it is placed on a turning wheel and more exactly shaped, trimmed and smoothed. Many articles, particularly of the finer and more ornamental kind, are cast in plaster-of-Paris molds, into which the liquid clay, or slip, is poured. Parts, such as handles and feet, are very often cast in separate molds.

The potter forms other articles partly by molds and partly on the wheel, or he uses molds for the outside and shapes the inside by hand or by tools. An innovation of the latter process is carried out on a jigging machine. This machine makes possible mass production of similar pieces. Regardless of technique, the working of pottery, so far described, has had the dramatic quality of swift and visible formation.

The next stage is slow, delicate and momentous. After drying, the article is fired in the biscuit (bisque) oven, and after this initial firing it is called bisque ware. Each article is placed so as not to touch another in a sagger, or clay box, which protects the ware from direct contact with the flame. The number of articles in each sagger varies according to their size, shape and importance. The saggers are piled on each other in a kiln, and spaces are left between

highest heat and then as gradually removed from this area. Whether the kilns are for bisque ware, glazing or decorating determines the size and control of the fire. Firing time may vary from a few hours to several days.

In some cases the bisque stage completes the ware or, as with some stoneware, the pieces are set unprotected in the kiln and glazed by throwing salt into the fire boxes. The volatilized salt, mixing with the silica

Lenox, Inc.

From the artist's design, the sculptor skillfully shapes in plaster of Paris or clay the master model.

each pile, or bung, of saggers to allow for heat circulation.

The simplest kiln, or oven, is such that it requires the stacking of ware in saggers, firing and then cooling for a period before it can be emptied. The advent of mass production prompted the construction of various types of kilns that could be maintained constantly at high temperatures and that would allow for the articles being fired to be brought gradually into the area of

in the surface paste, forms an even glaze over the whole article. Though of infinite variety, glazes essentially are layers of molten glass fused by firing to the clay's surface. A single baking suffices for sanitary tilings and stoneware jars, but more elaborate or ornamental vessels need repeated treatments.

Often a printed pattern is applied to pottery at the bisque stage. Transfer paper is used from which the porous bisque ware

FROM RAW CLAY TO CAST POTTERY

The clean, dried and partially refined clay is removed from the storage bin so that it may be weighed, ground and mixed.

The mixing cylinder grinds the raw clay and other ingredients to face-powder fineness. Water added to this forms slip.

Left: slip being poured into molds. The mold absorbs moisture from the slip, causing a shell of clay to adhere to the mold. Right: inside of a cup being jiggered (shaped) by a tool called a profile.

Finishing cast ware by removing mold seams and other irregularities caused by the casting process.

Photos, Lenox, Inc.

The fragile, chalklike cast ware going into the bisque kiln will come out vitreous and translucent.

THE FINISHED PRODUCT APPEARS

Ware after its first firing (bisque ware) is cleaned, or scoured, by sandblasting.

Photos, Lenox, Inc.

The ware is dipped into the glazing compound (liquid glass), after which it is again fired.

Gold-leaf designs are carefully transferred from specially prepared tissue paper to the chinaware.

absorbs the ink. The ware is placed in a low-heat kiln to fasten the colors and then dipped into a glaze composition, after which it is placed in the glost, or glaze, oven and fired a second time to harden the glazed surface. Care is exercised in dipping the ware evenly and in preventing any article from touching another in the glost oven. If the glazing is successful, the pattern underneath the glaze comes out clearly. Ordinary ware, with or without a pattern, is

The marvelous range of the potter's craft is one of its most striking characteristics. It begins with a stone ink bottle and ends with works of art that are priceless. There is considerable doubt as to the value of many examples existing over this vast range, since antiquity is always being copied and counterfeited.

The art of pottery lends itself in a remarkable degree to imitations; and in that respect English potters have been particu-

Syracuse China

Hand painting by an expert. Science, craftsmanship, artistry together create a piece of lasting beauty.

completed at this stage, after being fired at a lower temperature, more gradually and for a shorter length of time than in the bisque process.

Further stages involve decorating the ware over the glaze. The colors and decorations are printed or painted on the glazed surface, and a third firing is given, in the enamel kiln, at a still lower heat. The article increases rapidly in value through the work that is being done upon it, as well as the growing delicacy of its finish.

larly successful. In Wedgwood's day, the excavations of Pompeii were calling attention to the beauty of classical forms, and the great potter not only produced vases in the classical spirit, but copied with exactness the works of antiquity. The celebrated Barberini (now known as the Portland) vase, found near Rome, was lent him by the Duke of Portland, and fifty copies of it were made with a wonderful fidelity. Each one was a masterpiece.

See also Vol. 10, p. 282: "Ceramics."

AC Spark Plug Div., General Motors Corp.

Porcelain has many uses in industry. Above, it is used in manufacturing spark plugs for automotive engines. Right, high-voltage bushings and stand-off insulators are made of steatite and electrical porcelain.

General Ceramics Corp.

Jones & Laughlin Steel Corp.

Tile Council of America

Crouching workmen reline the bustle pipe of a blast furnace with heat-resistant brick.

Ceramic tiles are increasingly used in decoration to-day, as in the floor and wall of this attractive home.

Ceramic tile is used extensively in modern industrial buildings because it is both attractive and practical.

Tile Council of America

THE ASTONISHING GIANTS

Elephant, Hippopotamus, Rhinoceros, Giraffe

THERE was a time when giant mammals roamed over the forests and plains of the world. There were early members of the elephant family — the mastodon, the mammoth and the *Dinotherium*. There was the truly colossal *Baluchitherium,* an early hornless rhinoceros that stood eighteen feet at the shoulder. These creatures have vanished from the face of the earth, and we know of them only from their fossil remains, found in the rocks. We might well distrust the evidence of our own eyes as we examine these fossils were it not for the existence, in this twentieth century, of certain survivors of the age of monsters — particularly the elephant, the rhinoceros, the hippopotamus and the giraffe.

These animals are not closely related, but they form part of the large group of hoofed mammals, or ungulates — a group that includes the horses, cattle, tapirs, pigs, camels and so on. They show certain similarities in the form of the teeth and in the structure of the limbs; they feed almost entirely on vegetation. Elephants, rhinoceroses and hippopotamuses have the same type of skeleton. The backbone, to which are attached many ribs, serves as a girder balanced on the strong forelegs. The large, heavy head counterbalances the weight of the body. The hind legs provide the chief thrust in walking or running.

The elephant is among the most interesting of beasts. Besides its huge size and its well-developed brain, it has a most unusual elongated snout — the trunk — which is remarkably well adapted for collecting the quantities of food that the animal needs.

In the course of its evolution, the elephant grew larger and larger. It consumed more food and gradually grew larger teeth for grinding it. The study of the teeth, in fossil form, of extinct elephant species has furnished important evidence in reconstructing the animal's past.

The modern African and Indian elephants are two of a long line of animals that goes back some forty to fifty million years and that includes the mastodons, the mammoths and numerous more primitive elephantlike types. The earliest known member of this line was *Moeritherium* of Africa. A comparatively puny creature, it stood only two feet high. It possessed small tusks on both upper and lower jaws, and it probably had a short snout, like that of a tapir. From this type at least two branches of elephant development arose.

One branch led to the *Dinotherium* of Eurasia and Africa. It was a distinct type of elephant having no tusks on the upper jaw but possessing down-turned tusks on the lower jaw. It became extinct perhaps less than a million years ago.

The other branch led to the *Gomphotherium* of Africa, Eurasia and North America. This animal, which had elongated, tusk-bearing upper and lower jaws, flourished some 25,000,000 years ago. From this long-faced elephant came mastodons (of which some had a lower as well as an upper pair of tusks), mammoths and present-day elephants. The mammoths differed from the mastodons in having a much higher and shorter head and a short, tuskless lower jaw. Our modern elephants resemble the mammoths more closely than the mastodons.

Woolly mammoths were depicted quite frequently in Stone Age drawings. Well-preserved carcasses of these huge animals have been found embedded in the soil of Siberia. Apparently they had bogged down in deep mud, which later had become frozen.

In the early part of the last century, a mammoth was discovered in a marvelous state of preservation near the mouth of the Lena River, in Siberia. Its flesh and hair and internal organs were intact; in fact, its last meal of shoots and cones of pine and fir was still undigested in its stomach. Its brain was quite uninjured. The big animal had undergone astonishingly little change since it had met its doom.

A tasty meal
of mammoth

Dean William Buckland, an English divine and geologist (1784–1856), once had some guests of his partake of the flesh of a mammoth that had been unearthed in a Siberian marsh. The animal had sunk into this marsh hundreds of thousands of years before; yet its flesh was quite palatable! Siberia has been called a mortuary of mammoths. At one time the Russians exported considerable quantities of ivory derived from these huge animals.

Mammoths measuring thirteen feet in height have been found; but the average seems to have been about ten feet. These animals were comparable in height, therefore, to the elephants of the present day. Mammoth tusks were very long; in extreme cases, they may have extended over ten feet from the sockets in which they were imbedded. They were more slender, however, than those of modern elephants. The mammoth was covered with very long hair that almost reached the ground. The tail was short; a tuft at the end was provided with stiff bristles. The ears were small and they were covered with fur.

Apparently the mammoth was the only member of the elephant family that became adapted to conditions of extreme cold. In France, it maintained itself until toward the very end of the glacial period. It may have survived even longer in the tundras of northern Siberia.

The trunk of the elephant was a key factor in the animal's evolution. This powerful, muscular organ, which is actually an elongated snout and upper lip, serves as the chief tool of the ungainly creature in the struggle for survival.

The trunk developed through one of those experiments of nature whereby, as an animal changed in size or habits, it changed its means of getting food. Since the early elephants were browsing animals, it seems that as they grew in stature, they either had to increase the length of their necks or find another way to reach the ground with their mouths. The head was so heavy that it was impossible to grow a longer neck and the accompanying muscles to support it. Instead, the early elephants, such as *Gomphotherium,* for example, developed elongated lower jaws.

As the lower jaw lengthened, the nose and upper lip gradually grew beyond the upper jaw. But as the snout, or trunk, increased in length, the necessity for the long lower jaw disappeared. As a result, this jaw began to shorten and the trunk became the means for getting food. The two remaining incisor teeth of the elephant's upper jaw developed into continually growing tusks. The tusks were weapons and also served perhaps to dig up roots.

The trunk has become
a versatile organ

From an exaggerated, piglike snout, like that probably found in *Moeritherium,* the elephant fashioned a trunk that was capable of grasping objects. The powerful muscles needed to manipulate it arose chiefly from those parts of the facial muscles that in other mammals move the sides of the nose. As a consequence, the trunk of modern elephants is far more satisfactory than a long neck, for not only can it reach the ground but it can be moved upward, backward and sideways as well.

The elephant can do a great many things with its trunk. It uses the organ, for one thing, to draw in water. However, contrary to popular belief, it does not actually drink water through the trunk. Water is sucked into the organ and is then sprayed into the mouth. The trunk is also used to convey food to the mouth. By means of this useful organ, the elephant can feed on the vegetation on the ground and in the trees above its head. Among the big animal's favorite foods are the tender twigs and branches from the upper limbs of treees. Without the elephant's trunk, most of these delectable tidbits would be quite beyond its reach. Occasionally the desired twigs are too high for even the trunk to reach. In this case, the animal solves its feeding problem in a remarkably simple and practical way; it pushes the tree over, using its head as a battering ram.

The trunk of the elephant serves as a sort of fifth limb, and a very powerful one, too. With it, the animal has been known to lift as much as a thousand pounds. It does not carry heavy loads on its trunk, however. A ponderous log, say, would be lifted by the trunk to a position ac oss the tusks and then held in place by the trunk. With this organ, the elephant would test the weight of the log before lifting it to the tusks, where it would be carried, perhaps for a considerable distance.

Strong as the elephant's trunk is, it is capable of quite delicate work. On the end of the Indian elephant's trunk, there is a fingerlike lobe, which the animal can manipulate with quite amazing dexterity. (The African elephant has two such lobes.) Consider for a moment how the animal grasps the comparatively tiny peanuts that are fed to it by child visitors to the zoo. The lobe of the trunk is also a very sensitive organ of touch.

The trunk serves as the elephant's organ of smell. The animal's eyesight is poor and its hearing is not much better; hence it depends to a considerable extent on its sense of smell — one of the most acute in the animal kingdom — to keep it in touch with what is going on. If one were to observe an elephant in the wilds for any length of time, he would see that the trunk is constantly in motion sniffing the air and inspecting the objects in the immediate vicinity of the animal.

Elephants belong to the order Proboscidea and are the largest land animals in the world. At the present time they are represented by only two species — the African and the Indian. The African elephant is slightly larger than the Indian. African bulls normally reach a height of eleven feet, and a weight of six tons; occasionally,

individuals thirteen feet tall have been found. Females are about a foot and a half shorter, on the average, than the bulls. Indian bulls usually attain a height of nine or ten feet. One consequence of the African elephant's greater size is that it rarely lies down after reaching maturity. As the animal grows older and larger, it becomes more and more difficult for it to lift its great bulk to a standing position once it has reclined. It may spend the last thirty or forty years of its life on its feet — even while sleeping. Indian elephants, on the contrary, lie down quite regularly. Only old members of this species sleep on their feet.

There is a good deal of controversy in naturalist circles today concerning the intelligence of the elephant. There is no doubt that the animal is one of the most intelligent of the mammals (not including man, of course), but there is much disagreement as to how it compares with the dog, the horse and the great apes. In one respect, indeed — in its ability to learn from man — the elephant has few peers among the lower animals. Because of this ability, it has served humans in various ways for a good many centuries.

The big animal figures quite prominently in ancient warfare. When Alexander the Great invaded India in 326 B.C., the Indians sought to stop his advance by using elephants in their attacks. Apparently Alexander's troops were not intimidated. In the battle of Heraclea (280 B.C.), King Pyrrhus of Epirus hurled massed charges of elephants against the Roman legions and routed them. The famous Carthaginian general Hannibal employed elephants successfully in warfare. They played an important part in his famous victory over the Romans at Cannae in 218 B.C.

The elephant is one of the most reliable of show animals and among the mainstays of any circus. The first captive elephant to be brought to North America was a female, shipped from Calcutta, India, to New York in 1796. It became the feature attraction of the first American circus. Undoubtedly the most famous of circus elephants was Jumbo, which was originally exhibited in the Royal Zoological Gardens in London. It won renown because of its huge size; it stood ten feet nine inches at the shoulder and weighed six tons. The American showman Phineas T. Barnum bought the animal in 1882 for $10,000 and exhibited it in the Barnum and Bailey Circus for three years. Jumbo was killed by a railroad train in Canada in 1885. Its skeleton has been preserved for posterity at the American Museum of Natural History and its mounted skin at Tufts College, in Medford, Massachusetts.

Circus elephants can be trained to do many things. They can carry riders on their backs, play ball, stand on their hind legs or on their heads and bring a huge forepaw to within inches of the head of a trainer lying prostrate on the ground. They are gentle with children and graciously accept the peanuts that youngsters offer them. Only female elephants are used as performers; the bulls are apt to become unreliable at certain times.

Elephants are noted as work animals, performing remarkable services for man. Among other things, they have been used for centuries in logging operations in India and Burma. They played an important part in the construction of some of the outstanding architectural masterpieces of India. The big animals were used to carry and stack timber and to place masses of masonry in position.

The intelligence of elephants is attested in many ways. They have been taught to understand as many as twenty-five different commands. The big animals have been known to make tools for their own use. They remove branches, strip them of all but the topmost leaves and use them as fans to brush away flies. Sometimes they pull up stakes and use them as

Amer. Mus. of Nat. Hist.

African elephants with young at a training-station camp located in the Republic of the Congo.

back-scrapers. In passing spears to their riders, the animals have been trained to present them handle first. When traveling through a wooded area, an elephant will protect its rider from low-hanging twigs and branches.

The elephant leads a very interesting life in its natural habitat. It travels in herds of two hundred or more under the guidance of a big bull. The diet is strictly vegetarian; the herd moves from one place to another in search of fodder. The big animals lumber along, in single file, at about six miles an hour; no animal can travel farther in a single day.

Elephants may mate at any time during the year. At certain periods, the males go into a state of sexual excitation, called "musth." During such periods, a strong-smelling liquid runs from glands near the eyes, and the bulls are moody and ill-tempered. The period of gestation is about twenty-one months; the female bears a single calf at each birth. She may have six to eight young during her lifetime. Calves are covered with hair at birth; they stand about three feet high and weigh approximately two hundred pounds. They are suckled on a single pair of mammaries between the mother's forelegs and are not weaned until about the beginning of the third year.

The ungainly hippopotamus

The word "hippopotamus" comes from two Greek words meaning "river horse." The animal is a river-dweller, indeed; but the "horse" part of the name is not particularly appropriate, because the hippopotamus is more closely allied to the hog family. There are two living species: the common hippopotamus and the pigmy, or Liberian, variety. These animals belong to the family of the Hippopotamidae.

The common hippopotamus (also called the hippo) once ranged over much of Europe, the British Isles and Africa, but it is now restricted to a portion of the African continent, north of Zululand. As civilization advances and more and more areas are settled, the hippo becomes a problem because of the damage it does to crops. As a result, it is mercilessly hunted and its numbers have been greatly reduced.

The hippopotamus is the second largest land-living animal, by weight, in the world. The average specimen weighs about four tons; some individuals, as much as five. The massive head alone may weigh a ton. The animal is from twelve to fourteen feet in length and stands about four and a half feet at the shoulder. The huge, barrel-shaped body is supported by short, pillar-like legs. The hippo can open its huge mouth wider than any other animal, except the whales. The skin is very thick over most parts of the body, reaching a maximum thickness of two inches over some of the upper areas.

The internal organs of the hippo are of equally amazing dimensions. The stomach may reach a length of over ten feet. To fill this vast cavern, the animal may consume five or six bushels of vegetation in the course of a single day. The hippo's great appetite has proved useful to mankind, because it impels the animal to feed voraciously on river growths that would otherwise choke the streams.

Hippos live in herds of twenty to thirty individuals. Each herd keeps to the river most of the day, moving up and down the stream for considerable distances in search of food. Hippos can remain under water for as long as thirty minutes, if necessary; the normal interval, however, is from five to ten minutes. The animals can float on the surface or sink to the bottom at will. Nostrils and ears can be hermetically sealed, voluntarily, as the animals sink to the bottom of the river. Hippos are excellent swimmers; they can also walk along the river bottom at quite surprising speed — eight miles an hour, according to certain authorities.

At night, the members of the herd occasionally leave the river for short periods of time to feed on vegetation along river banks. A hippo will never go too far, and it will return before morning along the same path, following its own scent back to the river. Should a heavy rain wash away the scent, the animal will

Press-Inf. Bur., Govt. of India

Elephant plowing a field in the Naini Tal Terai forest area of India. Indian elephants, such as these, are somewhat smaller than the African species.

A sturdy pachyderm worker moving huge teakwood logs near Rangoon, in Burma. There, elephants play an all-important part in logging operations.

Ewing Galloway

be lost. On land, the animal can walk about as fast as a man and it can even gallop after a fashion. While it is on land, special skin glands pour forth an oily, reddish liquid that prevents the skin from drying out. The fact that this liquid resembles blood in appearance has given rise to the perfectly absurd belief that the hippopotamus sweats blood.

Hippos mate once a year. The bulls fight savagely at mating time, and their battles sometimes have a fatal outcome. The gestation period is about nine months; the female gives birth to a single calf. Even a baby hippo weighs a hundred pounds or so at birth and its weight increases rapidly. Usually it is suckled in the water. It travels from one place to another by riding on its mother's back.

The pigmy hippopotamus is found in Liberia and Sierra Leone, in West Africa. It stands only about three feet high at the shoulder, is about six feet long and weighs a mere four hundred pounds. Though far less bulky than the common hippopotamus, the pigmy resembles the larger animal in both appearance and habits.

A surly heavyweight — the rhinoceros

The rhinoceros is found today in the warm regions of Asia and Africa and also in Indonesia. This big, clumsy animal has a thick hide that makes it invulnerable to attack by even the largest flesh-eaters. The head, which is concave in front, is armed with one or two horns. These structures are outgrowths of the skin; they consist of compacted masses of hair. The horns continue to grow during the lifetime of the animal. The rhinoceros, or rhino as it is often called, is a plant-eater; it feeds on grasses and foliage of various kinds. It belongs to the family Rhinocerotidae, a name derived from the Greek and meaning "nose-horned animals."

The rhinoceros has a most unusual partnership with a bird called the tick bird, which spends its time removing ticks and insects from the rhino's hide. Thus it provides food for itself and contributes to the comfort of its big companion. At the

approach of an enemy such as man, the bird sounds a warning with loud screeching and much flapping of its wings. It is a most effective sentinel.

Most rhinoceroses have uncertain dispositions. They are not very intelligent and have poor eyesight and appear to be at a loss when confronted by a prospective enemy, such as man. They are likely to seek refuge in flight; but they are quite as apt to make a sudden charge. Since a rhinoceros cannot turn quickly, it is usually possible to side step its charge unless one yields to panic.

The commonest rhino of Africa is the black rhinoceros, which stands some five feet at the shoulder and weighs up to three thousand pounds. It has two horns, the front one being the larger of the two. Generally it is a solitary creature, except at mating time. Mating can take place at various times during the year. A single calf is born after a gestation period of some eighteen months. The mother suckles its young for two years or so.

The largest rhino is the white, or square-lipped variety, found in Central Africa. The animal is not really white, but a rather sooty gray in color. It weighs up to four tons and stands six and a half feet at the shoulder; like the black variety, it has two horns. Sometimes it lives in a small group.

The great Indian rhinoceros, which ranges over the plains of northern India, Nepal and Assam, has only one horn. Its hide is in the form of plates, hinged at the joints; hence it is sometimes called the "iron-plated rhino." The skin is dark gray in color. Other rhinos include the Javan, which has one horn, and the Sumatran, which has two.

The giraffe is the tallest land animal

We now come to a different type of giant, the giraffe, which is the tallest of all living land animals. The males average over fifteen feet in height and the females thirteen to fourteen; nineteen-foot specimens have been reported. The giraffe belongs to the family Giraffidae. It was

Right: the Indian rhinoceros (*Rhinoceros unicornis*). Note the deep folds of the rhino's skin.

Photos, N. Y. Zool. Soc.

THREE THICK-SKINNED GIANTS

Left: the black rhinoceros (*Diceros bicornis*). It is found in Africa, to the south of Ethiopia.

Right: the hippopotamus (*Hippopotamus amphibius*), one of the heaviest of the world's mammals.

415

called the cameleopard by the Romans, because it was supposed to be the offspring of a camel and a leopard. The giraffe travels in herds. Generally, each herd is made up of a full-grown male and several females with calves.

The forelegs are longer than the hind legs; hence the neck is thrust upward instead of forward. The neck itself, though it has the normal number of seven vertebrae characteristic of most mammals, is enormously elongated. The head can be placed on a line with the neck, so as to extend the reach still farther. The tongue, too, can be thrust out to an amazing extent.

The giraffe is one of the most specialized of animals. If the tree foliage on which it feeds were ever to become scarce, the animal would be hard put to it to find food. Ground-feeding is a difficult maneuver for the giraffe. It must straddle wide with its front legs in order to bring its mouth in contact with the earth. On the other hand, with its great reach and its flexible, prehensile lips and tongue, the animal can readily pluck the leaves of trees, avoiding thorns.

It has often been said that the giraffe is mute because it has no vocal cords. As a matter of fact, giraffes have been known to utter sounds. They groan when they are injured; the females call to their young when the latter stray. It is true that the vocal cords are quite undeveloped; yet

This striking photograph will give an excellent idea of the enormous mouth capacity of the hippopotamus.

Black Star

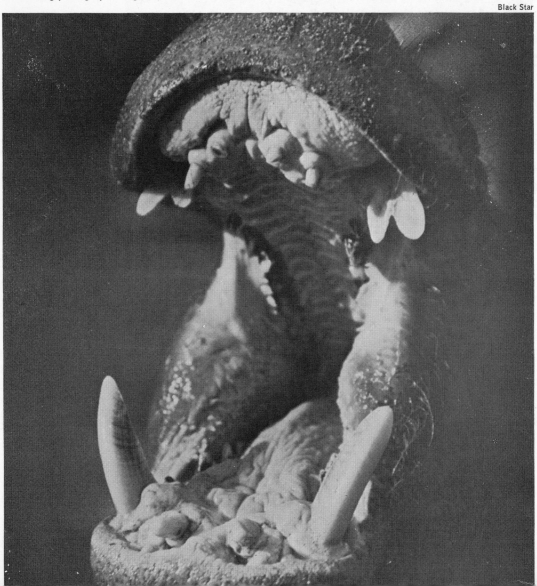

A fine specimen of Nubian giraffe in captivity. The giraffe is the tallest of the world's land animals.

somehow the giraffe can produce audible sounds without them.

Zoologists have long been puzzled by the giraffes' fantastic ability to withstand thirst. Restricted to Africa, south of the Sahara Desert, these animals are found, among other places, in the Kalahari Desert. In this arid region, as far as we know, they cannot find water for seven or eight months of the year. One must take into account, of course, the possibility that desert animals may know of water sources of which men are ignorant. Perhaps the Kalahari giraffe's supposed abstention from water may be somewhat exaggerated.

The giraffe is a most inoffensive animal and is even quite friendly in tourist areas. The horns are useless as weapons of attack, since they are covered with soft fur; however, the animal can deliver a powerful blow with its feet or its head.

A comparatively short-necked relative of the giraffe is the okapi, found in the upper Congo region in Africa. Its existence was not confirmed by zoologists until the turn of the present century. The male stands a little over five feet at the shoulder; its body is considerably more compact than that of the giraffe.

See also Vol. 10, p. 275: "Mammals."

Female Australian kangaroo, with her young in her pouch. Baby kangaroos, about an inch long when born, are suckled in the pouch. They remain there for some time after birth.

THE POUCHED MAMMALS

Kangaroos, Opossums and Other Marsupials

THE curious creatures known as the pouched mammals form a unique order of animal life. Their mode of reproduction, in particular, differs radically from that of the general run of mammals, such as moles, bats, mice, cats, apes and elephants. Most mammals, greatly as they may differ from one another in size and habits, reproduce in much the same way. The female produces a tiny egg cell that has very little yolk in it. This cell is fertilized within the body of the female, and after a time it comes to rest in her womb, or uterus. As the egg develops, membranes are formed around it. One of these membranes fuses with the wall of the uterus so that a firm connection is made between the developing young, or embryo, and its mother. This connection, which is called a placenta, is richly supplied with blood vessels. It serves as a passageway, by which nourishment and oxygen are passed from the mother to the embryo. After a certain length of time, varying with different species, the now well-developed young is forced from the womb and is born — that is, leaves the mother's body. The mammals that reproduce in this way are sometimes called placental mammals, or placentals.

The female pouched mammal brings forth its young in a different fashion. It produces an egg with a much larger supply of yolk than that of a typical placental egg. This egg is fertilized within the body of the female but no placenta is developed. (The rabbitlike bandicoot of Australia is an exceptional case; a primitive and only partly developed placenta is formed within its body.)

The embryo in the womb of the female pouched mammal is nourished on the yolk of the original egg cell; a nutritive solution, called uterine milk, in which the embryo is bathed, also furnishes a certain amount of food material. The embryo does not remain long in the mother's womb; it is forced out of the body at a very immature stage of development. The young of the opossum leaves the mother's body eight days after the fertilization of the egg; the young of the kangaroo remains in the womb for only thirty-nine days.

When the baby pouched mammal emerges from its mother's body, it is an almost shapeless bit of flesh. Hardly an inch in length, it is blind and naked. No external ears show; the hind legs and tail are mere buds. Surprisingly enough, however, the front legs are equipped with large clawed hands, whose nerve centers are well developed. By means of these hands the baby crawls up the mother's belly and finds its way to the pouch, or marsupium, which is located there. This pouch has given rise to the name marsupials, often applied to the group of pouched mammals.

The baby animal finds shelter and warmth in the marsupium. Within it are the mother's teats, through which the tiny little creature draws milk. The baby fastens on one of the teats, or nipples, with its mouth, so that the nipple is thrust far down into its throat. The upper and lower lips of the little animal grow together at the corners around the nipple, and the nipple itself swells. As a result the baby is securely fastened to its source of nourishment. Thus enclosed within its mother's pouch, the young marsupial feeds and grows. It leaves its snug shelter only when it is fairly well developed and capable of eating and digesting other kinds of food besides its mother's milk. But it does not leave its pouch sanctuary permanently, even at this stage of its development. As we shall see, it hops back into the pouch quite frequently, particularly if danger threatens.

In order to understand the relationship between the marsupials and the placental animals, we must go back to the early history of the mammal group. The earliest mammals, which were descended from mammallike reptiles, were in all likelihood small creatures that lived in trees. They had to remain inconspicuous in order to survive at a time when giant reptile carnivores were ruling the earth. The little mammals had sharp teeth, but because of their small size their prey was limited to worms and insects. They also fed upon eggs and, perhaps, fruits and buds.

From these early animals various mammal types evolved. Some of these flourished for a time and then became extinct. One type, which reproduced its kind by laying eggs, succeeded in maintaining itself; today it is represented by the platypus and the spiny anteater of Australia. Two other groups of mammals managed to survive, and became the ancestors of our modern placentals and marsupials. At the early stages of their development, both these types of mammals were small; they ate insects and various other kinds of food.

The early marsupials were opossumlike. They had small brains, but they were far more intelligent than any of the reptiles that were then living. Gradually marsupials spread throughout the tropical and temperate regions of the world. In the course of time they came into competition with the placentals and almost invariably succumbed to their rivals. Gradually the marsupials were driven out of Eurasia, Africa and North America. They continued to thrive in South America and Australia.

Various marsupials and herbivorous placentals had migrated to South America by way of the Isthmus of Panama. This intercontinental bridge had been broken off in the course of time and South America had become an island continent. The marsupials flourished there and evolved into quite dissimilar types. Some of them were rather short-legged, large-skulled carnivores, as large as bears; they preyed on other marsupials and on herbivorous placentals. One marsupial flesh eater had long piercing tusks, like those of the saber-toothed tiger. Other marsupials remained more or less opossumlike.

About a million years ago, the land bridge to North America was re-established. Great numbers of highly developed placental mammals pushed their way into the South American continent. The marsupials could not withstand this onslaught, and most of them perished. Today a few scattered species, including the opossums, are the only surviving representatives in the two Americas of what

Australian News & Inf. Office

A lively group of kangaroos on a New South Wales plain. These animals are often a pest.

was once a flourishing population of marsupials of many different kinds.

The marsupials of Australia fared much better than their American cousins. A number of opossumlike animals had invaded Australia when it was still connected to the Asiatic mainland. About 70,000,000 years ago, Australia became isolated from the rest of the world. Fortunately for the marsupials, no placental mammals, except for bats and several species of rats, had entered the now isolated continent. The marsupials, therefore, dominated this vast land, in which they increased their numbers and evolved into many different types.

The development of
different types of marsupials

One group became carnivorous. There were catlike marsupials (today called native cats), wolflike creatures (represented by the modern Tasmanian wolf) and a wolverinelike animal, from which the modern Tasmanian devil descended. Another group was insectivorous; it included marsupial anteaters with long snouts, marsupial moles with reduced eyes and well-developed fore limbs, and pouched mice, resembling the shrews of other parts of the world. A third group, including phalangers, kangaroos, wallabies, wombats and koalas, became herbivorous.

The phalangers were tree dwellers; they had prehensile tails, by which they could cling to branches. Some even developed extensions of skin along the sides of their bodies, so that they could soar from one treetop to the next. Kangaroos and wallabies became ground-dwelling plant eaters; they hopped about on their hind legs. The wombats became tailless, stout-bodied animals with rodentlike grinding teeth. The koalas became strikingly bearlike.

Giant marsupials of
the Australian continent

As is quite frequently the case with an actively evolving group, some of the Australian marsupials became giants, though they were not comparable in size with the dinosaurs that lived in previous eras. There was a lionlike marsupial species

Australian News & Inf. Office

The tree kangaroo, found in New Guinea and Northern Queensland, Australia.

that had a skull a foot in length. A giant kangaroo stood almost ten feet high. *Diprotodon,* an extinct relative of the wombat, was about the size of a rhinoceros, reaching a length of twelve feet.

The disappearance of marsupial giants like these is a puzzling problem. Changes in climate, decreasing food supply, the spread of poisonous plants, the outbreak of disease — all these factors may have been involved. The British biologist A. Dendy proposed the following theory to account for the disappearance of giant marsupials. He observed that many groups of animals, in the course of their evolution, have shown a marked tendency to increase enormously in size and that this tendency has been accompanied by the development of grotesque and utterly useless bodily structures. He believed that such untoward developments might lead in the course of time to the utter destruction of the species.

It is well known that the growth of the bodies of animals is controlled by internal secretions, the products of various glands. Dendy held that in the absence of certain growth-inhibiting secretions, various organs of an animal would continue to grow far beyond the normal limits. When a useful organ begins to develop and assumes some new function for which an increase of size would be advantageous, natural selection would favor those individuals in which the organ grew most rapidly and attained the largest size in the individual's lifetime. If growth were normally inhibited by some specific secretion, natural selection would favor those individuals in which the glands producing the secretion were least developed or least efficient. This tendency toward greater size would be handed on, from generation to generation; in time, excessive size would no longer be an advantage to the animals in question.

Marsupials adapted to a variety of environments

The extinct giant marsupials are just one example of many animals that began their evolution as well-adapted organisms but became nonadapted in altered circumstances and so died out. Other marsupials of Australasia, however, were more successful than their giant relatives and have survived to modern times. They exploited the opportunities at hand and so spread out into a variety of habitats. The force in evolution whereby evolving animals enter new environments and become quite different from one another is called adaptive radiation. Some marsupials became predators and developed teeth for their flesh-eating habits; others took to the trees and developed organs for climbing; still others became burrowing animals. The marsupials of a particular way of life came to resemble the placentals of other lands who had the same kind of habits. This phenomenon in evolution, in which basically different animals with similar habits take on similar characteristics, is termed convergence.

In modern times the marsupials of Australasia have seriously decreased in numbers; some species have become extinct. The primary causes seem to be man and his introduced animals. The marsupials have been hunted by man for food and fur and wantonly killed for "sport." Man's agricultural and stock-raising activities have drastically taken away good grazing land from the herbivorous kangaroos and wallabies. The introduced animals, especially the dingo, the fox and the domestic cat, have destroyed thousands of marsupials. Since they reproduce slowly, usually having but one or two young a year, they cannot quickly replace their lost numbers. During a period of drought, many herbivorous marsupials and the introduced rabbits will starve. When favorable conditions return, the prolific rabbits soon re-establish themselves and consume much vegetation, whereas the marsupial population seldom recovers itself completely and must constantly compete against the voracious rabbit population for food.

The kangaroo family is a varied group of marsupials

Our survey of the Australasian marsupials begins with the kangaroo family (family Macropodidae), a varied group ranging in size from the kangaroos, which stand six feet, to the musk kangaroo, which is only rat-size. The kangaroos, wallaroos, wallabies and rat kangaroos all belong to this family. The front legs are relatively short and undeveloped; the hind legs are enlarged and muscular. The hind feet are elongated. The fourth toe is the longest; the short second and third toes are fused together; there is no first toe. These marsupials are all hopping animals.

There are no specific differences between kangaroos, wallaroos and wallabies except size and body build. All have front teeth specialized for clipping grasses, shoots and leaves; the molar teeth have ridges on their crowns for grinding this vegetable food. Canine teeth are lacking or merely rudimentary.

The kangaroos (genus *Macropus*) are large, powerfully built animals capable of bounding at speeds up to thirty miles an hour for short distances. When moving at a slow pace they spring a distance of four to six feet at a single bound; as they in-

SMALL MEMBERS OF
THE MARSUPIAL GROUP

Above: New Guinea phalanger.

Right: gray phalanger.

Below: vulpine phalanger.

Above: murine opossum.

crease their speed they may make leaps of twenty-six feet. While they are moving rapidly, the heavy, thick tail is held out almost horizontally to counterbalance the forward thrust of the body. The tail is also used as a rudder to swing the body when the animal wants to make a turn at full speed. When the kangaroo grazes, its arched body rests on all fours. To move, the animal pushes its tail against the ground, at the same time lifting its hind legs together and swinging them forward.

The red kangaroos are plains dwellers, whereas the great grey kangaroos inhabit brushy country and open forests. Both are gracefully-built animals with exceedingly long hind legs and thick but tapering tails. Wallaroos are stockily built kangaroos that live in rocky regions of the mountains. They have roughened soles on their hind feet which prevent slipping as they leap from rock to rock. Kangaroos and wallaroos have a special stomach sac in which the cellulose and other carbohydrates of their vegetable food is broken down by means of bacteria.

The mother kangaroo gives little aid to her newborn young

Kangaroos mate but once a year and have one or, rarely, two young. The only help the mother seems to give her tiny offspring in finding the pouch is to lick a path in her fur from the opening of the birth canal to the entrance of the pouch. If the infant embryo loses its way in her fur or drops to the ground, she does not aid it. The time taken for the young to make the journey to the pouch is judged to be about fifteen minutes with the great grey kangaroo and from five to thirty minutes with the red kangaroo. Once the weaned young leaves the maternal pouch, it does not stray far from its mother, jumping back into its retreat at the first sign of danger from fox, dingo or eagle. If a female, carrying a young one, is chased by dogs or a hunter on horseback, she will dump her offspring from the pouch. Whether she does this to lighten her load or allow the young to escape is not known. It is said that if she survives the chase, she returns to the spot where the young was dropped. Kangaroos can fight savagely when cornered, propping themselves on the tail while kicking with their hind legs. Their hands grip strongly.

Wallabies are of medium and small size

Wallabies are kangaroolike in habits and appearance but are of medium and small size. The smaller ones dig ground burrows that have an open exit some distance from the entrance. When frightened, they dash for the burrow from some lookout point close by the entrance. Larger wallabies seek refuge in ground hollows scooped out under bushes. The brush wallabies (*Protemnodon*) are large, gracefully built and often distinctly marked and handsomely colored. Scrub wallabies, or padamelons (*Thylogale*), are smaller and are thick-tailed. They live in dense scrub areas and have tunnellike runways through the heavy undergrowth. Rock wallabies (*Petrogale*) are skilled in climbing and bounding over the rocks in mountainous areas. Their foot pads, like those of the wallaroos, are roughened; this gives traction on slippery surfaces. Nail-tailed wallabies (*Onychogalea*) get their name from the horny projection at the tip of the tail; they have silky fur. The hare wallabies (*Lagorchestes*) are solitary animals, much like hares in habits and appearance.

The forest-dwelling tree kangaroos (*Dendrolagus*) differ somewhat from the rest of the family because of their tree-dwelling habits. The hind foot is comparatively shorter and broader; though not a grasping organ, it has curved, sharp claws and roughened soles. The fore limbs are more equal in length to the hind, and the broad hands serve admirably in climbing. The tail is long and slender but not prehensile. Tree kangaroos ascend trees to sleep, feed on leaves and fruit and escape from predators. Though compact of body, they climb agilely, using the tail as a prop and balancer. Usually they descend a tree tail first, but if hard pressed they leap downward twenty or thirty feet to another tree or to the ground where they easily alight on all fours.

The Australian koala, or native bear.

The rat kangaroos (subfamily Potoroinae), diminutive members of the kangaroo family, are agile, ground-dwelling marsupials with prehensile tails. They feed on grasses, fungi, tubers and refuse. Rat kangaroos dig burrows or shallow depressions sheltered by bushes or a fallen tree. They make nests of stringy bark or grass; these materials are carried in small bundles to the nest site in a loop of the prehensile tail. Some rat kangaroos dwell together in colonies, but others live in pairs or are solitary. The female produces young several times a year, one or two at a birth. With one species, the infants remain in the maternal pouch four months.

The musk kangaroo (*Hypsiprymnodon*) seems to be a link between the kangaroos and the phalangers, or "possums." The hind legs are not so disproportionately large; the tail is opossumlike, hairless and scaly. Unlike the kangaroos, the musk kangaroo possesses a big toe; this is not opposable to the other toes.

The phalangers (family Phalangeridae) are cat-size or smaller marsupials, specialized for tree living. They have sturdy limbs and curved claws. Unlike the kangaroos, they possess a first hind toe; this is opposable to the other toes and is used for gripping. The tail is either prehensile or, as with the gliding phalangers, is extravagantly furred, almost featherlike, and serves as a rudder. The phalanger's head is somewhat squirrellike; the well-formed maternal pouch opens toward the front. The second and third toes are joined together but have separate claws; this arrangement serves as a comb to preen the fur. Phalangers are night-roaming and feed on insects and vegetation.

The honey mouse, or honey possum (*Tarsipes*), is a tiny phalanger with an elongated, trunklike muzzle, which can be thrust deeply into flowers for nectar and pollen. The pygmy possums (*Cercaërtus*) are small phalangers that hibernate in cold weather. Other noteworthy phalangers include the striped possums (*Dactylopsila*), which are handsomely marked with black and white and distinguished by their long, probelike fourth fingers; the ring-tailed phalangers (*Pseudocheirus*), which get their name from the habit of curling the end of the tail in a ring; the common possums (*Trichosurus*), which are somewhat foxlike with a thickly bushed tail, large ears and a pointed muzzle; and the cuscuses (*Phalanger*), which have small ears and a prehensile tail with a naked and scaly tip. The gliding phalangers belong to three separate groups, but all possess gliding membranes along the sides of the body. The graceful glides of the greater flying phalanger (*Schoinobates*) sometimes cover 120 yards.

The quaint koala, or native bear (*Phascolarctos*), appears to be related to the phalangers, but it differs from them by having only a rudimentary tail and well-developed cheek pouches; the maternal pouch faces to the rear instead of forward. The koala is completely arboreal, climbing by means of its strong limbs and powerful claws. This robust marsupial, which has a woolly coat, feeds exclusively on the leaves of only a few species of eucalyptus. Koalas usually breed only once every two years. One young (sometimes two) is born. When the young abandons the pouch it climbs to the mother's back where she carries it until the cub is almost full grown. A curious habit of the female koala is to manufacture a pap, or soft food, of eucalyptus leaves in her body at the time when the young one is being weaned. The cub eats this pap, which contains no excrement, from the mother's anal opening; apparently this food greatly speeds up the young's growth.

Unlike the phalangers and koalas, the wombats are ground-dwelling marsupials, being heavy-bodied with short, stout legs. They are excellent burrowers and will sometimes excavate an underground tunnel extending a hundred feet in length. Their front feet carry powerful digging claws; their rodentlike front teeth gnaw through any roots that obstruct their tunneling activities. The tail is a mere nubbin, and the head large. These nocturnal animals live alone, pairing only at mating time. Grasses, fungi and the inner bark of some trees and shrubs make up their diet. The common wombat (*Phascolomis*) is distinguished by having short, rounded ears, coarse hair and a naked area on the muzzle. It is fairly large, measuring three and a half feet. The hairy-nosed wombat (*Lasiorhinus*) is smaller; it has soft, silky fur and longer, tapering ears.

The bandicoots (family Peramelidae) are ratlike or rabbitlike marsupials having

Photos, N. Y. Zool. Soc.

Tasmanian devil, a marsupial carnivore.

pointed muzzles and large ears. Their hind limbs are fairly long, with the second and third toes fused; the front limbs are small. These curious little animals are both herbivorous and carnivorous. They eat various kinds of vegetable matter; they also feast on insects, mice and lizards. Australia's bandicoots include the short-nosed, long-nosed, pig-footed and rabbit bandicoots.

The marsupial mole (*Notoryctes*) is outwardly completely molelike. Its body is small and compact, its muzzle cone-shaped; the tiny ears lie buried in the short fur of the head, the legs are short and stocky, the fingers have enlarged digging nails and the tail is a rudiment. The non-molelike feature, of course, is the pouch, which

The common wombat of Australia is a rodentlike marsupial specialized for excavating ground burrows.

opens to the rear. This little marsupial plows through the desert sand about three inches below the surface, coming above ground at intervals. It feeds on ants and probably other insects and worms.

Another insect eater, the banded anteater (*Myrmecobius*), feeds on ants, termites and their eggs and larvae. This marsupial is rat-size and has a long, bushy tail, prominent ears and a pointed muzzle. Its tongue, by which it captures insect prey, is long and thin. The female marsupial anteater lacks a pouch; the young are protected only by her fur as they cling firmly to her nipples.

lizards. This animal is short-legged, with a large head and strong jaws. Like the marsupial wolf, it is confined to Tasmania. The Tasmanian wolf (*Thylacinus*) lives only in the most inaccessible parts of its island home. The largest marsupial carnivore, it is strikingly doglike, running on its toes more than does any other marsupial. It has piercing canine teeth, flesh-shearing premolar teeth and bone-crushing molars. The animal is nocturnal, hunting alone or in pairs for wallabies and other marsupials, rats and birds.

Aside from the pouched mammals of Australasia, the only other marsupials in-

Female common opossum with young. Six to eight young make up a normal litter.

The pouched mice (subfamily Phascogalinae) are also insectivorous. They are similar to mice but have numerous needle-like teeth and more tapering muzzles; some species live in trees, while others are ground dwelling. The native cats (*Dasyurus*) supplement their insect diet with birds and birds' eggs. These white-spotted carnivorous marsupials are weasel- or marten-like and are mostly nocturnal and tree-living. The Tasmanian devil (*Sarcophilus*) is a fairly small though powerful flesh eater. It can climb but usually lives in rocky areas, where it hunts for the smaller wallabies, rat kangaroos, rats, birds and

habit the New World. These include the South American opossum rats (*Caenolestes*), which are ground-living and have no pouches, and the arboreal woolly, murine and common opossums (family Didelphidae) of South, Central and North America. Most possess naked, prehensile tails. The females bear many young at one time. Those that lack the pouch carry their young at first attached to the nipples. Later the infants climb to the mother's back, wrapping their tails around hers.

One species of opossum, the yapock, or water opossum, seems to be the only marsupial specialized for an aquatic life. It is found along the rivers and other bodies of water from Guatemala to southern Brazil. Above, it is ashy gray, with broad black cross bands; it is whitish below. This aquatic marsupial is about two and a half feet long. It has webbed feet, a marsupial pouch and a naked, prehensile tail. The yapock feeds upon aquatic animals; it is particularly fond of fish.

The most familiar of the opossums is the Virginian, or common, opossum (*Didelphis virginiana*). This is the animal to which the name "opossum" (from the language of the Virginia Indians) was first applied. It ranges from New York to Florida and from the Atlantic coast to the Great Lakes and south to Texas. It is about as large as a cat; its tail is some fifteen inches long. It has a long pointed snout, sharp teeth, prominent ears, a long, almost naked prehensile tail and a marsupial pouch. It has long, grizzled gray external hairs and short, whitish underfur. Its den is comfortably bedded with leaves and dry grass; there, five to fifteen young are born once or twice a year.

The fecundity of the opossum is one reason why it has survived since Mesozoic times, for it is the prey of many predatory animals, including horned owls, wildcats, foxes, wolves and bears. Man hunts the opossum, too, for its fur and its flesh.

N. Y. Zool. Soc.

The yapock (*Chironectes panamensis*).

The animal is particularly exposed to attack by its enemies when it seeks its food on the ground.

The varied diet of the opossum

Opossums are nearly omnivorous. They eat fruit, mushrooms and other vegetable matter; they hunt for the eggs and young of birds; they pounce on wild fowl and rabbits at night; they search smokehouses and even farm kitchens at night for rodents and insects; they raid chicken coops. According to one writer, "the opossum will catch a hen and drag the squawking victim off. It will not abandon its prey until the dogs that have been aroused by the clamor have overtaken the robber and have fastened upon it with their teeth."

The origin of the phrase "playing possum"

Often when this animal (or any of its tropical relatives) finds itself cornered, it will fall suddenly limp and apparently dead; whence the well-known phrase "playing possum." When this takes place, the opossum draws back its gums from its glittering teeth, acquiring a truly cadaverous appearance. It will not stir, even if it is picked up by the tail or rolled around with one's foot. If one turns one's back, however, the animal is likely to start up suddenly and run away.

Generally this "playing possum" is considered to be an instinctive feigning of death. It is true that some people have attributed it to "paralysis by fear," but this is not very likely. As a matter of fact, the opossum will not always resort to this stratagem; if it has a fair chance it will fight, particularly if it is a mother defending the young. It shows a courage and a ferocity in the use of its teeth that may be quite disconcerting.

The ruse of "playing possum" is not always a safeguard, since many animals will seize an opossum that is feigning death as readily as one that is active. The artifice seems to be a vestige of some very ancient habit of defense.

See also Vol. 10, p. 275: "Mammals."

Amer. Mus. of Nat. Hist.